8/12/97

The Oystercatcher
From individuals to populations

Oxford Ornithology Series

Edited by C. M. Perrins

1 *Bird Population Studies: Relevance to Conservation and Management* (1991) Edited by C. M. Perrins, J.-D. Lebreton, and G. J. M. Hirons

2 *Bird–Parasite Interactions: Ecology, Evolution, and Behaviour* (1991) Edited by J. E. Loye and M. Zuk

3 *Bird Migration: A General Survey* (1993) Peter Berthold

4 *The Snow Geese of La Pérouse Bay: Natural Selection in the Wild* (1995) Fred Cooke, Robert F. Rockwell, and David B. Lank

5 *The Zebra Finch: A Synthesis of Field and Laboratory Studies* (1996) Richard A. Zann

6 *Partnerships in Birds: The Study of Monogamy* (1996) Edited by Jeffrey M. Black

7 *The Oystercatcher: From Individuals to Populations* (1996) Edited by John D. Goss-Custard

The Oystercatcher

FROM INDIVIDUALS TO POPULATIONS

Edited by

JOHN D. GOSS-CUSTARD

Furzebrook Research Station
Wareham, Dorset

Oxford   New York   Tokyo
OXFORD UNIVERSITY PRESS
1996

Oxford University Press, Walton Street, Oxford OX2 6DP
Oxford   New York
Athens   Auckland   Bangkok   Bombay
Calcutta   Cape Town   Dar es Salaam   Delhi
Florence   Hong Kong   Istanbul   Karachi
Kuala Lumpur   Madras   Madrid   Melbourne
Mexico City   Nairobi   Paris   Singapore
Taipei   Tokyo   Toronto
and associated companies in
Berlin   Ibadan

Oxford is a trade mark of Oxford University Press

Published in the United States
by Oxford University Press Inc., New York

A catalogue record for this book is available from the British Library

Library of Congress Cataloging in Publication Data

The oystercatcher: from individuals to populations/edited by John D. Goss-Custard.
(Oxford ornithology series; 7)
Includes bibliographical references (p.   ) and index.
1. Oystercatcher—Behaviour.   2. Oystercatcher—Ecology.   3. Bird populations.
I. Goss-Custard, John.   II. Series.
QL696.C452096   1996   598.3'3—dc20   95–47806

ISBN 0 19 854647 5

Typeset by Footnote Graphics, Warminster, Wiltshire
Printed in Great Britain by
Bookcraft (Bath) Ltd, Midsomer Norton, Avon

We dedicate this book to Professor Rudi Drent
in appreciation of the great influence he has had
on our research over two decades.

John Goss-Custard also dedicates the book
to Professor David Jenkins, for his friendship,
professional insights and support over many years.

Acknowledgements

We are all extremely grateful to Dick Visser for converting our rough diagrams into his magnificent figures, to Klaus-Michael Exo for taking on the terrible job of compiling the references and to Jan van de Kam for providing his excellent photographs. The editor is very grateful to Eileen Barrett for her help in getting the manuscripts into shape.

Contents

List of Contributors   ix

Introduction   1
John D. Goss-Custard

Part I Individual adaptations

1 *Food and feeding behaviour*   7
Jan B. Hulscher

2 *Prey size selection and intake rate*   30
Leo Zwarts, John T. Cayford, Jan B. Hulscher, Marcel Kersten,
Patrick M. Meire and Patrick Triplet

3 *Specialization*   56
William J. Sutherland, Bruno J. Ens, John D. Goss-Custard and
Jan B. Hulscher

4 *Feeding with other Oystercatchers*   77
Bruno J. Ens and John T. Cayford

5 *Where to feed*   105
John D. Goss-Custard, Andrew D. West and William J. Sutherland

6 *How Oystercatchers survive the winter*   133
John D. Goss-Custard, Sarah E.A. le V. dit Durell, Cameron P.
Goater, Jan B. Hulscher, Rob H.D. Lambeck, Peter L. Meininger
and Jamil Urfi

7 *Why do Oystercatchers migrate?*   155
Jan B. Hulscher, Klaus-Michael Exo and Nigel A. Clark

8 *Life history decisions during the breeding season*   186
Bruno J. Ens, Kevin B. Briggs, Uriel N. Safriel and Cor J. Smit

9 *Rearing to independence*   219
Uriel N. Safriel, Bruno J. Ens and Alan Kaiser

10 Haematopus ostralegus *in perspective: comparisons with
other Oystercatchers*   251
Philip A.R. Hockey

Part II Population ecology

11 *Oystercatchers and man in the coastal zone*   289
Rob H.D. Lambeck, John D. Goss-Custard and Patrick Triplet

12 *The carrying capacity of coastal habitats for Oystercatchers*   327
John D. Goss-Custard, Andrew D. West, Ralph T. Clarke,
Richard W.G. Caldow and Sarah E.A. le V. dit Durell

13 *Population dynamics: predicting the consequences of habitat
change at the continental scale*   352
John D. Goss-Custard, Sarah E.A. le V. dit Durell, Ralph T.
Clarke, Albert J. Beintema, Richard W.G. Caldow, Peter L.
Meininger and Cor J. Smit

Conclusions:
From individuals to populations: progress in Oystercatchers   384
John D. Goss-Custard

Appendix 1:
The Exe estuary Oystercatcher–mussel model   390
Ralph T. Clarke and John D. Goss-Custard

References   394

Index   433

Contributors

ALBERT J. BEINTEMA, Institute for Forestry and Nature Research (IBN-DLO), Bosrandweg 20, P.O. Box 23, 6700 AA Wageningen, The Netherlands.

KEVIN B. BRIGGS, 2 Osborne Road, Farnborough, Hampshire GU14 6PT, England.

RICHARD W.G. CALDOW, Furzebrook Research Station, Wareham, Dorset BH20 5AS, England.

JOHN T. CAYFORD, Oak Cottage, Langton Green, Eye, Suffolk IP23 7HKL, England.

NIGEL A. CLARK, British Trust for Ornithology, The Nunnery, Nunnery Place, Thetford IP24 2PU, England.

RALPH T. CLARKE, Furzebrook Research Station, Wareham, Dorset BH20 5AS, England.

SARAH E.A. LE V. DIT DURELL, Furzebrook Research Station, Wareham, Dorset BH20 5AS, England.

BRUNO J. ENS, Institute for Forestry and Nature Research (IBN-DLO), P.O. Box 167, 1790 AD Den Burg, The Netherlands.

KLAUS-MICHAEL EXO, Institut für Vogelforschung, An der Vogelwarte 21, D-26386 Wilhelmshaven, Germany.

CAMERON P. GOATER, Department of Biological Sciences, University of Alberta, Edmonton, Alberta T6G 2E9, Canada.

JOHN D. GOSS-CUSTARD, Furzebrook Research Station, Wareham, Dorset BH20 5AS, England.

PHILIP A.R. HOCKEY, Percy FitzPatrick Institute of African Ornithology, University of Cape Town, Rondebosch 7700, South Africa.

JAN B. HULSCHER, Zoology Department, University of Groningen, Kerklaan 30, P.O. Box 14, 9750 AA Haren, The Netherlands.

ALAN KAISER, The Mitrani Center for Desert Ecology, The Jacob Blaustein Institute for Desert Research, Sede Boqer Campur, 84990 Israel.

MARCEL KERSTEN, Zoology Department, University of Groningen, Kerklaan 30, P.O. Box 14, 9750 AA Haren, The Netherlands.

ROB H.D. LAMBECK, Netherlands Institute of Ecology, Centre for Estuarine and Coastal Ecology, Vierstraat 28, 4401 EA Yerseke, The Netherlands.

PETER L. MEININGER, National Institute for Coastal and Marine Management/RIKZ, P.O. Box 8039, 4330 EA Middelburg, The Netherlands.

PATRICK M. MEIRE, Institute of Nature Conservation, Kliniekstraat 25, B-1070 Brussels, Belgium.

URIEL N. SAFRIEL, Department of Evolution, Systematics and Ecology, The Alexander Silberman Institute of Life Sciences, The Hebrew University of Jerusalem, 91904 Jerusalem, Israel.

COR J. SMIT, Institute for Forestry and Nature Research, P.O. Box 167, 1790 AD Den Burg, The Netherlands.

PATRICK TRIPLET, 40 Rue Carnot, 80550 Le Crotoy, France.

WILLIAM J. SUTHERLAND, Department of Biological Sciences, University of East Anglia, Norwick NR4 7TJ.

JAMIL URFI, Department of Environmental Biology, University of Delhi—South Camps, Delhi, India.

ANDREW D. WEST, Furzebrook Research Station, Wareham, Dorset BH20 5AS, England.

LEO ZWARTS, Rijkswaterstaat, P.O. Box 600, 8200 AP Lelystad, The Netherlands.

Introduction

John D. Goss-Custard

One of the major challenges facing the science of ecology is to predict how natural populations will respond to the many changes to which our planet is subject. This is not just because ecological systems are complex and that single-species populations, and multi-species communities of species populations, may be affected by changes in the intricate eco-system of which they are a part. The challenge arises from the need to make predictions to wholly new circumstances that lie beyond our direct experience and, in many cases, may have no evolutionary precedent. Simultaneous changes in climate, land-use, agricultural and industrial practice and recreational activities, for example, introduce novel combi-nations of factors into the environment. The required predictions are dif-ficult to make yet they must be reliable; policy decisions are increasingly influenced by ecological forecasts, and failures can be expensive in many ways. This increasing requirement to make predictions provides an addi-tional stimulus to the scientist's always fascinating task of understanding the natural world.

Although much of the work reviewed in this book is basic research, a high proportion of the studies have been influenced by the need to make such predictions. Throughout the research programme, however, there has been an awareness that the most reliable predictions are based on a thorough understanding of ecological and evolutionary processes. Unless we understand the basic underlying principles, our study species may respond in quite unforeseen ways in the new conditions, and so under-mine the basis of our prediction. Acquiring such understanding requires a fertile interplay between theoretical and empirical studies. The rapid progress made in Behavioural Ecology—the study of the evolved de-cision rules used by animals in responding to their biotic and abiotic environment—over the last two decades owes much to the close inter-action between theoreticians and empiricists. Theory has had a large influence on the design of many of the studies reviewed in this book, whether they were observational, experimental or both. We hope that the thinking of theoreticians will be similarly influenced by our findings. Indeed, one of the two main themes of this book is to explore the extent to which modern theory explains the behaviour of a well-studied bird in nature and how, in turn, field studies may show how that theory has to advance.

The second major theme is that understanding and predicting the behaviour of bird populations must be based on studies of the variation between individuals. After all, it is not a population that responds to a changing environment, but the individual animals themselves. The one thing that virtually everyone knows about Oystercatchers is that different individuals specialize on a particular prey species and in the way in which they deal with them, and that the young adopt the specialization of their parents. As one author said, 'It doesn't matter what I lecture on with Oystercatchers, almost the first question will be about individual specialization and its cultural transmission!' In fact, the book seriously questions this conventional view. Nonetheless, individuals vary enormously in the strategy they adopt, or are forced by competition from other Oystercatchers to adopt, when they make their decisions about how they should respond to the contradictory demands placed upon them by their environment. Without such individual variations being taken into account, neither the behavioural nor population ecology of the species could ever be fully understood.

A close and reciprocal relationship has developed in Oystercatcher research between basic and applied projects, insofar as they can ever be clearly distinguished. The major applied issue in Oystercatcher research 40 years ago was whether the species was a serious pest of commercial shellfisheries and, if so, whether culling large numbers would alleviate the problem. Subsequent research work was stimulated more by the worldwide concern as to how populations of waders (or shorebirds, in North America) might be affected were the quality and quantity of their habitats to be changed by increasing human activity. Although there are over 200 shorebird species, much work continued on the Oystercatcher because its activities are relatively easy to quantify. Being large and conspicuous, and not too timid, it can be easily watched on both the breeding and feeding grounds. Unlike almost all other shorebirds, whose small camouflaged chicks are so difficult to locate, Oystercatcher pulli can be followed with relative ease, so the birds' reproductive success can be measured accurately. Their foraging activities can be quantified with a precision that can seldom be achieved in other species. As a result, some sophisticated theoretical issues have been studied in the field in quite remarkable depth. The Oystercatcher has thus attracted many studies of behavioural, and recently physiological and morphological, adaptation, conducted simply for their own sake.

The book itself arose because many of us meet regularly at conferences on the ecology of wetland birds and waders organised by the International Waterfowl and Wetlands Research Bureau (IWRB) and the Wader Study Group (WSG). Although up to two hundred workers from around the world attend, the meetings are very informal and friendly. Over the years, they have helped to engender the spirit of cooperation that is so important for the study of animals that move so freely across national boundaries. Many personal and professional friendships have developed, along with numerous collaborative projects. This book is one such project, following workshops on Oystercatchers held at the joint IWRB/WSG meeting in Ribe, Denmark, in September 1989 and at the Norddeutsche Naturschutzakademie in Schneverdingen, Lower Saxony, Germany in March 1991. The meeting at Ribe showed a clear need to bring together under one cover a synthesis of all the research carried out over four decades, the results of much of which were still only available in internal research reports, often written in languages with which many were not familiar. Instructive comparisons were drawn between Oyster-catcher species worldwide and the European species on which the major-ity of us had worked. At Schneverdingen, each contributor developed the theme of a chapter and unpublished information was unselfishly exchanged, both at the time and subsequently. Happily, students of Oys-tercatchers have not followed the example of their study subject, where competition is the way of life. Instead, a spirit of cooperation has pre-vailed which has turned out to be the cornerstone of the collaborative success of this book.

The book is divided into two parts. The chapters in Part I consider the adaptations for survival and reproduction shown by individual Oyster-catchers. Although morphological and physiological characters are dis-cussed, most attention is given to the behavioural choices made by the birds. One reason for this emphasis is that the decision rules used by Oystercatchers when choosing between alternative foraging or breeding strategies has attracted much of the research attention over the last two decades. The other reason is that these decision rules are central to attempts to derive population-level phenomena from studies of individ-ual animals, the topic addressed in the second part of the book. The chapters at the beginning of Part I focus on the foraging decisions of how, on what, where and when to feed. As much of the research on these choices has been conducted on the wintering grounds, Part I begins with the nonbreeding season rather than the breeding season, which is the more usual practice in books on birds. Later chapters discuss the migration between wintering and breeding areas and the decisions made in spring on how, where, when and with whom to take up a breeding territory and raise chicks. Up to this point, all the chapters focus on the much studied Eurasian Oystercatcher. The final chapter in Part I pro-vides an overview of how the adaptations shown by *Haematopus ostralegus* compare or contrast with those shown by Oystercatcher species from other parts of the World.

Part II describes the research that has been done to understand how anthropogenic changes in the nonbreeding and breeding environments of Oystercatchers might affect the abundance of this and other shorebird species. The first chapter reviews the many changes that are taking place in the coastal habitats of Oystercatchers and evaluates their possible effect on the survival and reproduction of the birds. The next two chapters review the attempts that are being made to predict the effect on local and global population size of many of these environmental changes. Field studies of the variation between individual Oystercatchers in how they respond to each other and to their common, changing environment play an important role in developing the models needed to make these predictions. Our progress in making such population-level predictions from the study of individual Oystercatchers is briefly evaluated in the Conclusions to the book.

Part I

Individual adaptations

1 Food and feeding behaviour

Jan B. Hulscher

Introduction

Oystercatcher species throughout the world are bound to the sea-shore (Voous 1965). At the most, they breed 400 km inland and return to the coast immediately afterwards. In coastal regions, Oystercatchers chiefly feed in the intertidal zone. All species share the habit, unique amongst waders, of opening lamellibranchs with their stout, laterally compressed bills. As they only eat the flesh, they are able to feed on the larger-sized prey individuals. Other mollusc-eating birds swallow their prey whole, the dimensions of their throat restricting them to the smaller size classes of prey. As molluscs often make up the bulk of the benthic biomass in intertidal areas, especially in winter (Beukema 1976, 1982), selection may favour Oystercatchers because of their ability to exploit a large proportion of the biomass of some very abundant prey organisms that are inaccessible to other species. This too, no doubt largely explains why Oystercatchers are relatively numerous amongst carnivorous birds on intertidal flats. This chapter briefly details the habitats where Oystercatchers live and their main prey species. It then explores the formidable and subtle capabilities with which these sturdy birds are able to exploit their feeding grounds.

Feeding habitats

The main feeding habitats of Oystercatchers in Europe are estuaries, rocky shores, beaches and inland fields. Estuaries are exploited throughout the year. The birds feed in the intertidal zone, concentrating on sandflats (clay content less than 5%), and usually avoiding soft mud (clay content more than 20%) (Zwarts 1988). While rocky shores within feeding territories supply food in the breeding season, few Oystercatchers spend the winter there (Feare 1971; Goss-Custard and Durell 1983). Sandy beaches in general are unimportant as feeding grounds. However, Oystercatchers may feed on beaches after gales, when fresh food items are deposited ashore and when food is inaccessible in frozen, ice-covered estuaries. Inland breeding Oystercatchers mainly use fields with short vegetation; in The Netherlands, even grass verges alongside roads are used before egg-laying starts. Tilled fields, inland riverbeds, heather and bracken fields are also used on a small scale for feeding by breeding Oystercatchers. In winter, Oystercatchers regularly feed on pastures adjoining estuaries; for example, when prolonged bad weather makes mudflats inaccessible (Goss-Custard *et al.*, Chapter 6, this volume).

Main prey species

Although the European Oystercatcher has a large prey spectrum (Dare 1966), only the most important prey species are discussed here. The bivalves *Mytilus edulis* (the Edible Mussel) and *Cerastoderma edule* (the Edible Cockle) provide the staple food for the majority of Oystercatchers in their main estuarine areas (Fig. 1.1). The bivalve *Macoma balthica* (the Baltic Tellin) is widely eaten during the breeding season. Locally, other bivalves can be important foods, such as *Mya arenaria* and *Scrobicularia plana* in many places in winter, oysters (*Crassostrea gigas*) in France (Lunais 1975) and the Bloody Cockle (*Anadara senilis*) on the Banc d'Arguin and in Guinea-Bissau in North-west Africa (Zwarts 1985; Zwarts *et al.* 1990a; Swennen 1990). In northern estuaries, gastropods, such as Periwinkles (*Littorina* spp.) and polychaetes, especially the Ragworm *(Nereis diversicolor)* and the Lugworm (*Arenicola marina*) occur in the diet. Among crustaceans, the Shore Crab (*Carcinus maenas*) and the Shrimp (*Crangon crangon*) can be important, especially in summer. The crab *Uca tangeri* is taken along the North-west African coast (Zwarts *et al.* 1990a). Along rocky shores, Oystercatchers mainly take molluscs, such as mussels, and gastropods, such as the Limpet (*Patella* spp.), Dogwhelk (*Nucella lapillus*) and periwinkle, while on beaches they take Sandhoppers (*Talitrus saltator; Amphipoda, Crustacea*) and the polychaete worm *Nerine cirratulus*. In fields, they eat a number of species of earthworms (*Lumbricidae*), leatherjackets—the larvae of crane-flies *(Tipula* spp.)—and the caterpillars of moths.

Anatomical adaptations for feeding

Opening large bivalves requires force and the Oystercatcher is well-adapted to apply it. It has a sturdy build compared with other waders. In addition, the big cervical vertebrae with long spines provide attachment for the muscles which operate the heavy skull bearing the massive bill. The legs and particularly the toes are robust and counteract the strong forces exercised by head and bill when opening large prey. In contrast, the musculature of the gizzard is quite meagre compared with that of species which swallow bivalves whole and so crush the shell in the gut (Piersma *et al.* 1993b).

The bill, although long and deep and laterally compressed, is relatively compact. The ratio of its depth (upper plus lower mandible at the frontal-nasal hinge) to its length is about 1:5, as compared to 1:7 in the Black-tailed Godwit (*Limosa lapponica*) and 1:8.5 in the Curlew (*Numenius arquata*). Although both these species also probe for deeply-buried prey, including bivalves, they swallow them intact. The bone component in each jaw of the Oystercatcher forms a strong tube, triangular in cross-

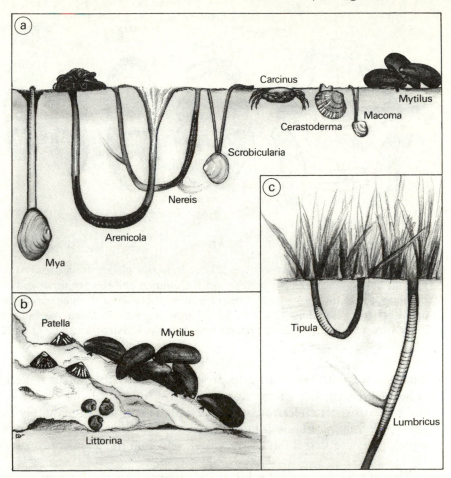

Fig. 1.1 The main prey species of Oystercatchers on (a) intertidal muddy and sandy shores; (b) intertidal rocks; and (c) fields. Adapted from Ens and Zwarts (1980) and Hulscher (1964b).

section and with strengthened corners, which encloses a central canal containing nerves and blood vessels (Fig. 1.2). It can thus withstand strong lateral stress when robust prey, such as mussels, are opened with a vigorous sideways levering action (Heppleston 1970; Hulscher and Ens 1992). As the vertical cross-section of the tip is formed as a blade, the bill can also resist strong dorso-ventral forces operating upon it, when the prey is either bitten or prized open (Bolze 1969). In males, the gonys (the underside of the mandible from the bill-tip to the point where its left and right branches meet in the mid-line) is shorter and the cross-section larger than in females. Males thus seem to be better adapted than females to opening sturdy prey (Hulscher and Ens 1992). The flat inner sides of both the upper and lower mandible at the bill-tip provide a large

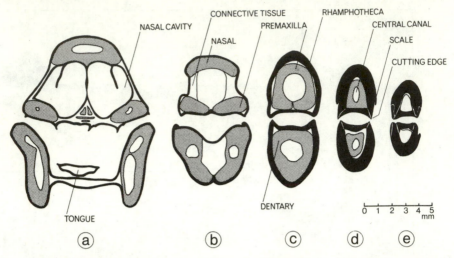

Fig. 1.2 Macroscopic cross-sections taken at various places along the Oystercatcher bill: (a) halfway between the feathers on the culmen and the proximal end of external nare; (b) at the proximal end of the gonys where left and right mandible-halves meet; (c) to (e) at about 30, 20 and 10 mm respectively from the bill-tip. Bone compartments stippled, horny rhamphotheca in black (Hulscher and Ens 1992).

surface area with which to grip prey. Finally, the tomia (the raised horny ridges on either side of the upper and lower jaw) of the distal 3 cms of the bill forms an effective cutting device. The cutis, immediately beneath the epidermis, contains numerous touch (Herbst) corpuscles in a layer enveloping the first 3 cm of the outer and inner sides of both mandibles (Bolze 1969).

The precise form of the bill-tip is adapted to the currently predominant food type in an individual's diet. In lateral view, the tip may be pointed or blunt or intermediate between these two extremes (Fig. 1.3). A pointed bill is associated with a diet of soft-bodied subterranean prey, such as worms and leatherjackets (Hulscher and Ens 1991); a blunt bill with thick-shelled bivalves, such as cockles and mussels, while an intermediate bill-tip is linked to thin-shelled prey, like clams and tellins, often in conjunction with polychaete worms. The width of the bill-tip varies between 0.6 and 2.8 mm, depending on the customary method used to open prey. Birds which open bivalves by stabbing the bill rapidly into a gaping shell or by forcing entry by prising the two valves apart have bill-tips of an average width of 1.2 mm. Those which break into bivalves by hammering a hole in the shell have bill-tips with an average width of 2.1 mm (Swennen *et al.* 1983; Hulscher and Ens 1991; Durell *et al.* 1993).

Oystercatchers often change their diet seasonally or periodically according to the circumstances (Sutherland *et al.*, Chapter 3, this volume). When this happens, both the form of the bill-tip and the length of

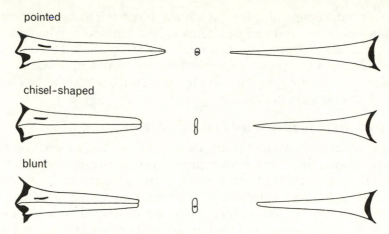

Fig. 1.3 Lateral, frontal and dorsal views of the three main bill-tip shapes of Oyster-catchers (Swennen *et al.* 1983).

the bill change according to the new technique used. There is a transition period of 10 to 20 days during which an interplay between the growth and abrasion of the rhamphotheca determines the resulting shape. At 0.44 mm (range 0.2–0.9 mm) per day, the mean growth rate of the rhamphotheca in Oystercatchers is high compared with other bird species (Hulscher 1985) so that a change in bill form can be quite rapid. For example, inland breeders have relatively long and pointed bills and predominantly feed on soft-bodied subterranean prey. When they return to the coast to winter, they switch to hard-shelled prey. Their bills then become blunt and the bill-length shortens (1.5–3mm) gradually (Suther-land *et al.*, Chapter 3, this volume). Birds with bills greater than 77.5 mm long, which are mostly females, abrade their bill-tips more (2.82 ± 3.03 SD, range 2.1–9.5 mm, *n* = 28) than do birds with bills less than 77.5 mm long (1.54 ± 1.98, range 3.5–5.0 mm, *n* = 30), which are mostly males. The greater rate of abrasion in the long-billed females can be attributed to their bill-tips being thinner than in males (Hulscher and Ens 1992). When the birds return inland the following spring, the reduced abrasion of the bill-tip allows the bill to grow longer again and the bill-tip to attain its previous pointed shape. The 1.5–3 mm longer and pointed bill may then be advantageous for feeding on deeply-buried sub-terranean prey.

Detecting and locating prey

Oystercatchers have no difficulty in detecting by sight those prey that are exposed on the substrate surface. Prey that are buried in soft sediments, and so not directly visible, usually maintain some contact with the sur-face for feeding and respiration and for expelling waste products. They thus produce visible traces which the birds might use to locate them by sight (Swennen and van der Baan 1959). However, when no visual cues are available, Oystercatchers must then rely on other senses to detect prey beneath the surface. In principle, the senses involved may be touch, taste, smell or hearing but, so far, only touch-detection has been studied in Oystercatchers in detail.

Motor patterns of head and bill in detecting prey

Four different movements are used to detect buried prey by touch or to locate individual prey whose presence has already been detected from visual cues: pecking, boring, sewing and ploughing. Each is adapted to deal with the particular life-style of the prey in the sediment and to the environmental circumstances. They are described in some detail to enable the mechanisms by which prey are detected and located to be better understood. A peck comprises a single movement of the head, in which the almost fully-closed bill penetrates the substrate at an angle of about 70°–85°, up to a depth of 0–5 cm. Between pecks, the head and bill slant down at an angle of 45°–60°. Pecks are directed in front of, or slightly to either side of, the bird. A bore consists of a series of rapidly executed, vertical up-and-down movements of the slightly opened bill on the same spot. The bill gradually penetrates the substrate up to its base. Sewing (or multiple pecking) is composed of distinct bouts of vertical probes made while the bird moves slowly forwards. The bill is opened 1–2 mm and moved straight up and down at a rate of 3–7 times per second. The bill penetrates the substrate to a depth of between 10 mm and fully up to its base. Between sewing bouts the bill is wholly retracted from the substrate and reinserted after the bird has moved forward a few centimetres, or much further. A sewing bout lasts between 1 and 5 secs. Ploughing is a variation of sewing. The bill is inserted obliquely, rather than vertically, to a depth of 1–2 cm and moved through the mud in straight lines. Shallow, vibrating movements are made with head and bill as the bird walks forward for several metres. The bill is then entirely withdrawn, inserted again and a new furrow started.

Detection by sight

Although Oystercatchers direct pecks towards any kind of irregularity on the surface, many of the buried prey provide clear cues of their presence beneath. For example, the deposit-feeding bivalves *Macoma* (Hulscher 1982) and *Scrobicularia* (Hughes 1969) make star-like tracts on the surface with their inhalent siphons. That Oystercatchers use such marks to locate prey has been demonstrated for *Macoma* (Hulscher 1982). Bouts of sewing on places with distinct surface marks resulted in a higher rate of successful prey retrieval than in the absence of such marks (Hulscher 1982). A captive bird was allowed to feed on a small area seeded only with *Macoma* where surface marks were either present or had been

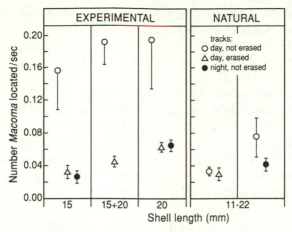

Fig. 1.4 The rate at which a captive Oystercatcher located prey by sewing when feeding on experimental populations of *Macoma*. In the left-hand three boxes, a one m^2 area was seeded with 50 *Macoma*, with the prey sizes varying between boxes. From left to right, they were; all small (15 mm), 25 small and 25 large (20 mm) and all large *Macoma*. The two right-hand boxes refer to a 1 m^2 (left box) or 2 m^2 (right box) respectively that contained natural *Macoma* populations. *Macoma* tracks were either present or absent. In the latter case during the day (as indicated), surface cues were simply erased by smoothing the substrate surface; at night the cues could not be seen. The means (± SE) of 30 minute-long observation periods are shown (Hulscher 1982).

erased by smoothing the mud. In all cases, the success rate of sewing was higher in daylight in areas where surface cues had not been erased and was lower at night, suggesting that the presence of cues helped the birds to locate *Macoma* (Fig. 1.4). Experiments conducted on a natural mudflat community, which included *Macoma*, also found that the success rate was higher where cues had not been erased, although the difference this time was not significant; perhaps the surface marks made by other prey species interfered with the recognition and capture of *Macoma*. Although the siphons of *Macoma* and *Scrobicularia* can be extended a long way from the burrow entrance when the substrate is very wet, for unknown reasons, Oystercatchers do not react to them, even though the siphons are moving and clearly visible to the human eye (Wanink and Zwarts 1985; J. B. Hulscher unpublished information). This is surprising as many other shorebirds do crop these siphons (F. Moreira, personal communication) and Oystercatchers eating another bivalve, *Mya*, specialize exclusively on them (personal observation); in fact, doing so is quite lucrative, since 40–50 % of the total meat content of the bivalves is located in the siphons (Zwarts and Wanink 1984, 1989).

Other prey species also make surface marks that Oystercatchers might use. Under water, the little round openings of the inhalent and exhalent siphons of the filter-feeding *Cerastoderma* are visible, level with the mud surface. *Mya* shows a single round entrance of the siphon burrow under water and often on the exposed mudflat. *Arenicola* produces characteristic toothpaste-shaped casts of faeces on the surface, while *Nereis* makes star-shaped feeding marks, resembling those of *Macoma* (Hulscher 1964c). Observations show that Oystercatchers locate *Cerastoderma*, *Mya*, and *Arenicola* in daylight by these surface marks, while those feeding on *Nereis* probably look for burrow entrances (Hoekstra 1988). The worm can be detected directly when its head extends from the burrow as it feeds on the surface. It can also be located indirectly when it produces a water current at the burrow entrance whilst feeding within its burrow (Esselink and Zwarts 1989). When *Arenicola* defecates, its tail remains at the surface for two seconds. Oystercatchers rush towards these casts as they are being made and catch the lugworm by its rear end, as has also been observed in Sooty Oystercatchers (*H. fuliginosus*) by Wells (1966). Bill marks left in the sand suggest that Oystercatchers equally often direct pecks at the funnel-like entrance of an *Arenicola* burrow (Hulscher 1964c), although with unknown success. Crabs under a film of sand in water may betray themselves by blowing bubbles when short of oxygen.

Having detected the presence of buried prey from surface marks, Oystercatchers either peck or sew in order to locate and retrieve them. When is either technique used? Prey that lie near the sediment surface, such as *Cerastoderma*, are located, as well as attacked, by pecking (Davidson 1967; Hulscher 1976c). Once a cockle has been detected, Oystercatchers probably gain precise information about its position in the sediment. Just before attacking, the birds reorientate either their head or their whole body to bring their bill into line with the persistent gape between the two valves (Hulscher 1976c). In contrast, sewing is used for prey that are buried at some depth. Pecking and sewing are combined in localizing terrestrial prey, such as earthworms, leatherjackets and caterpillars, which live in tubes with one or two openings to the surface. Oystercatchers seem to confine their searching to definite spots, chosen perhaps on the basis of cues visible on the surface (Safriel 1967; Heppleston 1971a; Veenstra 1977).

Localization by touch

Touch-hunting is directed at prey that are hidden in the substrate or, like some mussels (Ens 1982), covered by weed or mud. Oystercatchers are probably able to touch-locate such prey in darkness and in day-time in the absence of visual cues, by means of the concentrations of Herbst corpuscles in the bill-tip. On the assumption that the bill-tip contacts the prey by chance, Hulscher (1976c, 1982) developed a touch-detection model for Oystercatchers feeding on *Cerastoderma* and *Macoma*. The model was tested with a captive bird feeding on artificial and natural

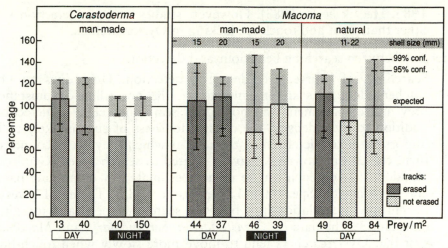

Fig. 1.5 The percentage of the prey detected by a captive Oystercatcher, shown by the columns, compared with the number expected to be found by touch by a randomly searching bird, expressed as 100%. The expected values are based on the observed times spent sewing by a captive Oystercatcher feeding on man-made populations of *Cerastoderma* and on man-made and natural populations of *Macoma*. The surface tracks of the prey were either erased or not erased, as indicated. Vertical bars and intermediate shading indicate 95 and 99% confidence limits of expected means (Hulscher 1976c).

populations of *Cerastoderma* and *Macoma*, both by day and at night (Fig. 1.5). All results with *Macoma* and those in daylight with *Cerastoderma* conformed to the predictions of the model. The proportion of prey taken coincided well with the numbers that were expected to be encountered by chance, given the cross-sectional area of the prey and of the tip of the bill, the depth and density of the prey in the substrate and the length of the bird's bill. However, a significantly smaller proportion of the cockles were taken at night than was expected by the model. Hulscher argued that, when hunting by touch on the very high densities of prey used in this experiment, the birds rejected a proportion of those encountered as being unsuitable. As Oystercatchers probably decide within a fraction of a second whether or not to attack any one prey they encounter, such rapid encounters would be missed by the human observer and would explain the discrepancy between the model predictions and these night observations. The model was also tested successfully in Oystercatchers hunting on *Scrobicularia* by touch (Wanink and Zwarts 1985), and is discussed further for a range of other prey species by Zwarts *et al.* (Chapter 2, this volume).

Localization by other senses

Prey discrimination by taste is widespread in birds (Berkhoudt 1985) and has also been found in four calidrine sandpiper species (Gerritsen *et al.*

1983; Heezik *et al.* 1983). However, studies on prey detection systems other than sight and touch are lacking in Oystercatchers.

Interplay between searching behaviour and detection

Searching, of course, precedes prey detection. How does an Oyster-catcher adapt its searching method to the prevailing circumstances of prey and environment and how does searching relate to detection? In addition to their density, the conspicuousness of cues is likely to be an important factor determining the rate at which prey are encountered. In turn, cue conspicuousness may be related to search speed (Gendron and Staddon 1983), which can be reliably estimated by combining pace rate and pace length (Speakman 1984; Grolle 1987). Table 1.1 gives some insight into the relationships between searching speed and foraging circumstances. The data on *Cerastoderma, Scrobicularia* and *Macoma* suggest that Oystercatchers tend to move more slowly when touch-hunting than when hunting visually. This is best illustrated by Oystercatchers feeding on the same area of *Cerastoderma* during the night as during the day. Oystercatchers always hunt cockles at night by touch by moving the bill at random through the substrate and presumably encountering every cockle. The number of cockles encountered this way is likely to be much larger than the number of those providing a visual cue in daytime. Thus, when feeding on this prey, the Oystercatcher encounters a higher density of cues at night and thus covers the ground more slowly than during the day. The negative relationship between cue density and the rate of moving may also apply when Oystercatchers search only for visual cues. Among mussel-feeders, for example, stabbers search faster than hammer-ers, apparently because the density of vulnerable prey for stabbing birds is much lower than for those that hammer. In general, *Nereis* cues are scarce and Oystercatchers taking these worms walk relatively quickly

Table 1.1 Moving speed (cm s^{-1}) of Oystercatchers searching for various prey types, either visually or by touch

Prey	Visual	Touch	Reference
Cerastoderma	22.3 (day)	16.7 (night)	Hulscher 1976c
Scrobicularia		14.7	Boates 1988
Macoma		10.1	Bunskoeke *et al.* 1996
Mytilus stabbing	25.0		Speakman 1984
	17.9		Boates 1988
	17.0		Cayford & Goss-Custard 1990
hammering:			
ventrally	8.5		Meire & Ervynck 1986
	8.3		Cayford & Goss-Custard 1990
dorsally	7.3		Cayford & Goss-Custard 1990
Nereis	28.4		Boates 1988
	21.3		Bunskoeke *et al.* 1996
Littorina	16.1		Boates 1988

and can increase their intake rate dramatically by speeding up their pacing rate (Hoekstra 1988). *Littorina*, on the other hand, occurs in high densities and Oystercatchers hunting them visually do so at a very low searching speed (Boates 1988).

Conspicuousness and cue-frequency probably also determine which detection method is chosen when both the visual and touch options are available. This was suggested by the experiments testing the random-encounter touch model. On areas of high cockle density, Oystercatchers always pecked in daytime. However, when the density fell below 40 m^{-2}, Oystercatchers alternatively fed by pecking and by sewing or ploughing. It seems likely that Oystercatchers hunting visually search for cockles that are gaping. As most cockles close their shells firmly to avoid dessication, gaping cockles are probably scarce at low cockle densities. The number of visual cues encountered in daylight might therefore drop below a critical minimum so that a situation comparable to the one occurring at night occurs. In darkness, Oystercatchers always plough, irrespective of cockle density, and then detect a higher number of cockles per unit time of sewing than when they are pecking during the day at the same densities. Hence, sewing seems to be more successful than pecking, so birds may turn to this method during the day when cockles are scarce and thus difficult to find visually. But if sewing is such an effective technique for locating cockles, why do birds not always use this method during the daylight, even when cockles are numerous? One possibility is that ploughing is energetically costly. Touching a cockle may also cause some of them to close, thus increasing the difficulty and risk of opening them. Indeed, the mean handling time of successfully opened cockles within density classes is significantly shorter in daylight than in darkness, implying easier access during the day (Fig. 1.6).

How Oystercatchers prepare and eat prey

As Dewar (1908, 1913) first reported, Oystercatchers use two methods for opening the same bivalve species: stabbing and hammering. Variations of both the stabbing and hammering techniques are also used to attack gastropods and shore crabs. This section asks: what is stabbing and hammering? Why and when do Oystercatchers use either method and what determines the choice?

Stabbing

The technique for stabbing mussels will be described in some detail. Essentially the same procedure is used for stabbing into other bivalves and crabs, the details having been investigated by combining field observations (Dewar 1908; Tinbergen and Norton-Griffiths 1964) with observations on captive birds (Hulscher 1982). Mussels clump together by byssus threads that emerge from the ventral surface of the valves. The shells

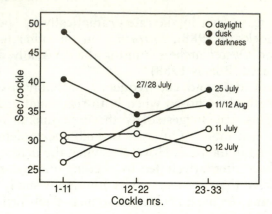

Fig. 1.6 Comparison of the handling times of cockles (range 25–36 mm, mean 31.4 mm) opened and eaten successfully in daylight and in darkness by a captive Oystercatcher feeding on a natural cockle bed with a density of cockles of 150 m^{-2}. Shown here are the mean handling times per group of eleven cockles in the sequence in which they were eaten over five days of observation (Hulscher 1976c).

gape slightly under water and are closed only loosely when the shells are still moist. As soon as such mussels are touched, the valves are pulled together very tightly. Oystercatchers feeding by stabbing move over moist mussel beds at a steady pace, tapping mussels as they go, mainly on the ventral margins of the two valves. In this way, they probably test whether shells are gaping. When the mussel is gaping widely, the bird makes a hard and rapid downward jab between the valves in the dorsal posterior region. The stab is directed at the posterior adductor muscle, which is then severed (Norton-Griffiths 1967). When the shell gapes only slightly, the bird drives the tip of its bill between the two valves, preferably at the ventral margin. Having inserted its bill, the Oystercatcher then brings it to a vertical position and, provided the shell gives way, rotates the shell so that the ventral margin faces upwards. At the same time, the bill is pushed vigorously downwards, thus penetrating deeply into the mussel. The bill usually enters the shell in the posterior half of the cleft, with the posterior end of the mussel turned towards the bird. Although the bird's upper mandible then becomes pinched between the two valves, the lower one is able to move freely in the shell, because the mussel gapes increasingly towards the rear. The sharp edges of the lower mandible are then used to sever the posterior adductor muscle (Hulscher 1988). The anterior adductor is then cut too before the bill is rotated through 90° and opened with force, prizing the two valves apart. Thus the idea, based on Dewar (1908) and since quoted many times in the literature, that a mussel is opened by turning the inserted bill and rupturing the adductor muscle has proved to be wrong. Were a bird to do this, it would run the risk of dislocating its mandibles (Tinbergen and

Norton-Griffiths 1964). Once opened, the flesh may be cleaned out from the shell *in situ* or carried first to another spot. The flesh is loosened from the shell by scissoring the mantle attachments all around before it is swallowed in one or two gulps. The valves are not damaged. Individual birds often use one or more specific sites for cleaning shells for several days on end, forming piles with several tens of empty shells.

Other prey are stabbed in a similar way to mussels. Cockles are stabbed at the persistent gape between the valves. Deeply-buried prey, such as *Macoma, Scrobicularia,* and *Mya,* can often be opened *in situ,* since they gape slightly (Hulscher 1982). Shore crabs are dealt with by stabbing Oystercatchers in the same way as bivalves (Hulscher 1964a; Tinbergen and Norton-Griffiths 1964; Baker 1974c). Once located by an Oystercatcher, a crab tries to defend itself by standing upright and stretching its pincers widely apart, ready for striking. The bird positions its body beside the crab, brings its bill in line with the pincers and stabs vertically downwards through the mouth. In doing so, the bill almost certainly cuts the supra-oesophageal ganglion, since the crab suddenly stretches out its appendages and becomes immobile. The Oystercatcher tears off the carapace, which falls to the ground with its inner side uppermost. The bird then removes the soft parts from the thorax, but usually leaves the gills.

Hammering

This technique is again first described for birds eating mussels, with ventral and dorsal hammering being distinguished. Ventral hammerers clasp the mussel between the mandibles, tear it from the mussel bed by pulling upwards until the byssus threads break and then carry it to a firm place. Accumulations of such empty shells can be used as anvils, especially in sandy areas (Davidson 1967; Heppleston 1971a). The mussel is orientated so that its flat, ventral surface is turned upwards and the hammer blows can be directed at the mid-ventral region. When hammering, a bird holds its bill vertically downwards with a stiff neck. Its body pivots from the pelvis and its toes grip tightly into the sand to maintain footing. The blows from the bill-tip cause a semicircular chip of the shell to be fractured from the ventral margin of one valve (Fig. 1.7). The Oystercatcher then inserts its bill, cuts the posterior and anterior adductor muscles and prizes the valves apart. The ventral margin is the most vulnerable part of the mussel, as demonstrated by Norton-Griffiths (1967). In dorsal hammering, the bird either attacks the mussel *in situ* from above, breaking the shell at the dorsal posterior margin, or at the side. In the latter case, it thus attacks the shell at the point at which the posterior adductor is attached to the shell, and where erosion, and so shell-thinning, often take place (Durell and Goss-Custard 1984).

Dewar (1913) suggested that the opposing valves of a mussel, which is not firmly closed, may rotate in the mesial plane when one valve receives

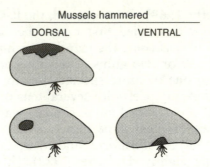

Fig. 1.7 Fractures (dark shading) of mussels hammered open on the ventral and dorsal sides by Oystercatchers (Ens 1982).

a blow from an Oystercatcher's bill, so leaving a cleft between them. This would allow the bill to enter without breaking the shell. Norton-Griffiths (1967) indeed showed that some mussels opened their valves 1 to 2 mm after three or four blows. The gaping reaction may also occur after just one firm stroke, especially in small mussels. Similarly, Hulscher (1982) observed a rotation of the valves in *Macoma* being tapped by Oystercatchers after they had been pulled out of the substrate. This might mean that at least some of the bivalves which are successfully attacked by dorsal and by ventral hammerers are actually opened without damaging the shell.

Other species of prey can also be opened by hammering. Drinnan (1957) and Baker (1974c) describe aspects of the opening of cockles by hammering and Hulscher (1982) considers Oystercatchers opening *Macoma*. When bivalves such as *Macoma*, *Scrobicularia* and *Mya* are not penetrated underground but are lifted to the surface, they are always opened by hammering. Limpets living in pools and on moist rocks do not draw their shell tightly over the foot. They leave a small gap between the shell and the substrate (Safriel 1967). Oystercatchers stalk these prey and rapidly insert the bill under the foot so the shell topples over. If this fails, fierce hammer blows are used to dislodge the limpet; thus, limpets and dogwhelks are dislodged by techniques that resemble both stabbing and hammering. There is as yet no consensus as to whether specific regions of a limpet are more vulnerable for attack than others and whether Oystercatchers adjust their blows to the size of the limpet (Safriel 1967; Feare 1971; Hockey 1981a). The soft contents are loosened with the tip of the bill, shaken out of the shell and swallowed. Intact opercula of dogwhelks and periwinkles are sometimes found in large numbers in the stomachs of Oystercatchers (Dare and Mercer 1973). More details on the way in which limpets and/or dogwhelks are handled are given by Dewar (1910, 1913), Safriel (1967), Feare (1971), Baker (1974c), Hartwick (1976) and Hockey (1981a).

Which prey are taken by the different techniques?

Individuals of one prey species are not equally likely to be taken by Oystercatchers employing different feeding techniques because the anti-predator defences of some prey are more vulnerable to some feeding methods than to others. For example, Speakman (1984) found that mussels opened by stabbers had, on average, smaller posterior muscle scars than the mussels of the same size that were on offer on the beds. If the size of the scar inside the shell reflects the size of the adductor muscle, it would appear that the birds only force entry into mussels with the weakest adductors. Similarly, ventral and dorsal hammering Oystercatchers preferentially take thin-shelled mussels (Durell and Goss-Custard 1984; Meire and Ervynck 1986; Sutherland and Ens 1987; Alting 1990; Cayford and Goss-Custard 1990; Meire 1993). Ventral hammerers select against shells that are thick on the ventral side, irrespective of the thickness of the dorsal shell, while dorsal hammerers do the reverse (Sutherland *et al.*, Chapter 3, this volume). Alting (1990) argued that selection against thick shells by ventral hammerers was by trial and error, as shells attacked unsuccessfully were equally as thick as those on offer. In contrast, Oystercatchers opening mussels by stabbing do not discriminate the thickness of shells (Durell and Goss-Custard 1984). Mussels also vary in the extent to which they are covered by barnacles and the degree to which their shells are eroded, both of which might indicate to hammering birds the thickness of the shell. Ventral hammerers appear to select for smooth, barnacle-free and non-eroded shells that are often brown rather than black, characteristics that are thought to be associated with rapid growth and a relatively thin shell (Ens 1982; Durell and Goss-Custard 1984). Dorsal hammerers strongly select for highly eroded shells and, perhaps, weakly for those with many barnacles; the degree of erosion may again be used as a clue to the dorsal shell being thin.

When and where are the different techniques used?

Individuals using the various feeding techniques are not mixed at random over the feeding grounds, either with respect to time during the exposure period or with respect to space. Rather, individuals feed at times and in places that favour their particular feeding technique. For example, on the Exe estuary, ventral hammerers begin feeding relatively late in the exposure period (J. D. Goss-Custard, personal communication). This might be explained by the observation that the intake rates of captive ventral hammerers eating cockles increase through the exposure period (Swennen *et al.* 1989), perhaps because the increasing resistance of the drying sediment makes it easier for the birds to open their prey. In contrast, stabbing birds start feeding at the beginning of the exposure period, when many of them also wade in the shallow water. As stabbers open gaping and lightly-closed mussels and cockles, their early arrival on the feeding grounds has been associated with a high availability of prey when the feeding area is wet (Zwarts and Drent 1981; Ens 1982; Sutherland 1982c; Goss-Custard *et al.* 1984). As the tide recedes and, where no pools remain, the substrate dries out, the bivalves close their valves and a decline occurs in the numbers of stabbing birds eating mussels (Zwarts and Drent 1981) and cockles (Swennen *et al.* 1983). Many stabbing birds then turn to forcing apart the valves of the more firmly closed, dry mussels which may allow them to maintain their intake rates over the low water period.

Spatial variations in the nature of the substrate are also often believed to partly underlie differences in the distribution of hammerers and stabbers over a mussel bed at any one time during the exposure period. In general, dorsal hammerers, which open the prey *in situ,* attack mussels on a firm substrate, although the support provided by the attachment to other mussels in a clump prevents the prey from being driven far into the underground in slightly soft areas. Ventral hammerers also congregate where the substrate is firm and there is a ready supply of anvils (Goss-Custard *et al.* 1982a). They may also prefer places where the attachment of mussels to the surface is weakest (Norton-Griffiths 1967). In contrast, birds that stab mussels often concentrate where the substrate is wet, perhaps because the prey there are more likely to gape; a soft substrate is therefore no hindrance to their feeding technique.

Changes in prey choice based on changes in the activity of the prey

Although individual birds may specialize for long periods on a single prey species, changes in diet do occur both in the short term, within one tidal cycle, and in the longer term, both within and between seasons. Why?

Changes during a tidal cycle

On Schiermonnikoog, *Nereis* is predominantly taken by breeding birds during the middle phase of the exposure period, while *Macoma* is taken at the beginning and end (Fig. 1.8) (de Vlas *et al.* 1996). The fraction of the diet consisting of a variety of other prey species remains more or less constant throughout. More worms are visible at the surface, and for longer, during the middle part of the exposure period (Esselink and Zwarts 1989). In contrast, *Macoma* becomes less active as the tide recedes, reaching a minimum around dead low water (Hummel 1985). When feeding on the receding tide, the shell of *Macoma* gapes open, allowing the siphons to be extruded. This makes *Macoma* both vulnerable and profitable to Oystercatchers because they can be easily and rapidly opened *in situ.* The profitability to Oystercatchers of handling *Macoma* decreases over the low tide period because an increasing pro-

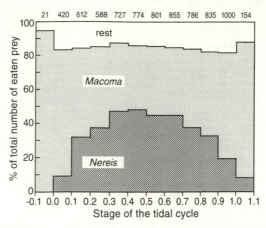

Fig. 1.8 Tidal trends in prey choice by adult Oystercatchers on Schiermonnikoog. The diet is expressed as the percentage of prey captures consisting of each prey type in scan observations. The stage in the exposure period is represented along a scale in which the time at which the area first exposed on the receding tide is represented by 0 and the time it covered on the advancing tide by 1 (de Vlas *et al.* 1996). Numbers on top refer to the total number of prey captures.

portion has to be lifted to the mud surface before they can be opened (E. J. Bunskoeke, personal communication) and this takes more time than opening a *Macoma in situ* (Hulscher *et al.* 1996). These data suggest that the change in diet during the course of the tidal cycle can be attributed to changes in the behaviour of the prey species over the exposure period which affect their relative vulnerability and profitability to Oystercatchers.

Changes within the breeding season

On Schiermonnikoog, *Macoma* (Fig. 1.9) predominates in the diet during the first half of the breeding season, until mid-June, whereas *Nereis* predominates in the second half (Bunskoeke *et al.* 1996; Hulscher *et al.* 1996). *Macoma* activity decreases notably from mid-June onwards, growth ceases (Beukema and de Bruin 1977) and burrow depth increases (Reading and McGrorty 1978; Zwarts and Wanink 1993). There is also a considerable decrease in the meat content over the same period (Zwarts 1991; Bunskoeke *et al.* 1996). In contrast, the mean burrow depth of *Nereis* is then at a minimum, coinciding with a peak in its feeding activity and hence in its catchability (Esselink and Zwarts 1989). The decline in the vulnerability and profitability of *Macoma* over the summer, coinciding with an increase in the vulnerability of *Nereis*, probably explains the within-season diet shift observed in these breeding Oystercatchers.

Changes between summer and winter and between years

Many prey species in the west European estuaries change their behaviour during the autumn in ways that affect their detectability and accessibility

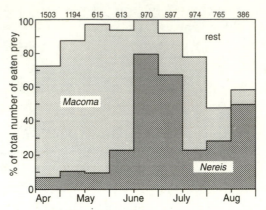

Fig. 1.9 Changes in the diet of adult Oystercatchers on Schiermonnikoog over half-monthly periods during three breeding seasons (1985–1987). The diet is expressed as the percentage of prey captures in scan observations consisting of each prey type (Bunskoeke *et al.* 1996). Numbers on top refer to the total number of prey captures.

to Oystercatchers. Crabs and shrimps migrate to deeper waters in winter and so are only eaten by Oystercatchers during summer (Hulscher 1964a; Dare and Mercer 1973). *Nereis, Macoma* and *Scrobicularia* in the Dutch Wadden Sea burrow deeper during the winter (Esselink and Zwarts 1989; Zwarts and Wanink 1989). *Mya* does the same, although to a lesser extent; as a result, this clam is proportionally more important to Oystercatchers in winter than in summer. However, changes in diet associated with changes in prey behaviour cannot always be so easily interpreted. Adult Oystercatchers on the Exe estuary showed a complete seasonal change in the diet by taking only *Nereis* during April–September and only *Scrobicularia* in December–March, even though the profitability of *Scrobicularia* remained the same throughout the study period. Indeed, the birds could have achieved much higher intake rates in summer by continuing to take *Scrobicularia* but, for unknown reasons, they did not do so (Boates and Goss-Custard 1989). Similarly, in the Wadden Sea, *Nereis* is important for breeders in both Denmark (Lind 1965), and at Schiermonnikoog (Hulscher and Ens 1992), yet non-breeders do not eat it (Blomert *et al.* 1983). Again, the reason for this is not known.

Some of the most important prey of Oystercatchers, notably *Mytilus* and *Cerastoderma*, remain accessible at or near the substrate surface throughout the year and are always important in the diet, when present. However, 'crashes' in cockle and mussel stocks do occur occasionally, especially when spat fails after a series of mild winters. The birds then turn to *Macoma*, suggesting that this prey species could previously have been exploited (Dare and Mercer 1973; Desprez *et al.* 1992), although with unknown consequences for Oystercatcher fitness.

Risks associated with feeding

Selecting particular bivalves may not only affect gross intake rate but may also, in different ways, influence the magnitude of a number of risks associated with feeding. One such risk is that of becoming parasitized. Birds, including the Oystercatcher, are often hosts in the life cycle of a large variety of bivalve parasites (Borgsteede *et al.* 1988). Although the effects on Oystercatchers of most of these parasites are not known, it may be advantageous to reduce the parasitic load to a minimum. Presumably because of this, Oystercatchers have been seen refusing *Macoma* parasitized by the trematode *Parvatrema affinis* (Hulscher 1982). A second risk arises when Oystercatchers handle large shells. For stabbing birds, getting the bill stuck between tightly clenched valves after failing to cut the adductors sometimes means death (Hulscher 1988). Failure to use the bill nimbly may also damage the tip, and thus the efficiency with which it can be used. The upper and lower mandibles of Oystercatchers caught alive for ringing may differ in length by up to 7 mm (J. B. Hulscher, unpublished information). This variation probably arises while birds are stabbing or hammering large bivalves, when relatively long pieces of a mandible may suddenly break off (Hulscher 1985), temporarily hampering efficient feeding (Sutherland *et al.*, Chapter 3, this volume). Such handicapped birds run a much higher risk of dying during bad winter weather (Swennen and Duiven 1983). Indeed, it may be the increased risk of damage to their longer and thinner bills when attacking robust bivalves that causes many females to feed on soft-bodied prey, such as worms, and the thin-shelled buried bivalves, such as *Scrobicularia* (Hulscher and Ens 1992; Durell *et al.* 1993). The risk of bill-damage in this species may prove to be a highly significant factor in both its behavioural and population ecology.

Despite the risk associated with eating large bivalves, Oystercatchers seldom take small hard-shelled prey. When they do, they eat them whole, with the shell intact. European Oystercatchers, swallowing whole *Macoma*, seem to be individual birds (J. B. Hulscher, unpublished information). Stomachs have contained whole mussels 5–12 mm long (Dewar 1908), cockles 8–9 mm long (Dare and Mercer 1973) and *Macoma* 5–6 mm long (Triplet 1989a), while *Littorina* shell fragments have been found in faeces (J. B. Hulscher, unpublished information). The American (Webster 1941a) and the African Black Oystercatchers (Hockey 1981a) eat intact small (< 12 mm) limpets while the Australian Pied Oystercatcher swallows whole bivalves smaller than 20 mm in length (Evans 1975). Probably it does not pay to open and clean small bivalve prey; indeed, it may even be physically impossible.

Night and day feeding

Coastal Oystercatchers feed extensively at night. However, it is not yet clear whether they are equally capable of collecting food at night as in day-time or whether they spend just as much time in feeding in the day as in darkness. The available data (Table 1.2) are far from sufficient. Field observations restricted to the first or last few minutes of the exposure period, when the birds forage near enough to land-based hides to be clearly seen (Sutherland 1982a; Goss-Custard and Durell 1987c), suggested that the intake rate at night is under half that during the day. However, other studies that cover the whole exposure period, either by using indirect methods in the field (Scheiffarth 1989; Kersten and Visser 1996b) or by performing experiments under controlled (Hulscher 1974), almost completely natural (Leopold *et al.* 1989) circumstances, show little difference between the rates of food intake during the day-time and at night.

As Zwarts *et al.* (Chapter 2, this volume) discuss, the rate at which the gut can efficiently process food, along with its limited capacity for storage, may set a ceiling to the total amount that can be processed in a single day-time or night-time exposure period that is below that which is required to maintain the bird over 24 hours. But while Oystercatchers must therefore feed on successive tides to obtain their daily energy requirements, it is not clear whether feeding is spread equally between night and day. During winter, food requirements are higher and there are less daylight hours available than in summer, so night feeding might be expected to be extensive at that time of year. In summer, by contrast, daylight feeding possibilities may not be fully exploited; thus, the more birds feed on one daylight exposure period, the less they feed during the next (Hulscher 1982). While this might imply that night feeding occurs at a constant rate, and at too low a level to make up for day-time intake rate deficiencies, this does not necessarily follow; the birds may simply vary their day-time and night-time feeding in parallel. It is therefore still unknown whether intertidal Oystercatchers prefer day-time to night-time feeding or vice versa or, indeed, whether there is any preference at all.

One factor that would be expected to alter the balance of advantage between night and day feeding is that the accessibility of prey is also quite likely to change between night and day. In some cases, the birds may even change their diet. The African Black Oystercatcher (*H. moquini*) has twice as many *Patella granularis*, a nocturnal forager, in its night-time diet as in its day-time diet (Hockey and Underhill 1984). Incubating European Oystercatchers feed to some extent during the night on *Nereis* (Kersten and Visser 1996b), apparently exploiting the nocturnal tendency of these worms to emerge from the burrow to feed (Vader 1964; Dugan 1981). The different feeding techniques used by Oystercatchers at night may also imply a difference between day-time and

Table 1.2 Review of the food intake of Oystercatchers feeding under comparable conditions during the day and during the night

Period	Prey	Parameter measured	Remark*	Day	Night	Ratio Night/day	Reference
Oct - Dec	Cerastoderma	Gut contents, ml(F)/bird	1	50	22.4	0.50	Drinnan 1958b
Winter	Mytilus	Number of mussels/5 min	2	0.789	0.382	0.48	Goss-Custard & Durell 1987c
Winter	Cerastoderma	Number of cockles/min	3	1	0.85	0.85	Davidson 1967
Winter	Cerastoderma	Number of cockles/10 mins	4	6.2	3.0	0.38	Sutherland 1982a,c
May	Mytilus	Number of mussels/10 mins	5	3.4	1.4	0.42	Zwarts & Drent 1981
Feb - March	Anadara	Number of cockles/hr	6	1.608	1.580	0.98	Swennen 1990
June 1987 (breeding)	Macoma and Nereis	g (FW)/min	7	0.93	1.09	1.17	Kersten & Visser 1996b
1989				0.66	0.53	0.80	
June (breeding)		% of low water period active, probably mostly feeding	8	22.7	19.1	0.84	Scheiffarth 1989
December	Mytilus	ml (F)/hr over total daylight and dark hours	9	16.2	8.1	0.50	Drinnan 1958b
December	Mytilus	g (DW)/hr over 12 hours daylight and 12 hours darkness	9	1.63	0.93	0.57	Heppleston 1971a
July	Mytilus	ml (F)/hr over 18 hours daylight and 6 hours darkness	9	6.40	5.44	0.85	Hulscher 1974
July	Mytilus	ml (F)/hr over 5-7.5 hours daylight and 5-7.5 hours darkness	9	10.80	15.35	1.42	Hulscher 1974
Summer	Cerastoderma	g (AFDW)/low water period	10	21.4	19.6	0.92	Leopold et al. 1989
	different	g (DW)/10 mins over 1360 minutes of daylight and 1380 minutes of darkness	11	0.57	0.56	1.00	Hulscher 1976c

F = fresh weight; FW = fresh weight; DW = dry weight; AFDW = ash-free dry weight.
*1. Netted birds 2. Observations during first hour of exposure period 3. Moonlit nights 4. Observations during last ten minutes of exposure period 5. During the night additional prey (not identified) were taken 6. Moonlit nights 7. Observations on feeding and weight increase (scale measurements) 8. Radio-tracked birds 9. Laboratory, food available 24 hours per day 10. Laboratory, fixed tidal schedule 11. Laboratory, semi-natural beds, 2-6 hours/low water period.

night-time feeding conditions. For example, there is a switch from local-ization by sight to touch, with consequences for the rate at which prey are found and handled. Whereas *Macoma* are found at night by touch at the same rate as they are found by sight during the day, *Cerastoderma* are found at an even higher rate at night than in day-time (Hulscher 1976c). Unless extra costs or risks associated with this method of feeding eliminate any benefits arising from the increased intake rate, this might imply that feeding at night is actually more profitable than day-time feeding. As in other shorebirds, much remains to be discovered about relationships between night-time and day-time feeding of Oystercatchers across diets, seasons and latitudes (McNeil *et al.* 1992).

Summary

Oystercatchers are unique amongst waders in that they are capable of opening the large and well-protected bivalve molluscs which contain so much of the intertidal macrobenthic biomass in European estuaries. Opening such a mollusc requires skill and force. Starting at the sensitive tip of the bill and moving along the massive bill, through the heavy skull, strong neck and body to its sturdy legs and robust toes, the anatomy and morphology of the Oystercatcher is well adapted to this role. But Oyster-catchers also take soft-bodied prey, such as worms and insect larvae. Surface-dwelling molluscs, such as mussels, are detected by sight and their vulnerability to attack tested by a peck. A mussel is vulnerable either because it has a relatively thin shell, through which the bird can hammer a hole, or because it is gaping slightly and/or has a weak adductor muscle and so can be stabbed or forced open. Most of the other prey species are buried in the substrate. They maintain contact with the sur-face to breath, feed and expel waste products. This provides cues for visually-hunting Oystercatchers to locate them. When such cues are not available, Oystercatchers are equally adept at localizing buried prey by touch. Depending on the nature of the substrate and behaviour of the prey, they do this using the random-contact techniques of sewing and ploughing the bill through the substrate. These may allow Oystercatchers to feed at night as fast, and perhaps even faster, as during the day. The diet is more varied than often believed, with both minor and major shifts occurring throughout the year. These can be related to tidal and seasonal changes in the behaviour of prey, which affects their vulnerability, and to seasonal changes in their body condition, which affects their value as prey. When an Oystercatcher changes diet for long periods, the bill-tip shape changes appropriately, through an interaction between growth and abrasion; the blunt bill-tip of a mussel-hammerer can grow into the pointed tip of a worm-eater within 10–20 days. Bill growth is also vital because of a not inconsiderable risk of bill damage when Oystercatchers hammer or stab into robust molluscs. Many of those that die in severe weather have damaged bills, this having apparently prevented them from building up the body reserves required during long periods of severe weather. Although the feeding adaptations of Oystercatchers give them unique access to a large prey biomass, the risks of exploiting it are nonetheless not small and may have important implications for both the behavioural and population ecology of the species.

Acknowledgements

This chapter benefited a lot from constructive comments on earlier drafts by Bruno Ens, John Goss-Custard and Leo Zwarts. I greatly acknowledge their help. I also thank Klaas Koopman for providing biometric data on individual Oystercatchers caught on the inland breeding grounds as well as in the wintering area.

2 Prey size selection and intake rate

Leo Zwarts, John T. Cayford, Jan B. Hulscher, Marcel Kersten, Patrick M. Meire and Patrick Triplet

Introduction

Do predators take all the individual prey belonging to one species that they encounter, and if not, to what degree are some prey under-represented in the diet and why? These questions are important in every study of the interaction between predators and their food supply. This chapter first analyses the degree to which the selection of prey sizes within one prey species by Oystercatchers can be understood as a passive process in which a randomly searching bird takes all the prey it encounters; this will be referred to as 'passive size selection'. As we shall see, however, Oystercatchers reject certain size classes of prey which they actually encounter, and hence there is also 'an active size selection'. The rules that Oystercatchers obey when selecting size classes actively, and whether these can be derived from optimal foraging models, are then discussed. Finally, the chapter discusses how Oystercatchers maximize their intake rate and how the rate at which food can be processed sets a limit to the food intake.

The analyses are restricted to five bivalve species: the Edible Mussel (*Mytilus edulis*) and Edible Cockle (*Cerastoderma edule*) and the three clams: *Macoma balthica*, *Scrobicularia plana* and *Mya arenaria*. There are three main methods for obtaining data on size selection by Oystercatchers eating these prey species: recovering opened and emptied shells, direct visual observations and size-specific depletion. Shell recovery is particularly convenient and is possible because, having eaten the flesh, the birds leave the opened shell behind. These are recognizable as prey by the damage done to the shell and/or by its position in the substrate. Recently emptied bivalves also contain some flesh, sometimes along the mantle edge and always where the adductors are attached on the valves. The collection of these shells is therefore an easy, reliable method of obtaining a frequency distribution of all the size classes taken by Oystercatchers in a certain area, although there may be a sampling bias in the case of mussels (Ens 1982; Cayford 1988b). It is thus no coincidence that there are more papers on prey size selection in Oystercatchers than there are in all the other wader species put together.

Predicted 'passive size selection'

If Oystercatchers locate prey beneath the surface of the substrate by randomly probing, the probability of a prey being encountered can be calculated from prey density provided that two other factors are also taken into account (Hulscher, Chapter 1, this volume). Firstly, a proportion of the prey may live out of reach of the bill. Secondly, the probability that a prey is actually hit by the bill-tip is a function of its size, or more precisely, the surface area the prey presents from above; this is referred to here as the 'touch area'. The observed size selection may only be compared with the predictions of passive size selection when the calculated encounter rate with prey of different size classes takes both these considerations into account (Hulscher 1982).

The proportion of prey that lie buried beyond the reach of the bill depends on bill-length and on the depth to which Oystercatchers insert it. Bill-length in Oystercatchers varies between 6.5 and 9 cm. When Oystercatchers extract bivalves from the substrate, they can probe so deeply that the base of the bill is pushed 0.5 cm below the surface. Even then, large *Mya* live out of reach (Zwarts and Wanink 1984, 1989) and the majority of large *Scrobicularia* are also inaccessible, at least during the winter (Zwarts and Wanink 1989). But, in fact, Oystercatchers do not usually probe to the maximum depth. The precise probing depth differs between prey species, but is, on average, always less than the bill-length. Oystercatchers probe to a mean depth of 4 cm when searching for deep-living *Scrobicularia* (Wanink and Zwarts 1985), 3–4 cm when feeding on the more shallow-living *Macoma* (Hulscher 1982), and only 0–3 cm when taking cockles which are found very near to the surface (Hulscher 1976c). The fractions of each size class of each clam species living in the upper 4, 6 and 8 cm of the substrate are shown for the winter and summer period in Fig. 2.1. The data show, for example, that whereas most *Macoma* are accessible to Oystercatchers in summer, less than half are in winter. As cockles live in the upper 4 cm of the substrate, with the majority being in the upper 1–2 cm, they are within reach of the Oystercatcher's bill throughout the year.

For Oystercatchers probing vertically downwards, the 'touch area' of a bivalve is equivalent to its surface area, measured in the horizontal plane (Hulscher 1982). This surface area has been determined by photographing from above the bivalve in its natural position with the substrate removed (Hulscher 1982) or by pressing the bivalve vertically into modelling clay (Wanink and Zwarts 1985; Zwarts and Blomert 1992). The 'touch area' of all bivalve species is elliptical, with the cockle being the most circular and *Scrobicularia* the most slender, with *Macoma* and *Mya* lying between. The first estimates of the 'touch area' in cockles (Hulscher 1976c) and *Scrobicularia* (Wanink and Zwarts 1985) were given as a function of the squared length but, when calculated over a larger range of size classes, a better fit was obtained with exponents slightly larger than 2, since small shells are particularly slender (Zwarts and Blomert 1992). The details of this exponential increase of 'touch area' with shell size are given in Table 2.1.

The real, or 'effective', touch area for a probing Oystercatcher is

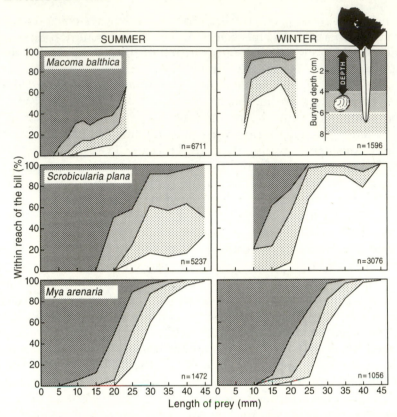

Fig. 2.1 Proportion of clams of three species living in the upper 4, 6 and 8 cm of the substrate as a function of shell size. The results are given separately for the summer months (July–September) and the winter period (November–March). The numbers of clams in the samples are shown (*n*) (Zwarts and Wanink 1989).

actually larger than the 'touch area' defined by the equations in Table 2.1, because the surface area of the bill-tip needs also to be taken into account (Hulscher 1982). Probing Oystercatchers leave behind imprints in the mud (see photo in Davidson 1967), enabling the bill-tip surface area to be measured. With the small space between the slightly opened upper and lower mandibles included, the probing surface equals a rectangle measuring 11 × 1.4 mm (Hulscher 1982). As the touch area depends on the surface area of the shell and not on its shape, for simplicity it is treated as a circle. The 'effective touch area' is then equivalent to the 'touch area' (πr^2) enlarged by the surface area of the bill-tip. Table 2.1 shows how to calculate 'effective touch area' from 'touch area' and gives, as an example, the surface areas for two size classes.

Mussel-eating Oystercatchers occasionally probe by touch, but usually appear to feed by sight (Hulscher, Chapter 1, this volume). This makes it impossible to use the method discussed above to measure the probability that prey will be detected at random. Ens (1982) argues that prey

Table 2.1. The 'touch area' of the shell (mm²) as a function of its length (mm). The 'touch area' is defined as the surface of a bivalve in its natural position, measured in a horizontal plane; from Zwarts and Blomert (1992). The corresponding surface area of the 'effective touch area' is equivalent to the surface area (set to a circle: πr^2) enlarged with the surface area of the bill-tip itself (11 mm × 1.4 mm = 15.4 mm²; Hulscher 1982) and the combined effect of the surface areas of bill-tip and bivalve (11 × 2π + 1.4 × 2π = 24.8π mm²); from Habekotté (1987) and Zwarts and Blomert (1992). As an example, the surface areas of the 'touch area' and 'effective touch area' (mm²) of two size classes (10 and 25 mm long) are given.

Species	Equation	Touch area		Effective touch area	
		10 mm	25 mm	10 mm	25 mm
Macoma balthica	0.151 mm$^{2.16}$	22	87	158	334
Scrobicularia plana	0.154 mm$^{2.09}$	19	129	80	287
Mya arenaria	0.125 mm$^{2.04}$	14	89	66	221
Cerastoderma edule	0.340 mm$^{2.07}$	40	266	128	494

encounter rate in visually hunting Oystercatchers may also be a function of the surface detection area of the prey, and thus about equivalent to the squared length of the shell. Other papers on size selection of mussel-eating Oystercatchers have followed the same assumption (Meire and Ervynck 1986; Sutherland and Ens 1987; Cayford and Goss-Custard 1990) and it will also be used here. Studies on mussel size selection have tended to focus on hammering Oystercatchers because it is possible to calculate the available fraction, i.e. those shells that are not covered with barnacles and which are thin enough to break (Fig. 2.2). Since mussels often grow on top each other, an unknown proportion of the mussels are

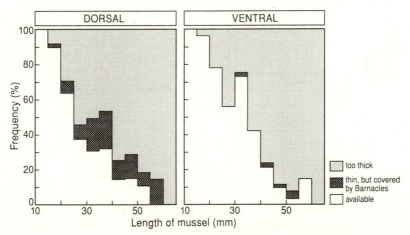

Fig. 2.2 Larger mussels are less available than small ones to Oystercatchers hammering on the dorsal or ventral side of the shell, either because the shell is covered by barnacles or too thick to break (Cayford and Goss-Custard 1990).

not actually visible to sight-hunting Oystercatchers (Meire 1991, 1996c; Goss-Custard *et al.* 1993). This may affect the observed size selection, since large mussels tend to cover small mussels (Hulscher 1964b; Norton-Griffiths 1967; Goss-Custard *et al.* 1993).

Observed size selection

Macoma balthica

Oystercatchers feeding on *Macoma* detect them by touch. Size selection in birds feeding on this clam has been studied on many occasions. To construct Fig. 2.3, the frequency distributions of the mm classes taken and of those on offer in the substrate were combined from all the studies because the size class distributions taken and on offer were similar, apart from the very large variation in the density of prey smaller than 10 mm long. However, these small *Macoma* were never selected, regardless of their contribution to the total prey population. On average, 81% of the prey taken were 15 to 20 mm long, compared with only 33% on offer at all depths in the substrate (Fig. 2.3a). In fact, Oystercatchers take *Macoma* only from the upper 4 cm of the substrate (Hulscher 1982). When the frequency distribution of the size classes actually within reach of the bill is calculated (Fig. 2.3b), the selection for larger size classes is even more pronounced because all small prey, but only a small proportion of the large prey, are accessible. But when the increase in effective

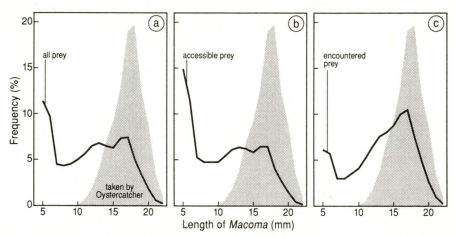

Fig. 2.3 Size classes of *Macoma balthica* selected by Oystercatcher (shaded) compared with the distribution of size classes in the prey population (solid line): (a) total prey population; (b) prey population in the upper 4 cm (see Fig. 2.1); (c) prey population in upper 4 cm, corrected for 'effective touch area' (see Table 2.1) and so actually encountered. The graphs give the averages of 26 studies in which prey on offer and those taken were compared (Goss-Custard *et al.* 1977b; Hulscher 1982 and unpublished information; Blomert *et al.* 1983; Habekotté 1987; Triplet 1989a).

touch area with size (Table 2.1) is also taken into account, the probability that small prey will be encountered dropped to a third of the previous value. Even so, there still remains a remarkable deviation between the observed size selection and that expected of a randomly probing Oystercatcher (Fig. 2.3c).

In all cases studied, Oystercatchers rejected *Macoma* less than 11 mm long, while the size classes 11–15 mm were taken much less frequently than would be expected on the basis of random searching. The relative risk of the medium-sized *Macoma* being taken appeared to depend on the density of the larger ones, 18–22 mm long (Fig. 2.4a), but not on their own density. There was also a difference in size selection between summer and the rest of the year (Fig. 2.4b); any confounding influence of the variable density of large *Macoma* was removed by restricting the analysis to studies where the density of large *Macoma* was less than 30 specimens m⁻². In conclusion, Oystercatchers concentrate their feeding on the largest *Macoma*, but when these are less common, they take relatively more of the medium-sized ones, especially in autumn and winter. Small prey are always rejected, however.

Scrobicularia plana

In contrast to *Macoma*, where different year classes regularly occur together, the frequency distribution of *Scrobicularia* size classes varies a

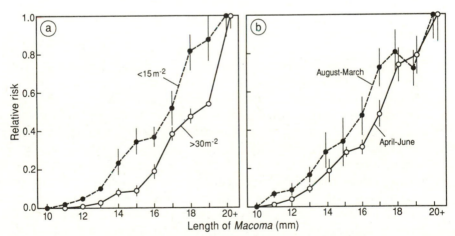

Fig. 2.4 The relative risk of *Macoma balthica* of different lengths being taken by Oystercatchers, calculated as the ratio per size class of the relative number of prey taken to the relative number in the prey population; data from Fig. 2.3. (a) Relative risk (± SE) when the density of large *Macoma* (18–22 mm long) was lower than 15 m⁻² (*n* = 10) or higher than 30 m⁻² (*n* = 4); (b) relative risk (± SE) in April-June (*n* = 10) and in August-March (*n* = 6). To rule out the confounding effect of prey density shown in (a), a selection is made for studies where the density of *Macoma* 18–22 mm long is less than 30 m⁻².

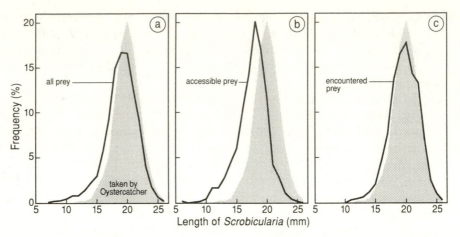

Fig. 2.5 Size classes of *Scrobicularia plana* selected by Oystercatchers (shaded) compared to the size classes in the prey population (solid line): (a) total prey population; (b) prey population in upper 6 cm (Fig. 2.1); (c) prey population in upper 6 cm corrected for 'effective touch area' (Table 2.1) and so actually encountered (Habekotté 1987).

lot depending on how many years have passed since the last recruitment took place. Three studies have been made of size selection in Oystercatchers taking *Scrobicularia* by touch. Blomert *et al.* (1983) studied Oystercatchers feeding on *Scrobicularia* between 24 and 48 mm long. The birds hardly took any prey larger than 37 mm because they were inaccessible (Fig. 2.1). Prey smaller than 28 mm long were taken less than expected, whether or not a correction was made for the effective touch area and the fraction that was accessible. Hughes (1970) found that Oystercatchers rejected prey smaller than 20 mm long when prey 25–40 mm long were available. Habekotté (1987) showed that small *Scrobicularia* were taken when there were few large specimens. In his study, the birds rejected prey smaller than about 13 mm long and selected the largest size classes available more often than expected (Fig. 2.5). In conclusion, there is, as in Oystercatchers eating *Macoma*, a rejection minimum size threshold which cannot be explained by the random touch model. However, in contrast to *Macoma* (Fig. 2.4), the lower size threshold for *Scrobicularia* appears to vary greatly between 13 mm (Fig. 2.5) and 28 mm (Blomert *et al.* 1983), depending on the density of the large *Scrobicularia*.

Mya arenaria

Touch-feeding Oystercatchers select *Mya* 15–40 mm long (Fig. 2.6). They reject the size classes smaller than 15 mm, although all are accessible (Fig. 2.1). In contrast, all clams larger than 40 mm live out of reach of the bill (Fig. 2.1). Taking depth distribution and effective touch area

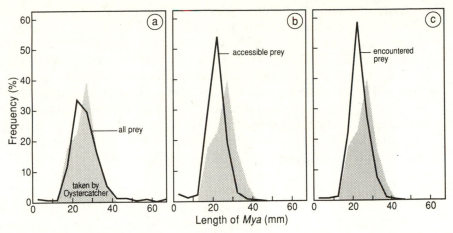

Fig. 2.6 Size classes of *Mya arenaria* selected by Oystercatchers (shaded) compared to the size classes in the prey population (solid line): (a) total prey population; (b) prey population in upper 6 cm (Fig. 2.1); (c) prey population in upper 6 cm corrected for 'effective touch area' (Table 2.1) and so actually encountered (Zwarts and Wanink 1984; Wanink and Zwarts 1996).

of the different size classes into account, Oystercatchers actively select *Mya* 20–30 mm long (Fig. 2.6c).

   Mya may also be located by sight since, when feeding, they extend their siphon up to the surface. The siphon holes are sometimes visible during calm weather and the first hours of the receding tide. For visually hunting birds, large *Mya* are easier to find than small ones, since the surface area of the siphon hole is approximately proportional to its length squared (Zwarts and Wanink 1989). As J. B. Hulscher (unpublished information) showed, Oystercatchers hunting visually for *Mya* directed their pecks at siphon holes on the mud surface. The birds took *Mya* 20–75 mm long, even though some of them lived out of reach of the bill. Such prey were pulled from the mud by grasping the big siphon. Prey 75–105 mm long also occurred in the study area, but apparently lived too deep to be lifted since captive Oystercatchers were able to eat these size classes when the prey were buried just beneath the surface (Zwarts and Wanink 1984).

Cerastoderma edule

Oystercatchers feeding on cockles either make distinct pecks or move the bill through the substrate (Hulscher, Chapter 1, this volume). Pecking Oystercatchers taking cockles use visual cues and do not peck at random (Hulscher 1976c). This limits the validity of the random touch model even though the effective touch area (Table 2.1) correctly measures the expected encounter rate. When the bill makes straight furrows through the mud, another random touch model is more appropriate (Hulscher 1976c).

Oystercatchers do not take first year cockles less than 10 mm long, when older cockles, 20–40 mm long, are available (Drinnan 1957; Brown and O'Connor 1974; Goss-Custard *et al.* 1977b; Sutherland 1982a; Triplet 1989b; J. B. Hulscher, unpublished information). When only first-year cockles are present, Oystercatchers may feed on them (Meire 1996b; Zwarts and Wanink 1996; T. Piersma, unpublished information), but this rarely occurs. When only second year, and older, cockles are available, there is a reasonable coincidence between the observed size selection and the selection predicted on the basis of random searching (Hulscher 1976c; Leopold *et al.* 1989).

Mytilus edulis

Oystercatchers use three techniques to open mussels (Hulscher, Chapter 1, this volume). Briefly, ventral hammerers tear the mussel from the bed, turn it over and hammer a hole on the weak ventral side (Norton-Griffiths 1967). Dorsal hammerers attack the mussel *in situ* on its dorsal side (Ens 1982), while stabbers stab, or force, the bill between the valves. Each technique requires different measurements to define the fraction of prey actually available to Oystercatchers.

Stabbing Oystercatchers reject mussels smaller than 20–25 mm long but take all larger size classes (Fig. 2.7a). Mussels larger than the apparent lower threshold of 20–25 mm are taken in proportions that conform to the assumption that the likelihood of a mussel being attacked is pro-

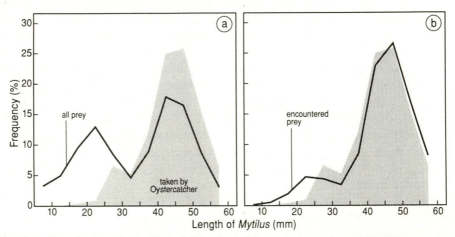

Fig. 2.7 Size classes of *Mytilus edulis* selected by Oystercatchers (shaded) compared to the size classes in the prey population (solid line): (a) total prey population; (b) prey population corrected for 'effective touch area' (i.e. squared length) and so actually encountered. The data are from 9 studies where Oystercatchers used the stabbing technique to open mussels (Koene 1978; Zwarts and Drent 1981; Blomert *et al.* 1983; Polman 1988; Ens *et al.* 1996b; J. B. Hulscher, unpublished information; L. Zwarts, unpublished information).

portional to its surface area, i.e. length squared (Fig. 2.7b). However, the rejection threshold differs between studies. When mussels about 50 mm long are numerous, stabbing Oystercatchers may even reject mussels as long as 40 mm (Zwarts and Drent 1981).

Hammering Oystercatchers reject thick-shelled mussels covered by barnacles (Durell and Goss-Custard 1984; Meire and Ervynck 1986; Cayford and Goss-Custard 1990). This affects the frequency distribution of the size classes actually available to hammerers, since the larger mussels are often encrusted by barnacles and many are too thick-shelled to be opened (Fig. 2.2). Like stabbing Oystercatchers, hammering birds reject mussels below 20–25 mm long, but, in contrast to stabbers, mussels larger than 60 mm are taken only infrequently. When a correction is made for the prey fraction that is unavailable due to shell thickness and barnacle cover (Fig. 2.2.), the proportion of each large prey class taken roughly coincides with their available density (Meire and Ervynck 1986), or explains at least a part of the deviation between observed size selection and frequency distribution of size classes on offer (Cayford and Goss-Custard 1990).

Size selection and optimal foraging

Predicted 'active size selection'

The random touch model tests the assumption that birds take prey at random. In fact, as the results in the previous section show, the observed prey selection often deviates from the predictions of the model. Oyster-catchers apparently prefer some size classes to others. Why? Prey size selection in Oystercatcher is analysed here within the framework of optimal foraging theory (Emlen 1966; MacArthur and Pianka 1966; Krebs and Kacelnik 1991). The basic assumption is that predators are able to rank prey according to their profitability, defined as the intake rate while prey are being handled. They are predicted to reject prey for which the profitability is below the current average intake rate over both handling and searching combined. The decision rule governing the rejection threshold is therefore based on the relative profitability of handling compared with continuing to search, i.e. whether the bird can achieve a higher net intake rate by continuing to search for more profitable prey than it could achieve by handling a less profitable, although more frequently encountered, prey. This leads to the prediction that, when the profitability of the prey remains the same, the rejection threshold should increase as the intake rate increases. The rejection threshold should also increase when the intake rate remains the same but the profitability of all prey types decreases; for instance, because the prey are lean. The rejection threshold should therefore be flexible within clearly defined limits, as illustrated in Fig. 2.8.

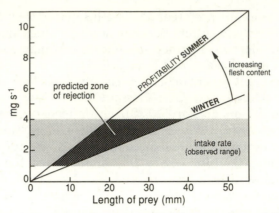

Fig. 2.8 The optimal prey size selection model. The two slopes delimit the seasonal variation in profitability (mg s^{-1} handling) as a function of prey size. Prey for which the profitability is below the intake rate (mg s^{-1} feeding, i.e. during both searching and handling) should be rejected. Which prey should be rejected thus depends on the level of the intake rate as well as on the length/profitability slope. The dark shaded field shows the expected range within which the lower acceptance threshold should be found when the intake rate varies between 1 and 4 mg s^{-1} feeding.

Small prey are less profitable

Even after a correction has been made for the small 'effective touch area' of small prey, Oystercatchers appear to take fewer of them than would be expected on the basis of the frequency with which they are encountered (Figs 2.2–2.7). The birds always completely reject prey less than about 10 mm long in *Macoma* and *Cerastoderma* and about 15–20 mm long in *Scrobicularia*, *Mya* and *Mytilus*. There might be a very simple explanation for this. Oystercatchers are specialized to open hard-shelled prey before they eat the flesh, in contrast to Knot (*Calidris canutus*) or Bar-tailed Godwits (*Limosa lapponica*) which swallow the prey whole and crush them in the stomach. There must be a size below which Oystercatchers are hardly able to separate the flesh from the shell. Whether or not this is close to the observed rejection threshold has still to be tested.

On the other hand, several papers have used the optimality approach to explain the size rejection threshold in Oystercatchers. The prediction is that, below a certain size threshold, prey are simply not worth taking since their energy value is too low given the time required to handle them; i.e. to open them and eat the flesh (Zwarts and Drent 1981; Ens 1982; Hulscher 1982; Sutherland 1982a; Zwarts and Wanink 1984; Meire and Ervynck 1986; Sutherland and Ens 1987; Cayford and Goss-Custard 1990). Handling times have therefore been measured in both captivity and in the field. When determining the relationship between handling time and prey size in the field, prey sizes have to be estimated by eye against something of known length, normally bill-length or the size of a colour ring. Fortunately, observers are quite consistent in their estimates and simple corrections are sufficient to give an accurate estimate of the size taken (Ens 1982; Blomert *et al.* 1983; Goss-Custard *et al.* 1987; Boates and Goss-Custard 1989).

The measurements of handling time in the five prey species are summarized in Fig. 2.9. Only times when prey were successfully taken are shown. Handling times for mussels are for stabbing Oystercatchers. Depending on the prey size, it takes an Oystercatchers 3–300 s to open and eat the flesh in a bivalve. The flesh weight increases from 5 mg to over 1000 mg over the range of size classes studied. In all prey species, flesh weight increases exponentially with size, with an exponent of about 3. Handling time is also an exponential function of prey size, but the exponent is much lower (Fig. 2.9).

Dividing prey weight by handling time gives the profitability, i.e. the intake rate in mg dry flesh s^{-1} while handling the prey (Fig. 2.10). When all data are lumped, profitability is more variable than handling time. This is due to the considerable variation in the flesh content of the prey; for instance, in *Scrobicularia* and *Cerastoderma*, winter weight is almost half that in early summer (Zwarts 1991). But, despite this, there is a highly significant relationship between prey profitability and size in all five bivalve species. This means that profitability can be ranked simply according to prey size.

Large prey are difficult and sometimes even dangerous to handle

Not all prey that are attacked are actually taken, so time is spent in failed attacks. When calculating the profitability of a particular size class, this wasted handling time has to be taken into account when calculating the average time needed to eat them. The difference this can make to the calculation of profitability can be illustrated by hammering birds. Oystercatchers successfully hammering into mussels spend more time in breaking into a large one than into a small one, but this is worthwhile as the flesh content is disproportionately greater. But if the wasted handling times are included in the calculation, the profitability of hammered mussels actually decreases in mussels larger than 50 mm long (Fig. 2.11).

Another factor which falls outside the scope of the simple optimality model being discussed here is the potential risk to the bill of attacking large prey. Both Sutherland (1982a) and Triplet (1989a) found that larger cockles are refused more often than small ones. As the time lost is insignificant, the birds may be reducing the risk that the bill-tip will be damaged. After being stabbed, mussels may also close their valves firmly on the bill, so that an Oystercatcher may eventually die due to starvation (Hulscher 1988). But the risk appears to be small and, with exception of mussels being hammered, the amount of time spent in wasted handling

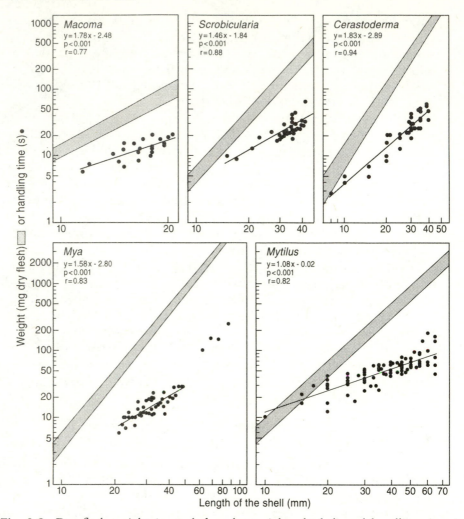

Fig. 2.9 Dry flesh weight (mg ash-free dry weight; shaded) and handling time (s; symbols) as a function of shell length (mm) in different studies. The seasonal variation in prey weights are from Cayford and Goss-Custard (1990) and Zwarts (1991). Data are from the following sources: *Macoma*: Blomert *et al.* (1983) and Hulscher (1982 and unpublished information); *Scrobicularia*: Blomert *et al.* (1983), Wanink and Zwarts (1985, 1996), Habekotté (1987), Boates and Goss-Custard (1989); *Cerastoderma*: Hulscher (1982 and unpublished information), Sutherland (1982b), Swennen *et al.* (1989), Triplet (1990), Ens *et al.* (1996b, c), Meire (1996b); *Mya*: Zwarts and Wanink (1984), Wanink and Zwarts (1996); *Mytilus* (stabbers): Koene (1978), Blomert *et al.* (1983), Speakman (1984), Linders (1985), Sutherland and Ens (1987), Cayford and Goss-Custard (1990), Ens *et al.* 1996c, Hulscher (unpublished information). The allometric relations are given: $y = \ln(\text{handling time, s})$, $x = \ln(\text{shell length, mm})$.

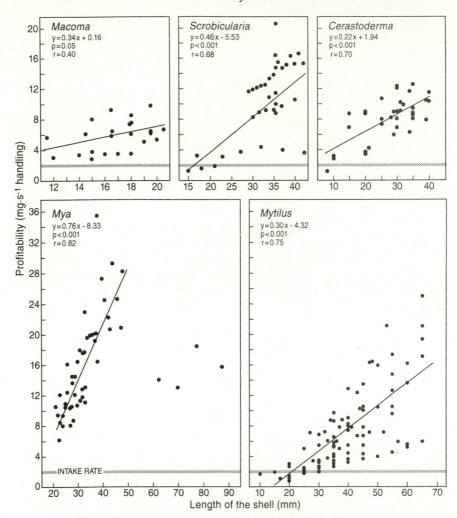

Fig. 2.10 Profitability (mg ash-free dry flesh s^{-1} handling) as a function of shell length; same data as in Fig. 2.9. The intake rate during feeding, averaged for all studies, is indicated by the horizontal line. The functions are given: y = profitability (mg s^{-1}), x = shell length (mm). The functions for *Mya* refer to shells less than 50 mm long.

time is not large. The general conclusion from the previous section that the prey size predicts profitability is largely unaffected.

Oystercatcher can vary encounter rate

Another important consideration in calculating whether a particular size class of prey should be taken is the rate at which prey are encountered. Oystercatchers can control encounter rates with prey through changes in their search behaviour. For example, when Oystercatchers switch from

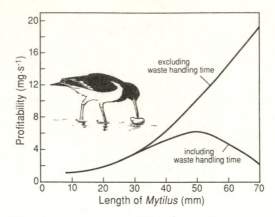

Fig. 2.11 Profitability (mg ash-free dry flesh s^{-1} handling) as a function of length for mussels which are hammered on the ventral side. Upper line refers to mussels that were actually eaten so the time cost was just the handling time for those mussels. The lower line also includes all the time wasted on other mussels of the same size class which were unsuccessfully hammered (Meire and Ervynck 1986).

touch to visual hunting (Hulscher 1976c, 1982), they change from randomly probing into the mud to searching on the surface for signs of the prey beneath (Hulscher, Chapter 1, this volume). This means that the encounter rate with potential prey has to be defined according to the feeding technique used. Again, Oystercatchers visually hunting for tracks or inspecting bivalves on the surface may vary the encounter rate with different prey types by varying search speed (Cayford and Goss-Custard 1990). The speed at which a foraging animal searches has been described as an adaptation to the crypticity of the prey (Goss-Custard 1977a; Gendron 1986; Zwarts and Esselink 1989) and observations on the walking speed of Oystercatchers feeding on different prey types is consistent with this idea (Ens *et al.* 1996d).

Although randomly probing Oystercatchers provide a good opportunity to measure encounter rates, it would be wrong to assume that they are fixed for a given prey density and depth distribution. This is because Oystercatchers can modify probing depth. As already described, for example, Oystercatchers probe twice as deeply when searching for the deep-living *Scrobicularia* than when they feed on a shallow-living prey, such as the cockle. Oystercatchers must therefore also make the decision on 'how deep to probe' and this too must be taken into consideration when the economics of foraging are being calculated.

It has been shown experimentally that a captive Oystercatcher decreased its probing depth from 7 to 3 cm when the density of the prey on offer, *Scrobicularia* 36–37 mm long, increased from 24 to 350 prey m^{-2} (Wanink and Zwarts 1985). It took more time to handle deep-living prey (Fig. 2.12a), hence prey profitability decreased with depth. More-

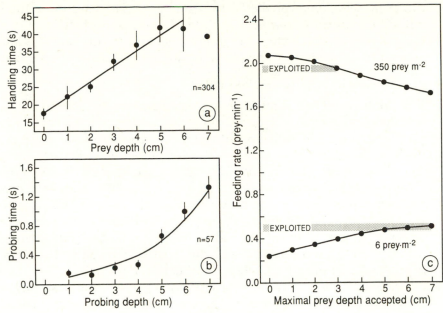

Fig. 2.12 Experiment to show why an Oystercatcher ignored deep, less profitable prey when the intake rate increased because of high prey density. (a) Time needed to handle *Scrobicularia* 37–38 mm long which are taken from different depths; (b) time needed to probe the bill to different depths (s ± SE); (c) predicted feeding rate (*Scrobicularia* min^{-1}) at two prey densities when an Oystercatcher took prey from the upper 1, 2, ... 7 cm of the substrate. Actual observed depth selection is indicated with the horizontal bars. As predicted, the bird took prey from all depths when the prey density was low but took only shallow prey when the prey density was high. However, the bird did this at a much higher rate than predicted (explanation given in the text). Data from Wanink and Zwarts (1985).

over, it also took more time in the first place to locate a prey at greater depths (Fig. 2.12b). The encounter rate, defined in optimal foraging models as the inverse of the searching time, for prey at different depths could be calculated from the effective touch area and the probing time at each probing depth. Since the encounter rate and the handling time were both known for each depth class, the optimal set of depth classes, which should be included in the diet to maximize the intake rate, could be predicted exactly with a multi-species functional response equation (Charnov 1976) (Fig. 2.12c). The depth selection made by the bird was very close to that which was predicted. The Oystercatcher took prey from all depths when the density was low but rejected the deep, less profitable prey when prey density was high; by doing so, it increased its intake rate at the high prey density by eating only the most profitable, shallow prey. It may be thought from Fig. 2.12c that the bird could have done slightly better by selecting prey only from the upper 1 cm, and not from the upper 3 cm, as it actually did. But, in fact, the Oystercatcher did even better than predicted by using a second selection criterion, as will be explained in the next section. The conclusion from this experiment is that Oystercatchers are indeed able to vary the rate at which they encounter different prey types and, in doing so, increase their intake rate.

Rejection of prey to increase intake rate

The previous section showed that Oystercatchers are able to adjust their searching behaviour in order to increase their encounter rate with more profitable prey. The next decision Oystercatchers have to make is which of the encountered prey they should take and which they should ignore. As noted above, the captive Oystercatcher studied by Wanink and Zwarts (1985) was able to attain an intake rate 50% above the predicted rate when prey density was high (Fig. 2.12c). Their explanation for this was that, at the high prey density, the bird only took prey from the upper 3 cm that were gaping and so could be stabbed immediately. As a result, the prey were opened and the flesh swallowed in less than 15 s, nearly twice as quickly as when a typical prey from the upper 3 cm was taken (Fig. 2.12a). The hypothesis that Oystercatchers may ignore closed, and thus less profitable, bivalves had already been proposed by Hulscher (1976c) who observed that Oystercatchers spent less time handling *Cerastoderma* when prey density increased. By rejecting closed bivalves when prey are abundant, Oystercatchers are able to increase their intake rate substantially.

There seem to be large differences generally in the profitability of prey of similar size due to the large variation in handling time. In the same way that it takes less time to eat shallow-living and/or gaping bivalves (Hulscher 1976c; Wanink and Zwarts 1985), there are also large variations in the profitability of hammered prey (Durell and Goss-Custard 1984). Despite the variation in profitability within a size class, size and prey profitability remain highly correlated (Fig. 2.10). This allows us now to focus on the question of why the small, less profitable prey are usually refused, but sometimes taken.

The optimal diet model predicts that birds should refuse prey for which the profitability is lower than the average current intake rate. Taking all studies together (Fig. 2.13), the average intake rate during feeding was 1.97 mg dry flesh s^{-1} (SD = 0.84; n = 197). The average intake rate differed significantly between the 12 prey species (P = 0.006, R^2 = 0.137) but there were also seasonal differences. The mean intake rate in 'summer' (April–October; 2.15 mg s^{-1}; SD = 0.87; n = 126) differed significantly from that in the 'winter' (November–March; 1.64 mg s^{-1}, SD = 0.67; n = 71; P < 0.001; R^2 = 0.086). Twenty-eight of the 31 studies recording an intake rate above 3 mg s^{-1} were made in summer, compared with only 19 of the 37 studies with an intake rate below 1.5 mg s^{-1}. This difference was due to the seasonal variation in flesh content of prey (e.g. Zwarts 1991; Zwarts and Wanink 1993), which was higher

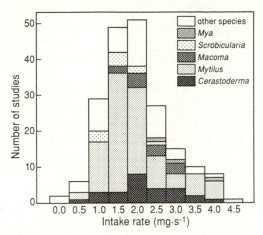

Fig. 2.13 Frequency distribution of intake rates (mg ash-free dry flesh s⁻¹ feeding) measured in 197 studies, given separately for five bivalve species; data for seven other prey species are lumped. The sources for the five bivalve species are given in the legend to Fig. 2.14. Other sources: Ragworms (*Nereis diversicolor*): Boates and Goss-Custard (1989), Triplet (1989b), Bunskoeke *et al.* (1996), Durell *et al.* 1996a and unpublished information); Lugworms (*Arenicola marina*): Bunskoeke (1988); earthworms: Heppleston (1971b), Hosper (1978); Leatherjackets (*Tipula paludosa*): Hosper (1978); Zwarts and Blomert (1996); Winkles (*Littorina littorea*): Boates and Goss-Custard (1992); Giant Bloody Cockle (*Anadara senilis*): Swennen (1990); Fiddler Crabs (*Uca tangeri*): Ens *et al.* (1993a).

in summer, making the prey more profitable than in winter. But taking 2 mg s⁻¹ as an overall average for intake rate, the predicted average lower size threshold for prey may be read from the profitability curves given in Fig. 2.10. The average rejection threshold should be about 10 mm for *Macoma* and *Cerastoderma*, 15 mm for *Scrobicularia* and *Mya*, and 20 mm for *Mytilus*. When these figures are compared with the actual selection of prey, there is a reasonable fit for *Macoma* (Fig. 2.3), *Cerastoderma* (see text on observed size selection), *Mya* (Fig. 2.6) and *Mytilus* (Fig. 2.7). On average, the optimal foraging model predicts the observed lower prey size thresholds quite well, although the size classes just above the predicted lower size threshold are under-represented in the diet.

Particular studies, however, have found some departures from prediction. Since the flesh content of bivalves may vary by a factor of two during the course of the year (Cayford and Goss-Custard 1990; Zwarts 1991; Zwarts and Wanink 1993), there is an opportunity to test whether seasonal changes in the profitability curves accurately predict a shift in size selection. Cayford and Goss-Custard (1990), for example, analysed the seasonal variation in the size selection for mussels on the Exe estuary, South-west England. The birds actively selected those size classes predicted to maximize intake rate for most months of the year. However, in spring, the birds fed on smaller mussels than were predicted from their profitability and availability. Cayford and Goss-Custard (1990) suggested that size selection may be influenced at certain times of the year by still unmeasured differences between size classes in prey quality and energy content.

However, other apparently contradictory results can be reconciled more easily with the optimal foraging model. Handling time does not increase as the flesh content of a bivalve increases because the time taken actually to eat the flesh is only a small fraction of the total (Wanink and Zwarts 1985). Most of the handling time is spent in lifting the bivalve to the surface, stabbing or forcing the bill between the valves, or hammering a hole in the shell. Hence the profitability of a given prey size may be as much as twice as high in early summer as in winter. The rejection threshold, therefore, should be lower in summer than during the rest of the year (Fig. 2.8). In fact, in *Macoma*, the reverse was found (Fig. 2.4b), as Oystercatchers were more selective in summer than at other times of the year. However, all data on size selection in early summer refer to breeding Oystercatchers which have the rather high intake rate of 3 mg s^{-1} (Ens *et al.* 1992; Bunskoeke *et al.* 1996). Such a high rate should have the effect of raising the rejection threshold for *Macoma*, thus countering the opposing effect on the rejection threshold of an increased prey condition (Fig. 2.8). Thus, in this case, two tendencies may be working in opposite directions.

The intake rates of Oystercatchers, measured over periods of several hours usually vary between 1 and 3 mg dry flesh s^{-1} (Fig. 2.13). When they feed at a rate of 1 mg s^{-1}, they would be predicted to accept all *Scrobicularia* and *Mytilus* larger than 15 mm long, whereas when the intake rate is three times higher, the rejection threshold should be raised to 20 and 25 mm, respectively (Fig. 2.10). The simple way to test these predictions would be to compare observed rejection thresholds with the lower thresholds predicted from the intake rate. There is, however, a methodological problem. In the field studies summarized in Fig. 2.13, intake rate is closely correlated with the size of the prey taken (see below); the expected dependence of the lower size threshold on intake rate may thus be attributed simply to size selection itself.

The intake rate is determined by a combination of three variables: the searching time, the handling time and the dry flesh weight of the prey taken. The search time primarily depends on prey density (Hulscher 1976c, 1982; Sutherland 1982b; Wanink and Zwarts 1985). Both handling time and prey weight increase with prey size (Fig. 2.9), whereas there is a seasonal variation in the flesh content of individual prey. The major factor determining intake rate in the reviewed studies was the size of the prey (Fig. 2.14); for instance, 59% of the variance in intake rate on *Scrobicularia* could be explained by prey size. A covariance analysis performed on the 197 studies which measured intake rate (Zwarts *et al.* 1996a)

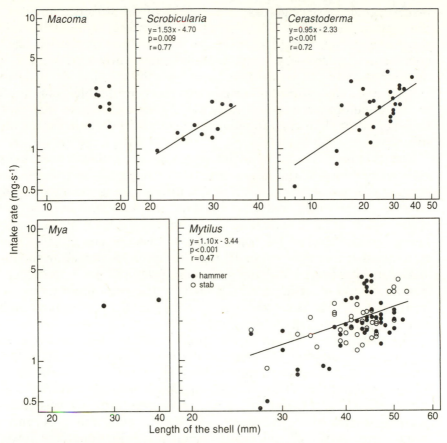

Fig. 2.14 Intake rate (mg ash-free dry flesh s⁻¹ feeding) as a function of average length of bivalves taken. Data are from the following sources: *Macoma*: Hulscher (1982 and unpublished information), Blomert *et al.* (1983), Bunskoeke *et al.* (1996), Hulscher *et al.* (1996); *Scrobicularia*: Blomert *et al.* (1983), Habekotté (1987), Boates and Goss-Custard (1989), Wanink and Zwarts (1996), Hulscher (unpublished information); *Cerastoderma*: Drinnan (1957), Davidson (1967), Brown and O'Connor (1974), Hulscher (1976c and unpublished information), Goss-Custard (1977b), Sutherland (1982a, b), Leopold *et al.* (1989), Ens *et al.* (1996b, c), Meire (1996b); *Mya*: Zwarts and Wanink (1984), Bunskoeke (1988), Wanink and Zwarts (1996), Hulscher (unpublished information); *Mytilus*—(○) stabbers and (●) hammerers: Drinnan (1958a,b), Heppleston (1971b), Koene (1978), Zwarts and Drent (1981), Blomert *et al.* (1983), Ens and Goss-Custard (1984), Speakman (1984), Meire and Ervynck (1986), Sutherland and Ens (1987), Cayford and Goss-Custard (1990), Boates and Goss-Custard (1992), Maagaard and Jensen (1994), Ens *et al.* (1996b), Meire (1996b), Goss-Custard (unpublished information), Hulscher (unpublished information). The allometric relations are given: $y = \ln(\text{handling time, s})$, $x = \ln(\text{shell length, mm})$.

revealed that prey weight explained 16% of the variance and that both the prey species and season had significant effects (R^2 = 0.24 and 0.19, respectively). When prey of different species, but of similar size, were compared, *Cerastoderma* and *Macoma* were shown to yield a higher intake rate than the other bivalves. The reason is that, for a given size class, *Cerastoderma* and *Macoma* are more profitable. This is due to their more globular shape (Zwarts and Blomert 1992), so that they contain more flesh than corresponding size classes in the other three bivalves (Fig. 2.10).

The increase in intake rate with prey size (Fig. 2.14) is thus due to the high profitability of the larger size classes (Fig. 2.10). The yield of the smallest prey taken (Fig. 2.10) is below the average intake rate of 2 mg s^{-1}. So even if the birds could eat these small prey continuously, without spending any time in searching for them, the average intake rate would fall to the low level of 1–2 mg s^{-1}. In contrast, the larger size classes can be handled at a rate of about 10 mg s^{-1} (Fig. 2.10), enabling the birds to attain an intake rate of 2 mg s^{-1}, even though as much as 80% of the feeding time is spent in searching. In fact, the searching time must have been less than this, as Oystercatchers taking large bivalves have an intake rate of 3–4 mg s^{-1} (Fig. 2.14).

The finding that the intake rate predominantly depends on size selection makes it difficult to interpret the field data on size selection. The prediction of the optimal foraging model is that the threshold size of acceptable prey should depend on the intake rate. However, the threshold level itself affects the intake rate. One way to break this circle is to examine studies where the density of the most profitable prey, and so the intake rate, varied considerably. When this was done, the lower size threshold changed in the direction predicted by the model; for example, Oystercatchers were more selective when the density of large *Macoma* was high (Fig. 2.4a). Similarly, Sutherland (1982a) found that Oystercatchers took fewer small cockles as the abundance of the large ones increased. Such findings allow the direction of causality to be established. On the assumption that these findings can be applied across all studies, we conclude that, as predicted by the optimal foraging model, the minimum size rejection threshold is determined by the intake rate which itself depends predominantly on the abundance of the larger and most profitable size classes.

Intake rate, processing rate, feeding time and daily consumption

The central tenet of the classical optimal diet model is that predators attempt to maximize their intake rate. Given a fixed food requirement, intake rate determines the time required for feeding. Maximizing intake rate therefore implies minimizing the time spent feeding, assuming that birds are working towards a predetermined amount of food. Under thermo-neutral conditions, an Oystercatcher is able to maintain its body weight when its daily consumption is about 33 g of dry food pellets or dry flesh (Hulscher 1974; Kersten and Piersma 1987; Swennen *et al.* 1989; Goede 1993). Since the energetic values of food pellets and dry flesh of bivalves are both 22 kJ g⁻¹ ash-free dry weight, and does not vary much (e.g. Brey *et al.* 1988; Zwarts and Wanink 1993), this is equivalent to 730 kJ. About 15% of the ingested energy is not digested (Kersten and Visser 1996a), so the daily metabolized energy of an Oystercatcher is 620 kJ, or about 2.5 times the basal metabolic rate (*BMR*) of 250 kJ (Kersten and Piersma 1987). However, when the air temperature drops below 10°C, the daily food requirements increase from 33 g per day to 50 g at an air temperature of 0°C (Kersten and Piersma 1987; Goede 1993). The same high level may be reached when Oystercatchers increase their body weight before migration. These data refer to Oystercatchers held in out-door cages. The requirements in wild birds are somewhat higher. First, free-living birds fly for 10 to 30 minutes per day. Assuming that flying costs are 12 times *BMR* (Masman and Klaassen 1987), this represents an expenditure of 20–40 kJ, equivalent to 3–9% of daily energy expenditure. Second, the lean body weight in captive birds was in most studies below the normal range observed in free-living birds. Since the energy requirement varies with the lean body weight, the average cost of living in the field must be some per cent above the estimated daily energy requirements in captive birds (Zwarts *et al.* 1996b). The daily food requirements of free-living Oystercatchers has therefore been estimated at 36 g.

Since tidal feeding areas are covered for half the time over high water, the daily feeding period is 12 hours if the birds feed at night. Given a daily consumption of at least 36 g dry flesh, Oystercatchers in the wild must consume 0.85 mg dry flesh s⁻¹, assuming they feed continuously over low water and at the same rate by day and by night (Hulscher, Chapter 1, this volume). An intake rate of about 0.85 mg dry flesh s⁻¹ is thus the absolute minimum required to sustain energy balance, unless Oystercatchers can extend the foraging period over high water in day-time by feeding in fields. Feeding at high tide rarely occurs in summer, but is more common in the energetically more expensive winter, especially when, due to gales and floods, the tidal feeding areas are only exposed for a short time (Heppleston 1971b; Daan and Koene 1981; Goss-Custard and Durell 1984).

The observed range of intake rates (Fig. 2.13) exceeds the minimum rate of 0.85 mg s⁻¹ in 93.4% of studies. One of the exceptions was a rate of 0.5 mg s⁻¹ recorded in the Eastern Scheldt, South-west Netherlands (Meire 1996b). On this occasion, the tidal food resources were so poor that Oystercatchers had to rely on cockle spat 8 mm long, which in all other studies were rejected because of their low profitability.

When Oystercatchers feed at the highest observed rate of 4 mg s⁻¹, it will only take them 2.5 hours each day to obtain the minimum daily consumption of 36 g. In contrast, when the intake rate is at the low level of 1 mg s^{-1} and the daily energy requirements are at their highest level of 50 g, the birds will need to feed for 14 h. Thus, although the average feeding time would be expected to be only 5 h a day (intake rate 2 mg s^{-1} and daily consumption of 36 g), the large variation in both intake rates and daily energy requirements mean that the actual feeding time may vary between 2.5 and 14 h a day.

In some circumstances, then, Oystercatchers would be expected to collect food as fast as possible, notably when the tidal exposure period is reduced and feeding at high tide is not possible (Goss-Custard *et al.*, Chapter 6, this volume). However, Oystercatchers cannot exceed the limit set by their digestive system. Kersten and Visser (1996a) studied the rate at which Oystercatchers process the flesh of bivalves. They found that the food processing rate could be described by three parameters: (1) a full oesophagus contains 80 g wet (or 12 g dry) flesh; (2) the birds start to defecate 0.5 h after the beginning of the feeding period; and (3) the processing rate is 4.4 mg wet (or 0.66 mg dry) flesh s^{-1}. The processing rate is thus much slower than the range of intake rates observed in the field (1–4 mg s^{-1}). Indeed, Oystercatchers in captivity can achieve even higher intake rates. The highest rate of 16 mg s^{-1} was observed by J. B. Hulscher (unpublished information) in captive birds eating mussel flesh. At that rate, a digestive pause would be necessary after only 13 minutes of feeding, which was indeed the actual length of the feeding period in Hulscher's birds.

The finding that the processing rate is usually much lower than the intake rate during feeding has several implications. At least in the long run, birds must interrupt their feeding when their intake rate exceeds their processing rate (Zwarts and Dirksen 1990). The overriding effect of this digestive bottleneck on the daily consumption of the Oystercatcher is shown in Fig. 2.15. The maximum intake during a feeding period is set by the three parameters given by Kersten and Visser (1996a). At the very low average intake rate of 1 mg s^{-1}, the gut fills and forces the bird to reduce its intake rate to a level equal to, or below, the processing rate after 530 min of continuous feeding. But with intake rates of 2, 3 or 4 mg s^{-1}, the alimentary tract is full after only 135, 77 or 54 min, respectively.

The length of time an Oystercatcher has to spend on the feeding area, including digestive pauses, to achieve a given food consumption, is also shown by Fig. 2.15. The only assumption here is that the bird starts to feed with an empty stomach and leaves the feeding area with a full oesophagus. With one feeding period a day, the bird would have to feed for 10 hours to achieve the minimum daily energy requirements of 36 g (Fig. 2.15a). Remarkably, the average intake rate while feeding does not matter, as long as it exceeds 1 mg s^{-1}. Furthermore, if Oystercatchers leave the feeding area with a full gut, it takes just over 5 hours to process

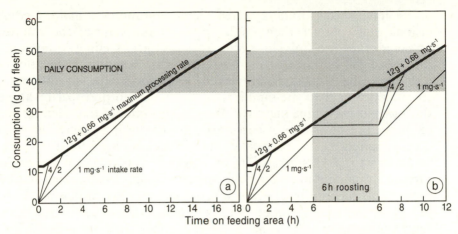

Fig. 2.15 Daily consumption as function of the time spent on the feeding area, when there is (a) one feeding period per day or (b) two feeding periods with a roosting period of 6 hours in between. The intake rate determines how long it takes before the bird is forced to pause or reduce its intake rate due to the digestive bottleneck, this being set by the three parameters: (1) the oesophagus may contain 12 g ash-free dry flesh; (2) defecation starts 0.5 h after the beginning of the feeding period, and (3) the processing rate is 0.66 mg s^{-1} (Kersten and Visser 1996a).

all the food, or nearly as long as the usual high water roosting period. Consequently, the total consumption is hardly affected when both the feeding and roosting periods are six hours long (Fig. 2.15b). The figure points to three important conclusions: (1) as long as the intake rate during feeding exceeds 1.21 mg s^{-1}, the gut processing rate determines the overall consumption during a low water period; (2) Oystercatchers cannot eat more than 26.2 g dry flesh during a low water period of 6 hours; so that (3) one low water feeding period of 6 hours is not long enough to obtain their food requirements.

What do birds do on the feeding grounds when their gut is full? Do they reduce their intake rate to the level of the processing rate of 0.66 mg s^{-1} or do they stop feeding and start again when the alimentary tract is partly or completely empty? Apparently Oystercatchers do not lower their intake rate when the alimentary tract is full, since intake rates as low as 0.66 mg s^{-1} are rarely recorded. Instead, Oystercatchers rest for a considerable part of the low water feeding period, just like Whimbrels (*Numenius phaeopus*) eating Fiddler Crabs *(Uca tangeri)* (Zwarts and Dirksen 1990). Active foraging is often restricted to the first and last hours of the low water feeding period (Brown and O'Connor 1974; Swennen *et al.* 1989). The higher intake rate on the incoming tide recorded in captive birds by Swennen *et al.* (1989) and in the field by Zwarts and Drent (1981) and Goss-Custard *et al.* (1984) probably guarantees that birds arrive at the high water roost with a full gut.

We have shown that Oystercatchers continuously make feeding decisions that enhance their intake rate. At the start of the chapter, we assumed they do this in order to minimize feeding time and thus maximize the time they can spend on other activities such as preening and aggressive behaviour. However, when the processing rate is usually so much lower than the intake rate, the birds can preen or be aggressive during their inevitable digestive pauses. This implies that feeding and, for example, preening are not necessarily competing activities. The question, already raised by Kersten and Visser (1996a) is this: why should we continue to expect Oystercatchers to always try to maximize their intake rate and thus minimize the time spent feeding? Maximizing intake rate would only seem to be relevant when the birds have difficulties in achieving an intake rate of about 1 mg s^{-1}; only in these circumstances is the total consumption determined by the intake rate itself (Fig. 2.15). It seems unlikely to be important for Oystercatchers to attempt to increase their intake rate when it has already reached levels of 3 or 4 mg s^{-1} and when the only apparent consequence of doing so is that they must pause earlier to allow for digestion.

While these arguments seem to apply to many situations in winter, they do not do so in the breeding season. The amount of food brought to the young often limits reproductive success (Ens *et al.* 1992). Breeding birds have to feed in a hurry to return to their nest and/or defend their territory or young, and the rate of provisioning is not limited by the capacity of their own gut to process food. Indeed, breeding birds feed at higher average rates than nonbreeding Oystercatchers, which are not time-stressed (Zwarts and Drent 1981; Hulscher 1982; Ens *et al.* 1996b). Moreover, Oystercatchers were able to handle prey faster and to increase their intake rates when they were experimentally forced to collect their food over short feeding periods (Swennen *et al.* 1989). It seems that Oystercatchers do not normally feed at the highest possible rate they can attain, even though they usually seem to make the most profitable choices in the sizes of prey they select. The task now is to investigate the trade-offs that the birds presumably make under various conditions. One such trade-off might be to balance a higher intake rate against the need to minimize the risk of damage to the bill when breaking into their large and well-defended prey (Hulscher, Chapter 1, this volume). This would mean that the long-term advantage of an undamaged bill would be set against the short-term goal of a higher food intake rate and reduced feeding time. Another trade-off might be between a lower intake rate but a reduced chance of being infected by parasites (Goss-Custard *et al.*, Chapter 6, this volume). What is clear, however, is that a model in which maximizing the intake rate is the only consideration is not adequate. More sophisticated models, which include trade-offs between a variety of changing influences, including the state of the bird's body reserves (Stephens and Krebs 1986), are now required.

Summary

Oystercatchers ignore small bivalves. These prey have a lower probability of being encountered, but this alone is not sufficient to explain why they are not taken. Small bivalves are not profitable given the time needed to open and eat them, and consequently they allow only low rates of return of energy. The size rejection threshold is not fixed, but varies according to the average intake rate during feeding. Oystercatchers increase their intake rate in two ways: they reject unprofitable prey and they adjust their searching behaviour in order to increase the encounter rate with the more profitable prey. The intake rate depends predominantly on the prey size taken. When large prey are present, the intake rate is high (3–4 mg dry flesh s^{-1}) and small and medium-sized bivalves are rejected, due to their lower profitability. When only small prey are available, the intake rate decreases to 1 mg s^{-1}. The daily consumption varies between 36 and 50 g dry flesh. This means that even when they feed at the average year-round rate of 2 mg s^{-1}, 5 h feeding a day is sufficient to meet their normal maximum daily energy expenditure. Ninety-two per cent of the intake rates observed in the field exceeded the rate at which the gut of Oystercatchers is able to process food (0.66 mg s^{-1}). Since processing rate forces the bird to stop feeding when their oesophagus is full (12 g), Oystercatchers must stop feeding for digestive pauses when their intake rate is high. Due to this digestive bottleneck, they have to feed 10 hours a day if there is one feeding period, or 8 hours a day if there are two feeding periods, digestive pauses included, to reach the minimum daily consumption of 36 g, required at thermo-neutrality.

Acknowledgement

We are grateful to Bruno Ens, Rudi Drent and John Goss-Custard for discussion and helpful comments on the manuscript.

3 Specialization

William J. Sutherland, Bruno J. Ens, John D. Goss-Custard and
Jan B. Hulscher

Introduction

Probably the best known fact about Oystercatcher biology is that individuals differ in the feeding technique they adopt. Niko Tinbergen even produced a film about Oystercatchers called *The specialist*. The suggestion that these birds learn which feeding technique to use from their parents has become a classic example of the cultural transmission of behaviour while the presence of individual specializations has numerous consequences for understanding many aspects of their physiology, behaviour and population ecology. Partridge and Green (1985) suggested that the three main factors that will result in differences in foraging behaviour between individuals are differences in morphology, frequency-dependent advantage and differences in habitat. In this chapter, we review the evidence for specialization, examine the three suggestions of Partridge and Green (1985) as to why it occurs, discuss briefly what the ecological consequences may be, and finally question the importance of cultural transmission in determining the specialization that individuals adopt.

Evidence for specialization

Specialization in diet

As described by Hulscher (Chapter 1, this volume), Oystercatchers feed on a wide range of prey species. By watching birds which have been marked with colour rings it has been shown that individual Oystercatchers often feed almost entirely upon one species, such as *Cerastoderma*, *Mytilus*, *Carcinus* or *Nereis*, for extensive periods of time (Norton-Griffiths 1968; Goss-Custard and Sutherland 1984; Hulscher and Ens 1991; Ens *et al.* 1996d), with the diet of individuals tending to be consistent between different breeding seasons (Table 3.1) as well as between winters (Goss-Custard and Durell 1983). It is clear that specializations often persist over many years.

The sequence of captures of different prey species by individual Oystercatchers on the Dutch Wadden Sea island of Schiermonnikoog provides an insight into prey specialization over a shorter time scale. Continuous observations of individually marked breeding adults were made for long periods, usually throughout the whole time for which the

Table 3.1 The correlation between percentage of each species in the diet for individual birds in successive years. In all years *Macoma* and *Nereis* were the staple foods. No mussels were available in the study area in 1986 (Hulscher and Ens 1991).

Years	*Nereis*	*Macoma*	*Mytilus*	*Mya*
1986 to 1987	0.72**	0.75**	-	0.37*
1987 to 1988	0.72**	0.47**	0.70**	0.16
1988 to 1989	0.81**	0.74**	0.02	0.32

*, $P < 0.05$; **, $P < 0.01$.

feeding grounds were exposed by the tide (Table 3.2). The chance that two successive prey captures belonged to the same species was highest for *Macoma* (94.9%) and lowest for *Arenicola* (47.3%). The chance that more than one species would be included in the diet increased slowly with the length of the observation period. Clearly, Oystercatchers in this study did not take all the potential prey types within periods of several minutes duration. Indeed, the two staple food items, *Nereis* and *Macoma*, were always present in each territory, yet particular individuals nearly always specialized on only one of them over the entire low water period (Ens *et al.* 1996d). On the other hand, prey specialization was generally not absolute, for most birds took more than one, usually two or three, species over the exposure period as a whole (Ens *et al.* 1996d).

Why ignore profitable prey?

An individual that specializes on a particular prey species thereby ignores prey that other individuals find profitable to feed on. One explanation for why they do this is that it is impossible to search for both prey effectively at the same time. How could this be? Gendron and Staddon (1983) propose that with increasing search rate (area searched per unit time), the probability of detecting prey declines. There is thus a trade-off between searching fast to encounter many prey in a short time and searching slowly to maximize the probability of actually detecting any prey that are encountered, leading to an optimal search rate at which the intake rate of prey is maximized. By definition, the probability of detecting a cryptic prey species at a given search rate will be lower than the probability of detecting a conspicuous prey species. As a result, the optimal search speed for cryptic prey will be lower than the optimal search speed for more conspicuous prey. Since animals can only search at one speed at a time, speed will be a compromise which may be so close to that which is optimal for one of the two prey that the other prey is effectively ignored.

This theory certainly fits the observations on Oystercatchers feeding on either *Macoma* or *Nereis*. The buried shells of *Macoma* are, under most circumstances, detected by touch (Hulscher 1982). *Nereis* also hide in burrows, but these burrows are generally too deep for the bill of the

Table 3.2 The correlation between the diet of successive prey captures (Ens *et al.* 1996d).

Previous prey (% of observations)	Arenicola marina	Macoma balthica	Mya arenaria	Mytilus edulis	Nereis diversicolor	Remaining prey	Number of observations	% of total
Arenicola marina	47.3	11.5	3.1	0.1	36.3	2.9	821	1.8
Macoma balthica	0.5	94.9	1.2	0.4	1.9	1.2	19855	43.4
Mya arenaria	2.3	20.2	52.5	5.8	13.5	5.7	1123	2.5
Mytilus edulis	0.1	4.6	2.9	87.6	2.9	2.5	2151	4.7
Nereis diversicolor	1.4	1.9	0.8	0.3	92.8	2.8	19875	43.5
Remaining prey	2.1	13.9	2.9	4.0	36.0	41.1	1619	3.5
Total							45744	100.0

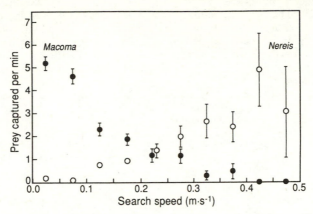

Fig. 3.1 The rate at which *Macoma* and *Nereis* are captured as a function of the rate at which the Oystercatcher searched, as inferred from the rate at which it paced. Based on short and detailed observations of Oystercatchers having one or both of these prey as staple food on Schiermonnikoog. Each observation ended after the bird had taken 50 steps during searching. Vertical bars represent 1 SE (Ens *et al.* 1996d).

birds to penetrate and the Oystercatchers probably catch the worms when they partly extend themselves from their burrows to feed on the surrounding mud. Detailed observations by Ens *et al.* (1996d) confirm that with increasing search speed, calculated from the pace rate, the rate at which the cryptic *Macoma* are captured declines, while the capture rate of the more conspicuous *Nereis* increases (Fig. 3.1). Thus, if an Oystercatcher chooses to capture *Macoma*, it should walk slowly while continuously probing the mud. If it chooses to hunt for *Nereis,* the bird should cover a large distance as quickly as possible to detect the occasional worm that has ventured out of its burrow. Since these are two incompatible searching techniques, the bird must choose between feeding on one of the prey or the other.

Specialization in technique

Not only do individuals specialize in their diet, but they also specialize in the techniques used to handle the prey. For example, amongst those individuals that specialize upon mussels, some Oystercatchers, referred to as stabbers (Hulscher, Chapter 1, this volume), feed by pushing their bill through the gaping valves, or by vigorously prising their way into closed shells, before chiselling out the flesh. Others, referred to as hammerers, hit one of the valves until it breaks and then insert the bill through the hole in order to chisel out the flesh. Furthermore, some hammerers almost always attack the dorsal surface of the mussel whilst others consistently attack the ventral surface (Table 3.3a). Closer examination reveals these specializations to be even more refined: amongst ventral hammerers, some largely attack the right-hand valve while others attack

Table 3.3 The feeding technique used by different birds. The different techniques are described in the text. In (a), the incidence of stealing mussels from other Oyster-catchers (kleptoparasitism) and of cleaning out mussels that had been opened and then lost by other birds (scavenge) are also shown (Sutherland and Ens 1987).

(a) Field studies

Bird	Stab	Dorsal hammer	Ventral hammer	Klepto-parasitism	Scavenge
KGBKKN	0	84	0	10	5
LGBKNN	220	37	0	49	18
LWAKBB	0	141	0	5	0
LBANNN	0	0	137	7	1
LBAKBN	55	163	0	7	1
ROAKKB	159	0	0	8	19
LGBNBB	0	65	0	0	4
ROABKN	0	66	0	1	0
RGBNNB	189	0	0	1	12
LBABBB	47	0	0	0	0

(b) Captive birds

Bird	Stab	Ventral hammer		
		Right	Left	Both
A	163	87	1996	1
B	109	729	7	0
C	43	233	310	0
D	35	796	4	0
E	49	221	75	0
F	50	84	151	1
G	93	0	0	0
H	1	44	24	0
I	42	0	0	0
J	14	0	0	0

the left (Table 3.3b). This degree of specialization is not restricted to birds feeding on mussels. Some individuals feeding on *Macoma* also preferentially attack the right valve while others concentrate on the left (Hulscher 1982).

However, although specialization is widespread and often intricate, it is not absolute and, in general, a bird is capable of using at least two feeding techniques on the same prey species (Hulscher 1982; Swennen *et al.* 1983; Goss-Custard and Durell 1988). For example, of 33 colour-ringed individuals feeding on *Mytilus* on the Exe estuary, 23 used two feeding techniques. Of 557 mussels consumed by these birds, 16.9% were opened using the minority technique, whether this was stabbing or hammering. Of 10 captive Oystercatchers feeding on *Mytilus*, only three

used the stabbing technique alone while five birds employed both stabbing and ventral hammering and two birds used all three techniques of
stabbing, ventral hammering and dorsal hammering (Table 3.3b).
Individuals with a preference for different valves revealed a similar
pattern: of six out of seven captive ventral hammerers that showed a
strong preference for attacking one valve, all attacked the minority valve
on some occasions (Table 3.3b). As with the diet, individuals clearly
often specialize in the strategy they use but they are nonetheless capable
of using an alternative under certain conditions.

Diet, feeding techniques, prey morphology and bill morphology

These studies of wild and captive birds raise the questions of why there
are a variety of techniques for opening prey and why individuals generally specialize on one of them, although not exclusively. One likely
explanation for the existence of a variety of feeding techniques is that the
prey of Oystercatchers, particularly mature bivalves, are difficult to
tackle and that birds with different feeding techniques concentrate on
exploiting different weaknesses in the prey. Durell and Goss-Custard
(1984) compared the shell thickness of *Mytilus* present on different mussel
beds with the shell thickness of mussels from these beds opened by birds
using various techniques (Fig. 3.2). The ventral hammerers selected those
mussels which were thinner on the ventral surface while the dorsal
hammerers selected mussels which were thinner on the dorsal surface. By
contrast, stabbers showed no preference for either dorsal or ventral shell
thickness (Durell and Goss-Custard 1984). Sutherland and Ens (1987)

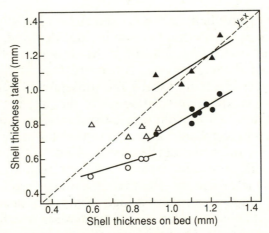

Fig. 3.2 The thickness of shells taken by Oystercatchers compared with the thickness of shells on the bed. Ventral thickness of shells opened by ventral hammering
(●); dorsal thickness of shells opened by dorsal hammering (○); ventral thickness of
shells opened by stabbing (▲); dorsal thickness of shells opened by stabbing (△)
(Durell and Goss-Custard 1984).

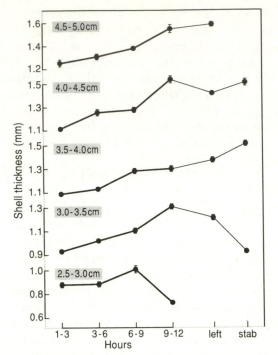

Fig. 3.3 The change in shell thickness of mussels of a given length taken by captive Oystercatchers over four consecutive three-hour periods. The shell thickness of those left (left) and those taken by stabbing birds (stab) are also shown (Sutherland and Ens 1987).

found that captive hammering Oystercatchers initially selected the mussels with the thinner shells but, as these became depleted, they selected mussels with thicker valves (Fig. 3.3); only the thickest shelled mussels were left at the end of the experiment. By contrast, stabbing individuals showed no preferences for thinner shells and took prey of average shell thickness.

The fact that different weaknesses of the prey require different feeding techniques does not explain why individuals specialize on only one technique. A hammerer that does not stab the stabbable mussels would seem to miss good opportunities to obtain food. The question, then, is why individuals do not flip rapidly between techniques as the occasion demands. One reason may be that individuals are 'committed' to a particular technique by virtue of a long-term, or basic, character. In Oystercatchers, a likely candidate is bill structure, which varies considerably between individuals. On Schiermonnikoog, for example, breeding adults were repeatedly caught on the nest and bill-tip shapes were classified in lateral view as being pointed, intermediate or blunt (Swennen et al., 1983; Hulscher 1985; Hulscher and Ens 1991). About two-thirds of the birds showed the same shape of the bill-tip on recapture (Table 3.4), and

Table 3.4 Relationship between bill-tip shape at first capture with the bill-tip shape when subsequently recaptured (Hulscher and Ens 1991)

Shape at recapture	Shape at first capture			
	n	Pointed (%)	Intermediate (%)	Blunt (%)
Pointed	26	65.4	34.6	0.0
Intermediate	131	12.2	66.4	21.4
Blunt	140	1.4	35.0	63.6

as may be expected, the majority of the changes were alterations in shape from intermediate into pointed or blunt or vice versa. Clearly, rapid and substantial changes in bill-tip shape are not usual. Insofar as bill-tip shape influences the efficiency with which alternative feeding techniques can be used, rapid changes in technique and diet would thus not be expected except, perhaps, under extreme circumstances, for instance during cold spells when some prey species are not available while others are.

On the other hand, large changes in bill-length and bill-tip shape can occur in individual birds over the longer term. How are these changes brought about? A sample of 56 individuals captured on the winter grounds and in the inland breeding area has shown that long-billed birds (mainly females) lose more of their tip in this process than do short-billed (mainly male) birds because the former have thinner bill-tips (Fig. 3.4). As described by Hulscher (this volume, Chapter 1), Oystercatcher bills grow at the rate of 0.44 mm per day, equivalent to one bill-length every six months. As a result, the bill-tip can change shape quite quickly, the mechanism apparently being abrasion. A pointed bill, which is associated with buried soft-bodied prey (Fig. 3.5) (Hulscher and Ens 1992; Durell *et al.* 1993), is 'moulded' through frictional forces as the bill is driven into the substrate. Abrasion of the tip continues until the cross-section, particularly the width, at the tip is sufficiently increased to withstand the sideways forces operating during probing for and extracting the prey. Similarly, the blunt and chisel-shaped bill-tips of Oystercatchers feeding on bivalves, and associated with hammering and stabbing respectively (Fig. 3.5) (Hulscher and Ens 1992; Durell *et al.* 1993), also seem to arise through abrasion (Hulscher 1985).

An intermediate bill-tip shape is associated with feeding on *Macoma* and *Nereis* and to a lesser extent on *Mya* (Fig. 3.5). *Macoma* is found moderately deep in the substrate while *Nereis* and *Mya* are both deep-dwelling species. Feeding on these prey types requires much probing and hence leads to abrasion of the outer corners of the two mandibles. *Macoma* is a small bivalve and it probably requires less force to open than is needed to open either a mussel or cockle; hence opening can be

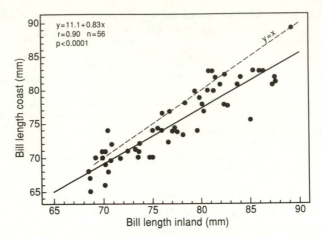

Fig. 3.4 Relationship between the bill-length of 56 adult Oystercatchers captured on the wintering grounds (Dutch Wadden Sea) and their length whilst on the inland breeding area. The $y = x$ line shows the relationship if bill-length remained constant (K. Koopman and J. B. Hulscher, unpublished information).

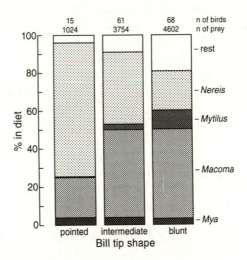

Fig. 3.5 The diet of adult male and female Oystercatchers on Schiermonnikoog during the summer (March–July) in relation to bill-tip shape. The numbers on top of the bars refer to the number of birds involved and the total number of prey eaten.

achieved with a finer bill-tip. The same applies to *Mya* whose shell always gapes, thus providing easy entrance for a pointed bill.

Bill-tip shape is clearly associated with particular food types, the apparent mechanism being abrasion. But, in the present context, the critical question is what happens when individual birds change diet or feeding technique? Unfortunately, direct field observations of concurrent changes in diet or feeding technique and bill-tip shapes are lacking and only indirect evidence from the field exists. Inland breeders in The Netherlands winter in the Wadden Sea where they predominantly feed on mussels and cockles and thus have intermediate and blunt bills. When they return to their breeding areas in March, they change to a diet of nearly 100% earthworms and, within two or three weeks, the bill-tips become pointed (Hulscher 1985). Experimental evidence for a change of bill shape upon a switch in food was found by Swennen *et al.* (1983). Individual captive birds feeding on an artificial mudflat containing *Nereis* had pointed bills but, when cockles or mussels were provided instead, the bill-tips became blunt; the reverse change in diet resulted in bill-tips changing from blunt to pointed. These changes in bill-tip were completed within 10 to 20 days, the relatively short transition period probably being due to the high and continuous growth and abrasion rates of the bill (Hulscher 1985). Although surprisingly rapid, such changes take too long to allow individuals to adjust their bill-tip shapes on a day to day basis, and this undoubtedly contributes to the high degree of specialization in both diet and feeding technique found in Oystercatchers.

As the bill-tip shape apparently follows the diet and feeding method, and not vice versa, it is necessary to ask why a particular diet or feeding technique is adopted in the first place. A favoured hypothesis is that the overall structure of the bill, as defined by its length and depth and hence strength, predisposes a bird to a particular diet and technique. Certainly, the diet not only varies with bill-shape but also with its length, with long-billed birds taking relatively more *Nereis* and *Mya*, and short-billed birds taking more *Macoma*, *Mytilus* and the other species (Fig. 3.6).

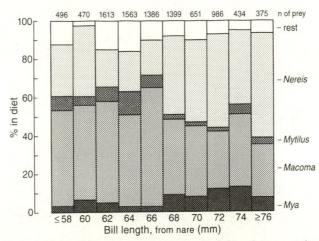

Fig. 3.6 The diet of adult Oystercatchers (males and females) on Schiermonnikoog during the summer in relation to bill-length (measured from the proximal end of the nostril). The total number of prey is shown along the top.

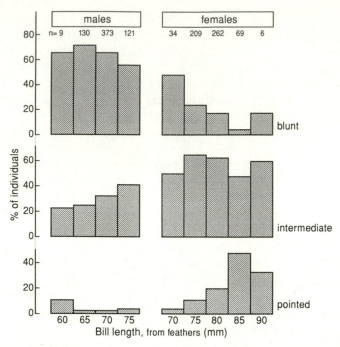

Fig. 3.7 Frequency histograms of the bill-lengths in 5mm classes (measured from the feather margin) of adult male and female Oystercatchers on Schiermonnikoog during the breeding season (May–June) according to the shape of the bill-tip. The total numbers (*n*) of birds in the sample across all bill-tip shapes are shown along the top.

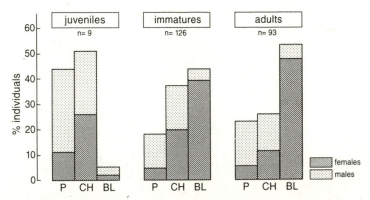

Fig. 3.8 The percentage of birds with pointed (P), chisel-shaped (CH) or blunt (BL) bill-tips in relation to age and sex (Durell *et al.* 1993). The number (*n*) of birds in each age class sample is shown along the top.

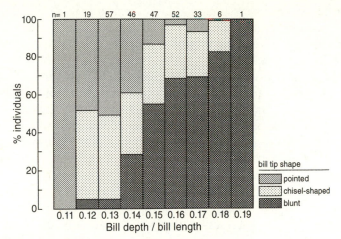

Fig. 3.9 The relationship between the percentage of birds with each bill-tip shape and the bill structure, expressed as the ratio of bill-depth to bill-length (Durell *et al.* 1993). The number (*n*) of birds in each sample is shown along the top.

Accordingly, the bill-tip shape also varies with bill-length (Fig. 3.7); short-billed birds have blunt bills while the long-billed birds, which feed on *Nereis*, *Macoma* and *Mya*, have pointed bills. Furthermore, Hulscher and Ens (1992) and Durell *et al.* (1993) showed that bill-depth also co-varies with diet and feeding method, these characteristics being related to sex (Zwarts *et al.* 1996c). Males tend to have blunt tips and to be mussel hammerers while females have pointed bills and feed more on worms (Fig. 3.8). These differences in diet and technique are related to the ratio of bill-depth to bill-length (Fig. 3.9); for example, a much higher propor-tion of the birds with relatively deep bills had blunt tips. The implication is that the diet and feeding technique depends upon the general structure of the bill, the currently favoured explanation being that the risk of breakage is probably greater in long, thin bills so that hammering is too risky (Hulscher and Ens 1992). Thus, although the bill-tip shape can be modified quite rapidly through growth and abrasion, the overall shape of the bill is less flexible and thus constrains the feeding options available to an individual.

Frequency dependent optimization

The overall form of the bill, however, does not on its own account for all the individual differences in specialization. Birds with comparable bill-forms do specialize on different prey and feeding techniques, and individ-uals can change both without any apparent change in overall bill structure. Other factors must affect the choice by individual birds. In the real world, Oystercatchers will not make choices independently of others; each bird is affected by, and affects, the choices made by others. Any gains from adopting a particular diet and technique are therefore likely to be frequency dependent. For example, only a certain proportion of mussel shells have thin ventral surfaces. As more birds specialize on ventral hammering, the intake rate of each individual is likely to decline as thin-shelled mussels become more scarce, and it will probably pay for some birds to adopt a different technique (Goss-Custard and Durell 1988). As an example of such a process, the removal of one sex of the downy woodpecker *Dicoides pubescens* resulted in the other altering its niche to encompass that of the other sex (Peters and Grubb 1983). How then can such frequency-dependent possibilities be tackled in Oyster-catchers?

Game theory provides an appropriate conceptual framework when alternative options can be identified, when the benefits from each are dependent on the frequency with which each option is adopted and when the relative frequency with which the options are used can change (May-nard Smith 1982). These criteria clearly apply to Oystercatcher special-izations. The expectation from simple game theory is that the intake rates should be the same for individuals adopting different techniques. If the gains are frequency dependent and individuals can switch techniques, as clearly they can, then the rewards should be equal across all tech-niques. However, five lines of evidence indicate that the various feeding specializations adopted by Oystercatchers are not equal alternatives.

First, adults specializing on *Nereis* and *Scrobicularia* on the Exe estuary feed in the fields at high water more often than adults specializing on mussels (Goss-Custard and Durell 1983). This suggests that mussel-feeders have less difficulty in meeting their daily energy demands. They may also suffer a lower risk of starvation during frost periods, when feeding in the fields is impossible. Indeed, of 14 corpses collected on the Exe after cold weather in Febuary 1991, all but one had pointed bills (Durell *et al.* 1993). Second, Boates and Goss-Custard (1992) compared the intake rate of adults specializing on four prey species: *Mytilus*, *Littorina*, *Nereis* and *Scrobicularia*. The greatest intake rate was achieved by birds feeding on *Scrobicularia*, even though this seems to be the least preferred prey. Amongst birds feeding on the mudflats, *Nereis* was preferred to *Scrobicularia* even though the intake was significantly lower on worms. Third, stabbing birds, especially the ineffient ones, are more likely to feed in coastal fields at high tide, suggesting that their intake rate on the estuary is lower than that of hammering birds (Goss-Custard and Durell 1988). Fourth, in all age classes examined, stabbing birds are signifi-cantly lighter than hammerers, in spite of the fact that the proportion of females is higher in stabbers than in hammerers and females tend to be heavier than males (S.E.A. le V. dit Durell, personal communication). Finally, across a sample of six beds, stabbing birds fed on average for ten per cent longer through the tidal cycle than hammerers (Goss-Custard *et al.* 1993).

We can think of four explanations as to why differences in intake rate persist between specializations.

Differences in morphology

Individuals may differ in morphology in ways which may be related to their age, sex, genotype or environment. Male Oystercatchers have relatively thick and short, and so probably stronger, bills and are thus the better sex at opening large sturdy prey, like *Mytilus* (Hulscher and Ens 1992; Durell *et al.* 1993). Female Oystercatchers have longer bills and thus are the better sex at reaching the deep-burrowing, and thus large (Zwarts and Wanink 1984), clams. Similarly, many Oystercatchers start to attack *Mytilus* relatively late in life, perhaps because the bones of the bill and skull and the musculature of the feet and head do not develop fully until a bird is several months, or even a year, old. If individuals vary because of such sex-related and age-related differences in morphology, individuals are no longer equally free to adopt any diet or feeding technique and so may differ in intake rate (Parker 1982).

Intake is an inappropriate measure to compare options

Although intake rate is clearly important, especially during periods of severe weather, it is unlikely to be the sole factor determining a bird's choice. One clear example of this is that, on the Exe, Oystercatchers have a strong aversion to foraging on areas of sloppy mud (Goss-Custard *et al.* 1992). As the energy cost of walking through mud is negligible (Speakman 1987), explanations other than increased energy expenditure are needed to explain the preference of Oystercatchers for a firm substrate.

Other reasons why intake rate may not be an appropriate measure of the return associated with different foraging options include the possibilities that: (i) different diets differ in variance (Sutherland and Anderson 1987); (ii) the intensity of parasite infection may be related to diet; for example, *Mytilus* has few parasites compared with *Cerastoderma* (Goater 1988); and (iii) digestability may vary between prey types. Zwarts and Blomert (1990) showed that the digestive efficiency of Whimbrel (*Numenius phaeocopus*) decreased as the amount of inorganic matter consumed increased. Birds feeding upon earthworms or polychaetes may also suffer a reduced digestive efficiency if the prey is coated with soil or mud. As captive Oystercatchers can increase their intake rate when the amount of time available for feeding is experimentally reduced (Swennen *et al.* 1989), field estimates of intake rate may more often reflect the current demand for energy or the rate at which the gut can process food (Kersten and Visser 1996a) than externally imposed constraints on birds that are always attempting to maximize their intake rate. The realization that foraging animals must trade-off a number of competing concerns, and so may not always feed at the maximum rate as early optimality models assumed (Stephens and Krebs 1986), has implications for interpreting field measurements of intake rate that have not yet been fully explored.

Diet and feeding technique is related to dominance

The feeding areas chosen by Oystercatchers when numbers are high is greatly influenced by the increased levels of interference experienced in the richest places (Goss-Custard et al., Chapter 5, this volume). As diet and feeding site are linked, it is possible that subdominant Oystercatchers are unable to feed at an adequate rate on mussel beds where the density of competitors is very high and instead feed on other prey, such as worms and clams, on the open flats where the density of competitors can be much lower (Boates and Goss-Custard 1992). Thus, many of the subdominant second- and third-winter birds leave the mussel beds of the Exe for the mudflats in autumn as the adults return from the breeding areas (Goss-Custard et al. 1982b). In the absence of competitors, mussel-hammering Oystercatchers are capable of achieving a higher intake than stabbing birds, but hammerers seem more susceptible to interference (Goss-Custard and Durell 1988). Feeding densities may be so high on the most preferred Mytilus beds that the intake rate of a hammerer of average foraging ability is depressed to the level of that of an average stabber. On the less preferred beds, interference is less important and the average intake of hammerers exceeds that of stabbers (Goss-Custard et al. 1984; Goss-Custard and Durell 1988). Although hammerers and stabbers do occur together on mussel beds, hammerers may, on average, be more dominant than stabbing birds and occur in the better feeding locations. If so, birds having equivalent morphology may become stabbers or hammerers according to their ability to dominate competitors.

Differences between specializations in ease of learning

Some diets and feeding techniques may be intrinsically more difficult to learn than others. This hypothesis seems most likely to apply to the ventral hammering specialization which is adopted relatively late during a bird's immaturity (Goss-Custard et al. 1993). Clearly, this is an insufficient explanation as many adults are stabbers but remains a possibility that has not yet been firmly rejected.

Conclusions

Although feeding specialization has been one of the most eye-catching aspects of the behavioural ecology of Oystercatchers, much remains to be discovered. For example, the degree of specialization shown by birds feeding inland during the summer is unknown, as is the relationship between winter and summer diets; the factors that determine the feeding specialization in winter may also have implications for the behaviour of

Oystercatchers during the breeding season. But some progress has been made in isolating the factors and processes that determine an individual's choice of specialization. Present evidence suggests that the gross morphology of the bill, including its length, breadth and depth, plays a major role in determining the diet and feeding technique adopted. The gradual hardening of the bill as birds mature may also have a considerable influence on the way in which feeding develops during the first few years; for example, hammering into *Mytilus* and *Cerastoderma* may involve too high a risk of damage for the juvenile bill with the result that most first-year birds either stab into shellfish or eat other, soft-bodied prey. But the morphology of the bill does not explain all the individual variation seen and questions remain that have not yet even been posed in the literature; for example, why does one bird choose dorsal hammering while another chooses ventral hammering? As discussed above, there are a number of hypotheses available to account for these differences, but the critical observations and experiments required to distinguish between them remain to be done.

Ecological consequences of specialization

Specialization has implications for almost every aspect of ecology, such as distribution and parasite burden. For example, niche width of a species can be wide (e.g. *Homo sapiens*) or narrow (e.g. the Sheep Nostril-fly (*Oestrus ovis*)). Within the range occupied by a species, all individuals can have a similar niche width or the niche may be partitioned between individuals. Roughgarden (1972, 1974) separated the within-phenotype component of niche width from the between-phenotype component in *Anolis* lizards and showed that the between-phenotype component contributed between 1.4% and 32.5% of the total niche width. Although such a precise measure for the components of niche width in Oystercatchers is not yet available, it is clear that the between-phenotype component is also likely to be important in this species.

In the absence of specialization, individuals would be expected to change their diet and feeding technique according to the relative profitability of the options available. However, when individuals do specialize, there may be a time delay, or hysteresis, in responding to changes in the prey population which could affect the fitness of individuals. In Oystercatchers, for example, changes in feeding conditions can be sudden and dramatic during cold spells, when the energy needs of the birds are maximal and the high-level intertidal mud flats may freeze over (Goss-Custard *et al.*, Chapter 6, this volume). Cockles are very sensitive to low temperatures and intertidal populations can be completely wiped out in severe winters (Hancock 1971). Other prey species may be much less affected, however; for example, *Mytilus* is moderately cold resistant while

Macoma is least affected by cold winters (Beukema *et al.* 1993). Snow cover and frost in coastal fields make it impossible for Oystercatchers to find earthworms. Because of all these changes, Oystercatchers are very vulnerable during cold spells and may emigrate temporarily to avoid the locally hostile conditions (Goss-Custard *et al.*, Chapter 6, this volume). However, in doing so, they may be forced to change diet or feeding technique which, in combination with unfamiliarity with the area, may further reduce survival (Hulscher 1989, 1990). Although there is little evidence, it is therefore quite likely that under such conditions selective mortality among food specialists may be expected and, in turn, affect their representation through the Oystercatcher population.

The acquisition of feeding specializations

An obvious and almost trivial explanation for differences between individuals in the prey species they take and the feeding techniques they use is that they occupy different habitats. Oystercatchers on rocky shores are inevitably more likely to specialize upon *Patella* while those on sandy shores are more likely to specialize on *Arenicola*. However, this begs the question as to which comes first—the habitat choice or the specialization. The more local the scale, the more difficult it is to determine whether the habitat determines the diet or vice versa, especially as there may be other constraints on habitat choice, such as a bird's dominance and the strength of its bill.

One way to start disentangling the process by which individuals acquire their specialization is through longitudinal studies of individuals or cohorts of maturing birds so that the sequence in which choices are made can be established. Many juvenile Oystercatchers on the Exe estuary feed on mussels when they first arrive on the estuary in late summer, and most of them use the stabbing technique. Studies of marked individuals show that, subsequently, many juveniles leave the mussel beds and feed over the winter on *Nereis* and *Scrobicularia* on mudflats or on earthworms in the fields. Few juveniles feed on mussels over the winter unless severely cold weather drives them off the more easily frozen fields and upper mudflats; in a cold spell in 1987, for example, some 80% of all juveniles returned temporarily to the mussel beds to feed (J.D. Goss-Custard, unpublished information). First-winter birds always return to the mussel beds in April where most remain for the spring and summer. Subsequently, some second-winter birds again leave the mussel beds on the return of the adults from the breeding grounds. The proportion of birds feeding on mussels gradually increases over the first few years of life, with an increasing proportion using one of the two hammering techniques (Goss-Custard *et al.* 1993). Clearly, the birds show a great deal of flexibility in their diet, and thus feeding method, as they mature and many changes in both diet and feeding technique occur.

The departure of young birds from the mussel beds of the Exe in autumn just as interference competition begins to intensify (Goss-Custard and Durell 1987a) and their return in April after the dominant adults have left might suggest that, over their first winter, the young birds feed on prey other than *Mytilus* because they are at a competitive disadvantage on mussel beds rather than because they are inexperienced at feeding on *Mytilus*. In a study of unmarked birds, Goss-Custard and Durell (1987a,b,c) compared the intake rates of mussel-feeding stabbing juveniles and adults in a series of matched-pair observations made at such low densities (< 100 Oystercatchers per ha) that interference was unlikely to have had much affect. Juveniles fed more slowly than adults from August through to October, but achieved parity between November and February. Thus, juveniles switch from mussels to other foods at the very time their intake rate achieves parity with adults. Although there are some indications that juveniles may not be so efficient at extracting the flesh even then (Norton-Griffiths 1968; J.D. Goss-Custard, unpublished information), the implication of these results is that their very low dominance, rather than poor foraging skill, causes juveniles to leave the mussel beds as the heavier demands of winter arrive. When juveniles return to the mussel beds in April, the level of interference is again low and, in any case, most of the competitively superior adults are absent. Although the profitabilities of the alternative winter foods to juveniles have not yet been established, the sequence of events does suggest that, during their first winter, individuals respond flexibly to the changing competitive circumstances by switching diet. Although more individuals were suspected of having done so, only six out of 30 colour-marked individuals of known age were observed feeding on mussels for the first time in their first winter. The majority start doing this in later years, with more birds beginning in summer than winter. Probably young Oystercatchers require a couple of years of increasing dominance and foraging experience to withstand the interference from older birds before they can embark on tackling a difficult prey like the mussel.

Culture

The hypothesis that feeding techniques in Oystercatchers are culturally determined is very well known; as an example of its fame, and although wrong in almost every important detail, the front cover of the book *The evolution of culture in animals* (Bonner 1980) pictures an Oystercatcher opening a mussel. It is probably this fact about Oystercatchers that attracts most interest from biologists in other fields.

Norton-Griffiths (1968) showed that both partners of a pair either stab or hammer; mixed pairings were not observed. In this study, hammering parents always produced young that hammered and stabbing parents always produced stabbing young. To determine whether the specializations had a cultural or genetic basis, he switched eggs between nests with parents of different feeding techniques. As chick survival is low, only three young survived until their feeding technique could be distinguished. All had the same technique as their foster parents, implying that the technique is learnt rather than genetically determined. Although the sample size of this neat experiment was small, it became a classic example of the cultural transmission of behaviour in birds. Furthermore, the implication that individuals could not adopt a technique without first learning it from their parents subsequently led to the belief that Oystercatcher specializations become fixed and so inflexible.

It now seems that this story is too simple (Sutherland 1987). First, as this chapter has shown, there is more flexibility amongst adults in diet and technique than previously thought; although both captive and wild birds have one favoured technique, they frequently use an alternative. Second, the diet and behaviour of Oystercatchers on the Exe estuary differs between birds of different ages, and many marked individuals have been seen to change as they mature; indeed, many young birds change during the course of their first few months. Third, some birds are reared inland but feed on the coast during the winter after leaving their parents and so they cannot learn their intertidal feeding techniques from their parents. The evidence thus points to there being much more complexity and flexibility in the diet and feeding techniques of Oystercatchers than implied by the interpretation that has been subsequently placed on Norton-Griffiths' experiments.

But although the story is more complex than originally thought, the possibility that young Oystercatchers learn by observing their parents feed remains a real one. Before becoming completely independent, the chicks often feed for themselves, while following their parents. They handle prey discarded by the parents, peck in burrows probed by the parents and clean out shells of molluscs consumed by the parents. At the very least it seems that early experience of how to feed on marine prey, which forms the winter food of almost all Oystercatchers (Hulscher, Chapter 1, this volume), might give young reared on the coast an advantage over young reared inland. Clearly, this area needs further research and it would be interesting to repeat Norton-Griffiths' experiment with a larger sample size. Our expectation is that there are advantages to specialization and that the specialization adopted by an individual will depend upon a suite of characters. In addition to the kind of prey species available, a minimal list of these would include the experience gained by watching the technique being used by the parents, the individual's own bill morphology, the techniques used by other Oystercatchers and its dominance and thus the interference experienced when feeding on each prey. At present, the relative importance of each of these processes, and the way in which they interact during an individual bird's development, is unknown.

Are Oystercatchers odd?

Despite their flexibility, Oystercatchers show a remarkable degree of specialization. As described earlier, some birds may even concentrate on hammering the right-hand or left-hand valve of a mussel. Why should this extraordinary degree of specialization be recorded in Oystercatchers but not in other species?

One possibility is that the prey of Oystercatchers are technically very difficult to handle compared with those of many other species. Individual prey differ in their weaknesses and so individual Oystercatchers must acquire the experience to overcome them. Accordingly, some birds, for example, become expert at locating *Mytilus* with thin right-hand valves and in breaking through at this point. Predator species that feed on prey that are technically easier may not require such individual specializations to be effective.

The second possibility is that such specializations are far more common amongst predator species than currently believed. If so, the main reason why specialization is so well documented in Oystercatchers is that they are easy to watch; for reasons outlined in the Introduction to this book, an army of researchers has studied, amongst other things, their feeding behaviour. Similar intensive studies of Great Tits (*Parus major*), for example, might reveal similar individual variation in feeding techniques.

Indeed, there is evidence for individual specialization in other species. Turnstones (*Arenaria interpres*) show six different feeding techniques: routing, turning stones, digging, probing, hammer-probing and surface-pecking (Whitfield 1990). In common with Oystercatchers, individual turnstone show a strong, but not absolute, prediliction for a particular specialization. Individual Curlew (*Numenius arquarta*) show feeding specializations that persist over many years and have a range of hunting and handling techniques adapted to specific prey and sizes of prey (Ens and Zwarts 1980). Similarly, Grant (1986) and Price (1987) have shown that differences in diet occur between individuals of the same Darwin's finches and that, as in Oystercatchers, some of this variation is related to variation in morphology. Wintering Black-capped Chickadees (*Parus atricapillus*) also show individual differences in feeding behaviour (van Buskirk and Smith 1989). Coal Tits (*Parus ater*) show individual differences in niche use which are again related to morphology and age (Gustafsson 1988). Indeed, specialization may apply very widely across many taxa. For example, individual Bumblebees (*Bombus vagans*) specialize over a short period of time on a limited range of flowers and, with practice, improve their ability to tackle a species morphology (Heinrich 1979). Although it is still difficult to state how frequently such specializations occur throughout the animal kingdom, it appears that in terms of their degree of specialization, Oystercatchers may not be so special as has so often been believed.

Summary

Individual Oystercatchers often feed almost entirely upon one species, such as *Cerastoderma*, *Mytilus*, *Carcinus* or *Nereis*, and use the same feeding technique for extensive periods of time, with specializations persisting over many years. But despite the Oystercatcher's reputation for inflexibility, specialization is not absolute and individuals can vary their diet and feeding technique over periods varying in length from one tidal exposure period to years. A likely explanation for the variety of feeding techniques employed is that the prey, particularly mature bivalves, are difficult to tackle and that birds with different feeding techniques concentrate on exploiting different weaknesses in them; for example, ventral hammerers attack *Mytilus* with especially thin shells on the ventral surface. However, that specialization may be profitable does not, in itself, explain why individuals do not flip rapidly between techniques as the occasion demands. This question is especially apt because, through a combination of rapid bill growth and the abrasion associated with particular diets and feeding techniques, the shape of the bill-tip in Oystercatchers can change within two weeks to that appropriate for a new diet or feeding technique. As the bill-tip shape apparently follows the diet and feeding method, and not vice versa, it is necessary to ask why a particular diet or feeding technique is adopted in the first place. The evidence suggests that the diet and feeding technique depends a great deal on the general structure of the bill. The risk of bill breakage is probably greater in long, thin bills so that certain feeding techniques and large shellfish can only be exploited by birds with stout bills. Thus males, which have relatively short and deep bills, can hammer into shellfish and so have blunt bill-tips while females have long and thin bills and feed more on worms and clams, and so have pointed bill-tips. Similarly, the immature feeding apparatus of immature birds may only allow them safely to stab into shellfish or take soft-bodied, buried prey. Although the overall shape of the bill is less flexible and thus constrains the feeding options available to an individual, it does not explain all the observed specializations and the possible role of other factors, such as frequency-dependent optimization in a competitive social environment, are discussed.

Despite their reputation for specialization and inflexibility, Oystercatchers are very adaptable. Rather than their diet and feeding technique being immutably determined by parental training, a variety of factors, including characteristics of the individual itself and the social and prey environment in which it matures and lives as an adult, may interact to determine the particular specialization a bird adopts in its own particular circumstances. And rather than being unusually specialized, Oystercatchers may merely be a well-studied example of a general tendency of many predators in a wide range of taxa to specialize in order to exploit the varying weaknesses in the anti-predator defences of their prey.

4 Feeding with other Oystercatchers

Bruno J. Ens and John T. Cayford

Introduction

Because Oystercatchers are conspicuous and feed on easily identifiable prey items in an open habitat, they are an ideal species for studying the two competitive processes among foraging predators of prey depletion and mutual interference. This chapter reviews studies of mutual interference, defined as the more or less immediate and reversible reduction in intake rate at high densities of conspecifics (Goss-Custard 1980); the effects of depletion are considered elsewhere (Goss-Custard *et al.*, Chapter 6, this volume). Early theoretical work (e.g. Hassell 1978), as well as empirical studies on shorebirds (Goss-Custard 1970; Zwarts 1978; Zwarts and Drent 1981), did not explicitly address differences between individuals. But field studies of marked birds revealed that individuals did not suffer equally from an increase in bird density. In common with Łomnicki (1978, 1982), who argued that, 'the assumption that all individuals are equally affected by increasing population density disagrees with empirical evidence and leads theoretical population biology into a blind alley', recent studies on shorebirds have focussed less on showing that interference occurs and more on measuring differences in susceptibility between individuals (Ens and Goss-Custard 1984; Goss-Custard *et al.* 1984; Goss-Custard and Durell 1987a,b,c, 1988; Whitfield 1985a). But whereas the aim of most studies on interference has been to understand feed-back processes at the population level (Goss-Custard 1980), this chapter explores how the optimal feeding and fighting decisions of individuals depend on competitor density, competitive ability, age and food type. Since the notion of competitive ability is central to theories on the distribution of unequal competitors (Sutherland and Parker 1985; Parker and Sutherland 1986), yet lacks a clear operational definition, we also examine this concept and how it might be measured in Oystercatchers. In doing so, the chapter reviews the wealth of important information on individual variation, while highlighting the major gaps. The argument is based for convenience on the mussel beds of the Exe estuary, with other work being reviewed for comparison.

What should a wintering Oystercatcher do to maximize its fitness? Oystercatchers separate from their partners and fledged chicks at the end of the breeding season so their major goal in winter seems simply to survive until the spring departure. Though small waders can be heavily predated at this time of year (Page and Whitacre 1975; Kus *et al.* 1984; Whitfield 1985b; Whitfield *et al.* 1988; Bijlsma 1990; Cresswell and Whitfield 1994), few Oystercatchers are taken, probably because of their large size (Whitfield *et al.* 1988). The best known cause of mortality is starvation during cold spells (Goss-Custard *et al.*, Chapter 6, this volume). Accordingly, and in the absence of information on the importance of timing and body condition in spring for breeding success, we assume throughout that wintering Oystercatchers aim to minimize the risk of starvation by maximizing their energy intake rate over the low tide exposure period.

Aggregation in Oystercatchers: do they feed in flocks?

On most estuaries, Oystercatchers are dispersed widely but aggregations do occur locally, usually on beds of mussels or cockles (Fig. 4.1b). Such aggregations have been viewed as feeding flocks (Goss-Custard 1970; Vines 1980; Rands and Barkham 1981), but this is probably incorrect. As Vines (1980) showed on the Ythan estuary, Oystercatchers feeding on a mussel bed are distributed non-randomly. Short nearest-neighbour distances were under-represented, suggesting avoidance, whereas intermediate ones were over-represented. Contrary to Vines' own view, the latter does not imply gregariousness; with some distances being under-represented, others have necessarily to be over-represented. Nor was there any indication that very long nearest-neighbour distances were also under-represented. Thus, there is no evidence that Oystercatchers actively seek others while feeding; rather, they appear only to aggregate on local concentrations of their food supply (Goss-Custard 1977b; Goss-Custard *et al.* 1981; Zwarts and Drent 1981; Sutherland 1982c,d).

Studies of marked individuals confirm that Oystercatchers on mussel beds, at least, do not flock in the sense in which the term is normally applied to wintering birds. Adults are long-lived and extremely faithful to their wintering grounds. On the mussel beds of the Exe estuary, many occupy well-defined home ranges to which they return year after year (Goss-Custard *et al.* 1980, 1982a). Others roam more widely, although even these birds usually have only a few favoured feeding areas and a particular route by which they reach them on the falling tide (Goss-Custard *et al.* 1981; B. J. Ens, unpublished information). The frequent independent movement of individual Oystercatchers through the exposure period show that they do not form feeding flocks with a more or less stable membership, as do Turnstones *(Arenaria interpres)*, for example (Whitfield 1985a); instead, individuals appear simply to return to specific sites which other birds also use. In fact, the absence of flocking at low water is thrown into stark relief by their dispersion at high water when they clearly do roost in tight flocks (Fig. 4.1a).

Finally, we may ask what Oystercatchers might gain by foraging in

(b)

Fig. 4.1 Photographs showing the spacing of Oystercatchers under different conditions. (a) Roosting Oystercatchers. (b) Oystercatchers feeding on cockles (front) and mussels (back) (J. van de Kam and B. J. Ens).

flocks. On the assumption that the primary advantage of flocking in waders is to reduce the risk of predation (Goss-Custard 1970; Myers 1984), Oystercatchers might be able to benefit, albeit weakly, from flocking at all stages of the tidal cycle. However, the risk of predation may be much reduced during low tide because avian predators seem to concentrate their hunting effort at high tide (Whitfield 1985b; Bijlsma 1990). Furthermore, as the rest of this chapter discusses, high rates of interference for many Oystercatchers may actively deter them from feeding close to others and thus foraging in flocks because of the energetic costs incurred.

Interference: the costs of feeding together

Measuring interference

Zwarts and Drent (1981) first capitalized on the special opportunities provided by the tide; each tide can be considered a *gratis* natural experiment. At the beginning and end of each exposure period, little feeding area is available and Oystercatchers necessarily feed at high densities compared to low water, when a much larger area is accessible (Fig. 4.2). By marking out squares and counting the birds periodically and measuring their intake rates, the interference hypothesis can be tested that intake rate decreases immediately as bird density increases. However, a number of factors may lead to spurious negative correlations between food intake and bird density: (1) the condition and growth rate of filter-feeding molluscs are poorer at the high shore-levels that expose first, leading to lower intake rates at the very stages in the exposure period when bird densities are highest (Goss-Custard and Durell 1987c); (2) as the tide recedes and bird density decreases, prey may become easier to attack by hammerers because the substrate dries out (Hulscher, Chapter 1, this volume), thus raising intake rate at a time when bird densities are generally low; (3) feeding motivation may vary through the exposure period, due to circatidal rhythms (Daan and Koene 1981; Swennen *et al.* 1989) or digestive bottle-necks (Zwarts and Dirksen 1990; Kersten and Visser 1996a); (4) as different individuals begin feeding at different times, there may be systematic changes through the exposure period in the food-finding abilities and/or dominance status of the sample of Oystercatchers being studied. However, by taking the effects of these confounding variables into account, and by testing for interference in individual birds, the evidence for interference has become convincing and very unlikely to arise only by spurious correlation.

The causes of interference

The early studies of interference were performed on unmarked individuals feeding on mussels (Koene 1978; Zwarts and Drent 1981; Sutherland and Koene 1982). They all reported a decrease in intake rate at high bird

Fig. 4.2 Photographs showing Oystercatchers feeding on a mussel bed at three stages of the tidal cycle. (a) The tide has just receded sufficiently to expose the first tops. (b) The mussel bed is exposed halfway. (c) The tide is fully out (B. J. Ens and J. van de Kam).

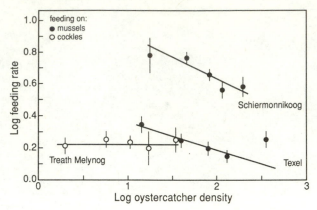

Fig. 4.3 Examples of evidence for interference from plots against bird density of the numbers of prey consumed per unit time (feeding rate) by samples of unmarked Oystercatchers (Zwarts and Drent 1981; Sutherland and Koene 1982). The prey were mussels on Schiermonnikoog and Texel but cockles at Traeth Melynog.

densities; two early examples are shown in Figure 4.3. Other studies on cockle-feeding birds that failed to show interference (Fig. 4.3) were probably conducted at bird densities too low to detect interference (Goss-Custard 1977b; Sutherland and Koene 1982). Interference in Oystercatchers may be due to one or more of three causes: (1) high densities of Oystercatchers may attract interspecific kleptoparasites which steal an increasing number of prey items; (2) more prey may be lost to conspecific kleptoparasites as competitor density rises; and (3) fewer prey may be captured as density rises because of increased amounts of prey depression, avoidance behaviour, displacement from good feeding spots or time lost in aggression (Goss-Custard 1980). The evidence for each of these mechanisms being responsible for interference is reviewed in turn.

Interspecific kleptoparasitism

The amount of food lost by Oystercatchers to interspecific kleptoparasites differs between studies. Oystercatchers feeding on Giant Bloody Cockles (*Anadara senilis*) on the Banc d'Arguin in Africa lost almost half their food to kleptoparasites, mainly Lesser Black-backed Gulls (*Larus fuscus*) (Swennen 1990). As Figure 4.4a,b illustrates, Herring Gulls (*Larus argentatus*) stole 12% of the mussels captured by Oystercatchers in one study (Koene 1978) while Common Gulls (*Larus canus*) and Herring Gulls each stole 4% of captured mussels in another (Zwarts and Drent 1981; L. Zwarts, personal communication). In contrast, Carrion Crows (*Corvus corone*) stole only 1.2% of the mussels captured by Oystercatchers on one mussel bed of the Exe (Ens and Goss-Custard

(a)

(b)

Fig. 4.4 Photographs of interspecific kleptoparasitism. (a) Herring Gull surprising an Oystercatcher just after it opened a mussel. (b) Common Gull chasing an Oystercatcher with a mussel (L. Zwarts).

1984), although more were stolen on another bed (Boates 1988). Little food is lost to kleptoparasites by Oystercatchers feeding on cockles and mussels in the Delta area of South-west Netherlands (P. Meire, personal communication). In many other studies, the amount of food lost to other species is not reported and probably negligible. Although the proportion stolen is very variable, kleptoparasites are clearly capable of stealing many of the prey captured by Oystercatchers.

Theoretically, interspecific kleptoparasitism of sufficient intensity could lead to interference if more prey are lost to kleptoparasites as Oystercatcher densities increase. Does this happen? Interspecific klepto-parasitism is increasingly common in birds as more of the following conditions are met: (1) The host species feeds in high densities; (2) the food occurs in large quantities; (3) the food items are large; (4) the food supply is predictable; (5) food is visible when captured by the host; and (6) the kleptoparasites suffer from food shortage (Brockmann and Barnard 1979). The first five conditions indicate profitable feeding conditions for the host species and clearly apply to Oystercatchers feeding on mussels. As for condition (6), Thompson (1986) has subsequently shown that the attack decisions of a kleptoparasite species are probably shaped solely by economic considerations which would apply whether or not the parasite was short of food. If so, species would be expected to steal from Oystercatchers simply when Oystercatcher densities, and so the numbers of potential victims and the profitability of stealing, are high.

On the Exe, the rate at which Oystercatchers lost mussels to Carrion Crows did not increase as Oystercatcher densities increased, but the rate of kleptoparasitism was low anyway (Ens and Goss-Custard 1984) and the crows found many mussels for themselves. But the Common Gulls stealing mussels (Zwarts and Drent 1981) and Lesser Black-backed Gulls stealing Giant Bloody Cockles (Swennen 1990) seemed to depend entirely on Oystercatchers. These kleptoparasites could not open the prey themselves and apparently defended a certain number of Oyster-catchers against other parasites, as was also noted by Thompson (1986) for Black-headed Gulls (*Larus ridibundus*) parasitizing gulls and plovers. The net effect of territorial behaviour among kleptoparasites could be that Oystercatchers actually lose fewer, rather than more, prey to klepto-parasites at high densities of conspecifics. Indeed, whether comparing sites or different stages in the exposure period, L. Zwarts (personal com-munication) found either constant or increasing numbers of Oyster-catchers per gull with increasing Oystercatcher densities. The evidence therefore suggests that, although Oystercatchers may under rare circum-stances lose much food to kleptoparasites, the rate of loss is probably not linked to short-term changes in the density of the Oystercatchers them-selves. Interspecific kleptoparasitism is therefore unlikely to be the cause of interference in Oystercatchers.

Intraspecific interactions and social dominance

If interference arises from intraspecific interactions, including kleptoparasitism, its severity is likely to vary between individuals according to their social status. This can be defined according to the outcome of the aggressive encounters between individuals that are particularly frequent on mussel beds. Two types of encounters are equally common and always end in clear victory: (1) in fights over feeding spots, one individual threatens another, characteristically adopting the diplomatist attitude (Fig. 4.5a). The victim usually moves away, so the attacking bird is scored as having won the encounter. However, sometimes the victim retaliates successfully and holds its ground and is then deemed to have won. Very often, retaliating birds performed the piping display (Fig. 4.5b). Whether the immediate benefit to the victor of these encounters is the acquisition of a good feeding spot is often unclear in the field, although Oystercatchers are known to defend feeding spots. Experimentally created rich patches of food, consisting of large *Mya arenaria* buried at shallow depth, were immediately defended by marked Oystercatchers, until they had depleted them after four days (B. J. Ens and L. Zwarts, unpublished information). Similarly, when Leopold *et al.* (1989) created a good feeding patch with large cockles and a poor feeding patch with small cockles in an experimental tidal ecosystem, the dominant bird 'excluded' the subdominants from the good feeding patch. Clearly, Oystercatchers do contest feeding spots and not just individual items. (2) In fights over food items, one individual usually rushes towards a potential victim as it opens a mussel. Quite often, there is no overt aggressive display. Though retaliations do occur, it is more likely that the victim will run away, with or without its prey. Even though the attacking bird is not successful in obtaining the prey—because the victim has carried it away—it is scored as having won the encounter. It is causing the victim to yield ground, rather than stealing the mussel itself, that defines the attacker as being successful. Similarly, the victim wins the encounter if it holds its ground, even if the mussel is stolen from it, or simply not relocated, after the dispute.

Operationally defined this way, wins and losses in encounters over food items or feeding spots can be used to calculate a dominance score which accurately predicts the outcome of aggressive encounters between marked individuals (Fig. 4.6a). The many empty cells in the matrix are due to the strong site fidelity of individual Oystercatchers; many birds were never seen to meet. Quite small differences in dominance score accurately predicted the outcome of an encounter, more or less irrespective of the absolute difference between the dominance scores of the birds involved (Fig. 4.6b). The argument is not circular, because only a minority of individuals was marked, so the dominance score is effectively based on encounters with unmarked individuals. Dominance score also

(a)

(b)

Fig. 4.5 Photographs of two important aggressive postures in Oystercatchers. (a) Diplomatist attitude. (b) Piping (J. van de Kam).

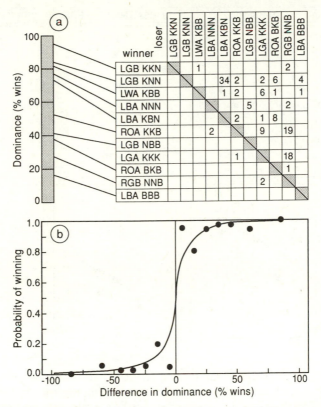

Fig. 4.6 (a) Dominance matrix showing how often marked Oystercatchers, ranked on the basis of their dominance score, won or lost encounters against each other (Ens and Goss-Custard 1984). (b) Probability that, in an encounter between two marked birds of known status, the subject bird beats its opponent as a function of the difference between them in their dominance score. The dominance score of each individual was measured from the outcomes of all recorded encounters against any Oystercatchers, whether individually-marked or not. In the left-hand part of the graph, the focal bird was the subdominant individual; the difference in dominance between the two birds is therefore negative. The left-hand part is the mirror image of the right-hand part, where the focal bird was the more dominant individual (Goss-Custard *et al.* 1995a).

correlates well with 'aggressiveness', the rate at which Oystercatchers initiate attacks (Goss-Custard *et al.* 1982a). It should be kept in mind, though, that this dominance score only applies to the site where the individual regularly feeds. In Turnstones, for instance, dominance is strongly site-dependent between birds from different 'home' areas, although not for birds from the same 'home' area (Whitfield 1985a).

Losing mussels or failing to find them?

The susceptibility of an individual to interference, defined as the slope of the curve relating its intake rate to bird density (Goss-Custard and Durell

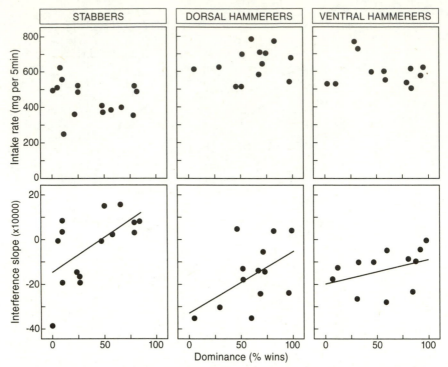

Fig. 4.7 Comparison of the feeding efficiency and the susceptibility to interference of individual Oystercatchers with different feeding methods in relation to their dominance. Top three graphs: the interference-free intake rate, measured at 50 birds per hectare. Bottom three graphs: susceptibility to interference, measured as the slope of the logarithm of intake rate against bird density (Goss-Custard and Durell 1988). Where significant, the regression lines are shown.

1987a), depends on its local dominance (Ens and Goss-Custard 1984; Goss-Custard and Durell 1988). The locally top-dominant individuals suffer least, if at all, from interference (Fig. 4.7). To understand why, the data from Ens and Goss-Custard (1984) on the two most dominant individuals were combined and compared with those from the remaining six subdominants. The dichotomy is not real, of course, as many individuals have intermediate status and intermediate behaviour (Ens and Goss-Custard 1986; Goss-Custard and Durell 1988), but it facilitates the analysis of the causes of interference in birds of low social status.

The intake rate of the dominants may not be affected, and may even slightly improve, as Oystercatcher density increases (Fig. 4.8a). More mussels are gained by dominants than lost at all feeding densities (Fig. 4.8a). Though dominants attack subdominants more when density increases (Fig. 4.8b), their net gain remains constant. It appears that dominants lose more as well as gain more mussels at high feeding densities, because even dominant birds are not free from being attacked (Fig. 4.8d). As also

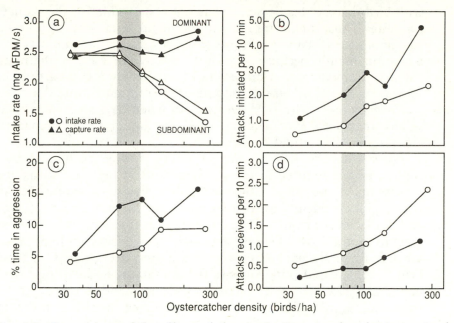

Fig. 4.8 Comparison of the effects of changes in Oystercatcher density on intake rates of dominant (closed symbols) and less dominant (open symbols) individuals (Ens and Goss-Custard 1984). The grey bar indicates the modal density of Oystercatchers. (a) Rate at which food was first captured (capture rate) and then ingested (intake rate). The difference between these two is the net rate at which food was gained in encounters with other Oystercatchers; the net result of food stolen from and lost to other birds. (b) Rate at which the focal individual initiated attacks against other Oystercatchers. (c) Percentage of time spent in aggression. (d) Rate at which the focal individual was attacked.

demonstrated by Boates (1988), an increasing amount of feeding time is lost in aggression as bird density increases but the effect on intake is small, because the rate at which dominants capture mussels is not affected (Fig. 4.8a).

In contrast, intake rate in subdominant individuals sharply declines with an increase in the density of conspecifics (Fig. 4.8a). They are attacked more frequently than dominants (Fig. 4.8d), while their tendency to initiate attacks themselves is reduced (Fig. 4.8b). Subdominants always lose more mussels than they gain, but this net loss increases with an increase in competitor density (Fig. 4.8a). Thus, part of the decline in intake rate of the subdominants is due to an increased loss of mussels. However, the rate at which mussels are captured from the mussel bed by the subdominants also declines (Fig. 4.8a). Indeed, in this example, 80% of the difference in intake rate between dominants and subdominants at high bird densities is due to the difference between them in their capture rate. As the time lost in aggression increases only slightly in subdominants with bird density (Fig. 4.8c), this is not responsible for the reduced capture rate. Thus, intake rate in subdominants decreases with an increase in competitor density partly because they lose more mussels to dominants but mainly because they capture fewer mussels in the first place. Though dominants gain more mussels than they lose and their capture rate is unaffected by bird density, their overall intake rate does not show a strong increase with bird density. So, when data from dominants and subdominants are averaged, the overall decline found in unmarked individuals (Fig. 4.3) would be expected.

The results also provide some insight into the mechanism by which capture rate is reduced in the subdominants at high bird densities. Prey depression is unlikely, otherwise all individuals would be affected equally, independently of their social status. Increased rates of avoidance and displacement from feeding spots at high competitor density therefore seem to be the most likely mechanisms. This raises two hypotheses, that are not mutually exclusive, to explain interference in Oystercatchers. First, the increased opportunities for dominants to steal food from subdominants as bird density increases induces avoidance behaviour in the subdominant individuals which, while reducing the risk of losing mussels from stealing, causes fewer prey to be captured. Second, as densities increase, subdominants are increasingly often displaced from good feeding spots, while the most dominant individuals are unaffected. As no data exist with which to test the second hypothesis, we now examine the evidence that stealing and avoidance provide a convincing, though perhaps only partial explanation, of the reduced capture rate at high densities of Oystercatchers.

Costs and benefits of intraspecific kleptoparasitism

When animals compete for resources, the costs and benefits of a decision by one individual depend on, and influence, the decisions taken by others. Game theory provides the conceptual framework within which the different optimizing decisions made by competing individuals can be explored and the evolutionary stable strategies (ESS) identified (Maynard Smith 1982). For kleptoparasitism, early ESS models predicted a stable ratio between the two incompatible strategies of 'scrounging' (stealing) and 'producing' (capturing) food (Barnard and Sibly 1981); clearly, it is logically impossible for all birds to survive only on stolen food. However, this simple producer-scrounger dichotomy does not apply to Oystercatchers (Fig. 4.6). Instead of a dichotomy, there is a dominance hierarchy along which the ratio between stealing and losing mussels gradually changes (Goss-Custard *et al.* 1982a; Ens and Goss-Custard 1984). In addition, robbing and searching are not mutually exclusive over the exposure period as a whole; stolen food contributed at most 15% to the total food intake in even top-dominant birds (Ens and Goss-Custard 1984). Clearly, there is a need for a model that recognizes that the same individual can both produce and scrounge, depending on the circumstances.

Recently, Vickery *et al.* (1991) developed an ESS model that does this by incorporating the idea that producing and scrounging may only be incompatible to a certain degree. It assumes that looking for potential victims to attack merely reduces the efficiency with which other prey are located. This model explains why stolen food contributes relatively little to total food intake but does not incorporate individual differences in dominance. It views robbing as the trait to be explored by game theory rather than the dominance hierachy which determines the profitability of robbing in the first place. Like Ens *et al.* (1990), we believe that the most pertinent question to be answered by game theory is how differences in social dominance can be evolutionarily stable. Once differences between individuals in social dominance are recognized, robbing behaviour can be viewed simply as a problem of prey choice (Kushlan 1978, 1979; Dunbrack 1979; Thompson 1986; Ens *et al.* 1990). That is, an individual should only initiate a robbing attempt if the expected gains from robbing exceed those from continuing to search (Thompson 1986). In a study of juvenile Oystercatchers, Cayford (1988a) indeed found that the tendency to attack conspecifics for prey almost disappeared as robbing became relatively less profitable during the course of the season (Fig. 4.9). However, contrary to prediction, robbing remained a profitable choice throughout in absolute terms. This may be explained if only the very dominant juveniles continued to steal food later in the season so that the estimate of the profitability of robbing became increasingly biased

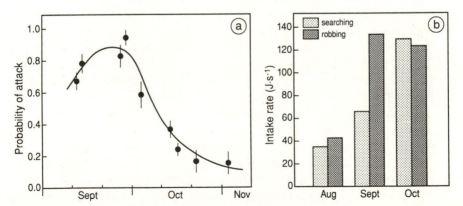

Fig. 4.9 Robbing by juvenile Oystercatchers during their first few months on the wintering grounds (Cayford 1988a). (a) The probability that juveniles attacked any birds they encountered which were handling mussels (values shown are means while bars indicate the SE). (b) The net energetic benefit of robbing another Oystercatcher compared with continuing to search for mussels on the bed.

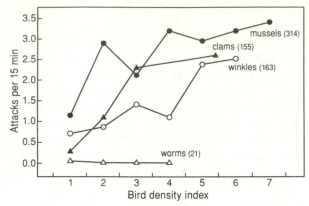

Fig. 4.10 Rate at which Oystercatchers specializing on different prey were attacked for prey at different bird densities, measured as an index and increasing from left to right (Boates 1988). The mean biomass (mg AFDM) of the prey consumed is shown in brackets.

upwards (Cayford 1988a). If so, this is an example of the difficulty of evaluating strategy options in unmarked birds that vary greatly.

The profitability of stealing depends on the amount of meat that potentially can be gained, the duration of the attack and the probability of success. Hence, predictions can be made to test the idea that robbers simply attempt to maximize their rate of gain by robbing. First, robbing should be directed at conspecifics handling large prey, since the potential gains and probability of success are higher. In support of this, Boates (1988) found in a study of Oystercatchers feeding on mussels, winkles, clams and worms that intraspecific kleptoparasitism at a given bird density was most common in birds eating large prey species (Fig. 4.10). In contrast, Cayford (1988a) did not find that the probability of attack increased with mussel size, possibly because of the small overall variation in the size of mussels captured. A second prediction is that robbers should attack before much meat has been consumed, a possibility that has not yet been tested. Third, attacks should be directed towards subdominants, as these are more likely to yield their prey, and this does appear to be the case (Ens and Goss-Custard 1986). Finally, attacks should not be initiated at too large a distance because of the greater chance of the attack being detected by the victim, which could then carry the mussel away. Vines (1980) showed that the probability of an attack for prey being successful decreased as the initial distance between attacker and victim increased. As the rate at which an individual encounters a less dominant individual handling prey within the threshold distance presumably increases as bird density increases, attack rate would be expected to increase with Oystercatcher density, as indeed it does (Fig. 4.8b). Clearly, the concept of a threshold attack distance, which is perhaps dependent on prey size and/or handling time, links the costs and benefits of robbing to bird density, and thus helps to explain the occurrence of interference in subdominant Oystercatchers. But for a complete understanding of the effect of density, the options that victims have for minimizing their losses must also be taken into account.

How to avoid losing mussels

What can subdominant individuals do to minimize the risk that they lose their mussels to dominants? This can be very important as Cayford (1988a) found that the probability of being attacked while handling a mussel approached 1 at very high bird densities. Basically, potential victims can minimize the opportunities for more dominant birds to launch a successful attack in the first place and maximize the probability that they can take effective evasive action when they are attacked. Thus, a distinction can be made between pre- and post-handling avoidance tactics. An Oystercatcher can alter its search path, feed elsewhere or simply stop searching. Alternatively, once a mussel is being handled, it can be transported to a safer place and the bird can regularly scan for robbers. In the following discussion, pre- and post-handling tactics are reviewed not in the sequence in which they take place but according to how severely they disrupt normal foraging.

Do subdominants scan for potential attackers while handling mussels? Cayford (1988a) showed that scanning rates, and therefore the time spent scanning, increased with Oystercatcher density (Fig. 4.11a) and that subdominants were generally more vigilant than dominants (Fig. 4.11b). He suggests that Oystercatchers handling mussels scan both to detect

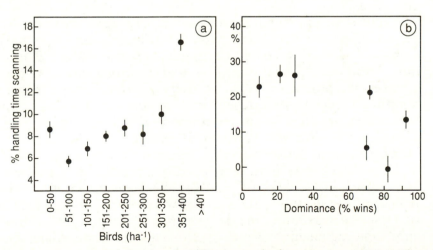

Fig. 4.11 Percentage of the time handling mussels that was spent in scanning: (a) averaged for different feeding densities of Oystercatchers; (b) averaged for individuals of different dominance status (Cayford 1988a). The vertical bars show 1 SE.

attacks and to continuously assess the changing risk of being attacked and so perhaps losing their mussel. On the basis of this assessment, Oystercatchers can then presumably decide whether to incur the cost of transporting the mussel to a safer place. But do subdominants carry their prey to a safer distance from nearby dominants? Mussel-feeders, which are very vulnerable to robbing, regularly transport their captured mussels (Boates 1988) and are more likely to do so when their nearest neighbour is close (Vines 1980). Furthermore, juveniles lose many mussels before learning to carry a mussel away (Cayford 1988a). There is thus some evidence that Oystercatchers do take steps to reduce the chances that mussels they have captured will be stolen from them.

Do Oystercatchers also change their search path to avoid coming close to more dominant individuals in the first place and, if so, does it lead to a reduction in capture rate, as argued earlier? The under-representation of short nearest-neighbour distances in Vines' (1980) study clearly indicates avoidance. By defining avoidance as a clear change in search path, or as a retreat with prey without overt aggression, Boates (1988) showed that avoidance rates did increase with bird density, except at very high bird densities. Although he could not measure any time cost associated with avoidance, he did find a clear time cost to aggressive encounters, for both attacker and victim, over and above the time spent on the encounter itself. For example, the search time of a winkle specialist that attacked another was increased by 10.3 seconds whereas the search time of the victim was increased by 32.7 seconds. While the evidence suggests that foraging Oystercatchers do take a number of steps to minimize the risk of having mussels stolen from them, finding out exactly how avoidance reduces the capture rate of subdominants remains a major challenge.

A final option for subdominants at the very high densities at which the probability of being attacked approaches 1 is for them simply to stop foraging and to wait for conditions to improve. Indeed, both Koene (1978) and Zwarts and Drent (1981) found a strong negative correlation between the proportion of birds feeding actively and bird density. In addition, model Oystercatchers in a standing posture had almost the same effect as real birds on the proportion of birds feeding (Fig. 4.12). However, the birds may have ceased feeding for reasons other than the increased risk of being attacked; Oystercatchers prefer to rest in flocks and the models may have preferentially attracted birds that wanted to rest. Indeed, Ens and Goss-Custard (1984) showed that it was the dominants, and not the subdominants, that were most inactive at high densities of Oystercatchers. Since the decision not to feed is quite a drastic one, subdominants might do better by moving away and feeding somewhere else; they could do this either by avoiding the high feeding densities early and late in the tidal cycle or by changing their feeding area on a longer-term basis, a possibility for which there is good evidence (Goss-Custard

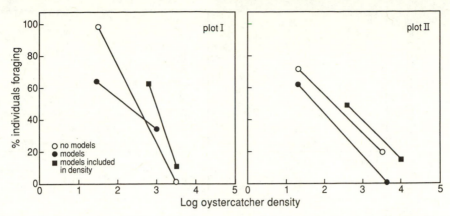

Fig. 4.12 Proportion of Oystercatchers engaged in foraging against the density of Oystercatchers for two study plots in the study of Koene (1978). Regression lines are shown for observations when there were no model Oystercatchers in the plot and for observations when they were present and either included, or not included, in the measure of bird density.

et al. 1982b). Instead of stopping feeding at very high densities, subdominants seem more likely to look for other places where the density of competitors is lower.

Dominance

Every time a subdominant is attacked it has the option to contest its current dominance position by standing its ground, but seldom does so. Although game theory will undoubtedly explain why subdominants nearly always give in, this section reveals that our knowledge of the costs and benefits of dominance are still insufficient to formulate a model.

Dominance behaviour

On the definition that a territory is a portion of an individual's range in which it has priority of access to one or more critical resources over others which themselves have priority elsewhere or at another time (Kaufmann 1983), some Oystercatchers on mussel beds defend territories. As these territories are non-exclusive, the term 'pseudo-territories' might be a better term (D. Harper, personal communication). Although in part a semantic matter, using this term may lead to a reorientation of our thinking, as it stresses that dominance is site dependent.

Dominants signal their status by piping, a display performed while either standing or running, and accompanied by a penetrating call (Makkink 1942; Cramp and Simmons 1983). Despite wide variations in intensity, the carpal joints are nearly always raised while the bill points downwards. Piping is performed solitarily, in which an individual looks

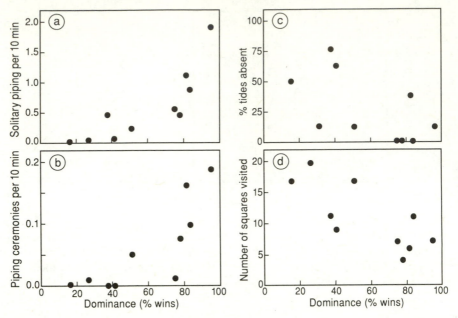

Fig. 4.13 Ranging and display behaviour of Oystercatchers in relation to their dominance (Ens and Goss-Custard 1984, 1986). (a) Rate at which the focal bird performed solitary piping displays. (b) Rate at which the focal individual engaged in piping ceremonies. (c) Percentage of tidal cycles during which the focal individual was not observed in its home range. (d) The size of the area over which the bird foraged, measured as the number of 25 × 25 m squares visited by the focal individual during 3.5 hours of foraging.

up, pipes and continues its activity, or in so-called 'piping ceremonies', in which two or more individuals pipe to each other over a short distance (Fig. 4.5b). These ceremonies usually develop when disputes over food items or feeding spots escalate because the victim retaliates. On the mussel beds of the Exe, the most dominant individuals most often engage in both solitary piping (Fig. 4.13a) and piping ceremonies (Fig. 4.13b), suggesting that piping signals an individual's dominance in a particular area (Ens and Goss-Custard 1986). The most dominant individuals also occupy the smallest home ranges (Fig. 4.13d), from which they are hardly ever absent (Fig. 4.13c). This strong site fidelity makes it difficult to test the prediction that, outside their pseudo-territories, dominants have no priority of access to food. However, dominants were less aggressive when they were chased from their territory by human observers (C.J. Smit, personal communication). Furthermore, when an artificial mussel bed was created on Schiermonnikoog in an area where breeding Oystercatchers defended mudflat feeding territories, they did not abandon their territories but no longer succeeded in evicting all intruders. The density of Oystercatchers increased more than tenfold and the social system very much resembled that on the Exe; the former territory owners became the top-dominants in pseudo-territories (B.J. Ens, unpublished information).

The costs and benefits of dominance

What are the benefits of dominance? Dominants steal food and so boost their intake rate and they may also locate good feeding spots by usurping those found by subdominants (Rohwer and Ewald 1981). Indeed, uncertainty surrounding the existence of competition for small-scale feeding sites may well be the major gap in current Oystercatcher studies. The rate at which individuals collect food in the absence of competitors, the 'interference-free intake rate', does not correlate with dominance (Fig. 4.7). Therefore, dominants would only be expected to have higher than average intake rates at medium to high bird densities, when subdominants would suffer from interference while some very dominant individuals may benefit slightly. In line with this prediction, there was a positive correlation between dominance and intake rate on the most preferred and densely populated mussel bed in the Exe estuary (Goss-Custard *et al.* 1984).

Although priority of access to food must be the major benefit of dominance, top-dominants do not have exclusive use of the resources in their pseudo-territory. We might expect dominant Oystercatchers to attempt to conserve a non-renewing food supply for use during harsh conditions, as may territorial Grey Plovers (*Pluvialis squatarola*) (Dugan 1982), yet not all intruders are excluded. Why? Based on the work Myers *et al.* (1979) on the costs of territory defence in Sanderlings (*Calidris alba*), and on the results of the experiment on Schiermonnikoog, we can hypothesise that top-dominants attempt to conserve their supply of mussels, but intruder pressure is so high that they cannot exclude other birds economically. Instead, they reduce the intake rate of these intruders, which may induce the intruders to feed somewhere else. An alternative hypothesis is that overwinter predation on mussels is so low that food remains abundant throughout so that the only benefit of dominance is the ability to steal food, and perhaps feeding spots, from other birds. Under the first hypothesis, the non-exclusivity of the pseudo-territory is viewed as the net result of the inter-dependent decisions of the top-dominant not going all out to evict all intruding subdominants and the decisions of the subdominants not to leave or to return quickly upon being chased (Kaufmann 1983). This hypothesis also implies that dominants do suffer somewhat from the presence of subdominants. Under the second hypothesis, dominants make no attempt to chase subdominants away permanently, but only displace them from good feeding spots. On this hypothesis, dominants ultimately benefit from the presence of subdominants because they help them to locate good feeding places.

If differences in dominance are simply determined by differences in fighting ability, there need not be a cost to dominance. Because we think this is unlikely (see later), we do expect dominance to have costs. What are these likely to be? Dominants devote more time than subdominants to aggression but without seriously impairing their rate of food capture. But dominants also spend more time in piping (Ens and Goss-Custard 1986) and this could be energetically expensive, like strutting in male Sage Grouse (*Centrocercus urophasianus*) (Vehrencamp *et al.* 1989). Regular presence on a small part of the mussel bed (Fig. 4.13c), presumably required to maintain the dominant's position, also means that dominants are less aware of changes in food supply elsewhere, a cost that depends on the variability of the environment. Thus the costs of dominance are less well understood than are the benefits.

Dominance and survival

Whatever the precise costs and benefits of dominance, there are unlikely to be no net fitness benefits to dominance. No existing data relate survival directly to dominance but some indirect evidence is available. Juveniles are on average subdominant to adults (Goss-Custard and Durell 1987b) and suffer a higher mortality (Goss-Custard *et al.*, Chapter 6, this volume). In winter, juveniles only achieve 59–82% of their potential intake rate on the mussel bed because of interference (Goss-Custard and Durell 1987c). Furthermore, juveniles feed more often than adults in fields over high water, suggesting difficulties in meeting energy demands (Goss-Custard and Durell 1983). Although this evidence suggests a causal link from low status via low intake rate to a high risk of starvation, several problems remain. First, feeding in fields may be an alternative, not an inferior strategy. Second, Oystercatchers feeding on cockles can increase their intake rate when feeding time is experimentally reduced (Swennen *et al.* 1989); therefore, a low intake rate does not necessarily imply food shortage (Zwarts *et al.*, Chapter 2, this volume). Third, because dominance is site dependent, and sites vary in quality, the comparison between the overwinter survival of dominants and subdominants should be restricted to one habitat. Finally, as Goss-Custard *et al.* (1982c) argued that most adult mortality occurs away from the Exe, perhaps on migration, body condition at the start of spring migration may be a more apt way to assess the significance of dominance than overwinter survival, especially as it may also have important consequences for subsequent reproduction (Ankney and MacInnes 1978; Ebbinge 1989). For the Oystercatcher, the critical time may be when they prepare for the spring migration rather than during the winter itself.

Correlates of dominance and status signalling

Compared to passerines (Piper and Wiley 1989), the determinants of dominance in Oystercatchers are poorly known. As yet, the only clear correlate of dominance is age; as in so many animals, adults on average dominate juveniles and immatures. It is therefore remarkable that when juveniles first arrive on the mussel beds of the Exe in early autumn, they

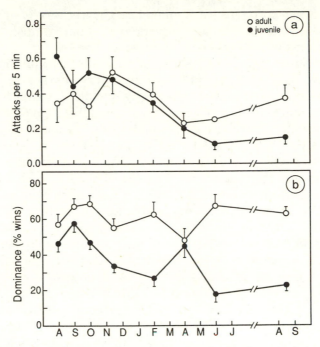

Fig. 4.14 Aggressive behaviour of matched pairs of juvenile and adult Oystercatchers over one year (Goss-Custard and Durell 1987b). (a) Rate at which birds attacked others. (b) Probability of winning an encounter with other Oystercatchers.

are rather aggressive and attack conspecifics at a much higher rate than adults (Fig. 4.14a) and also win more than 50% of their encounters. But by February, their dominance drops to less than 30% (Fig. 4.14b) and they initiate fewer attacks. Their initial high success may be due to bluff (Cayford 1988a). Most attacks by juveniles are attempts to steal mussels in which the bird apparently relies on surprise and a rapid approach to snatch the mussel away. Apparently, this strategy works, but not for long. For juveniles, the profitability of robbing declines during late autumn and winter, primarily because they become increasingly unsuccessful at winning encounters (Cayford 1988a). Indeed, the estimated 30% average dominance score in late winter may even be an overestimate, as most juveniles have left the mussel beds by then and the least aggressive ones may leave first (Goss-Custard *et al.* 1984). This highlights a major problem with most studies of juvenile Oystercatchers which are difficult to catch in sufficient numbers to follow individually-marked birds; how to distinguish real changes in behaviour from changes in the composition of the sample under study.

Age is not the only determinant of dominance as the scores of some adults fall well below the average for the juveniles. But nothing is known of the other correlates of dominance. Clearly, sex is one candidate but, as male and female Oystercatchers are virtually indistinguishable in the field, no study has yet related dominance to sex.

As reliable signals of fighting ability allow contestants to settle disputes without fighting and so risking injury (Parker 1974), and as Oystercatchers possess such a potentially damaging weapon as their bill, we would expect to be able to identify a reliable signal of fighting ability. Piping, which correlates so strongly with dominance (Fig. 4.13a,b), is one possibility but, to date, no studies have been done. Morphological features, such as plumage colour, may also indicate fighting ability (Rohwer and Rohwer 1978). The size of the white neck collar in wintering Oystercatchers could be such a signal although, as it declines with age (Dare and Mercer 1974a), it would have to be a reciprocal measure. However, Ens and Goss-Custard (1986) found no correlation between dominance and collar size among nine adults displaying a wide range of collar sizes. An alternative hypothesis, advocated by Shields (1977) and corroborated by Whitfield (1988) for Turnstones, is that this plumage variability serves to facilitate individual recognition. Though this may explain collar size variation in Oystercatchers (Ens and Goss-Custard 1986), it does not explain why collar size should change with age; what benefit might immatures gain by signalling that they belong to a subdominant class?

At first sight, the notion developed earlier in this chapter that dominance is site dependent is clearly at variance with the idea that dominance is a function of individual characteristics, such as age, sex and plumage, that vary irrespective of locality. However, dominance in the well-studied wintering migrant White-throated Sparrows (*Zonotrichia albicollis*) has both intrinsic (age, sex) and situational (site-dependent) components (Piper and Wiley 1989), and the same seems likely to apply to Oystercatchers. Identifying the contribution of intrinsic features, such as age, sex and fighting ability, and situational components in the dominance of Oystercatchers is an important goal for future Oystercatcher studies.

Social careers and competitive ability

This chapter has viewed interference as the result of adaptive robbing decisions made by individual Oystercatchers which differ in dominance. Based on this idea, this section develops an alternative to the concept of competitive ability as used by Sutherland and Parker (1985) and Parker and Sutherland (1986) to model the distribution of unequal competitors over habitats of different quality. In these models, individuals have high competitive ability if they are efficient at food finding or if their intake rate is little affected by the presence of competitors. However, the models assume that individual ranking by competitive abilities is fixed and that individuals are free to move without cost. In fact, both assumptions are likely to be false in the Oystercatcher for the same reason: the site-dependent nature of dominance. An Oystercatcher that moves from its pseudo-territory is likely to decline in dominance and so suffers more from interference; its competitive ability has therefore declined and so moving has a cost.

Any theory to explain the distribution of Oystercatchers over different mussel beds should recognize that competitive ability is likely to depend on the site to which the individual has become attached. It seems reasonable to assume that the fitness of a top-dominant on a high quality bed probably exceeds the fitness of birds of lower dominance, as well as the fitness of top-dominants on lower quality beds. Following Ens *et al.* (Chapter 8, this volume) on how nonbreeding Oystercatchers queue for breeding territories of differing quality, we suggest that wintering birds of lower dominance that regularly feed in the area can be thought of as queuing, albeit competitively with individuals of adjacent rank, to take the position of more dominant birds when they disappear. As birds move up in the hierarchy, they become increasingly committed to a smaller and smaller area, probably because it is uneconomical to defend status over a large area. In this way, individuals gradually commit their competitive activities to a particular site and their dominance rises in parallel.

This hypothesis of local competitive queues implies stable dominance ranking and a slow increase in rank with age. The first implication is borne out by the observation that the dominance ranking of 10 birds studied in 1980 was the same as in 1979 (Goss-Custard *et al.* 1982a). A slow rise in dominance over a number of years was shown in adults by Caldow and Goss-Custard (1996); dominance rarely increased by more than 10% from one year to the next (Fig. 4.15a) and the rate of change declined with initial dominance. This relationship can be used to calculate how dominance may change with age (Fig. 4.15b). With an initial dominance of 30% in the first year, it takes over 20 years to reach a dominance of 80%. Even when the rate of change is set to 20% in the first year, it still takes at least 10 years to reach a dominance of 80%. It is thus likely to take a great many years to become a top-dominant.

Such slow improvement of dominance is another argument against the view that social status is primarily determined by fighting ability. Though fighting ability may improve over the first few years of life, it seems unlikely to improve for ten or more years in a species that does not grow morphologically beyond one year of age. Nonetheless, it would at first sight be surprising if a very strong immature did not select a weaker top-dominant and beat it to become more dominant itself. However, there are several possible reasons why this might not be a successful strategy. First, beating the top-dominant alone may not be enough; the other individuals in the local queue have also to be beaten. In Turnstones, the selective aggression and unprovoked attacks by birds on con-

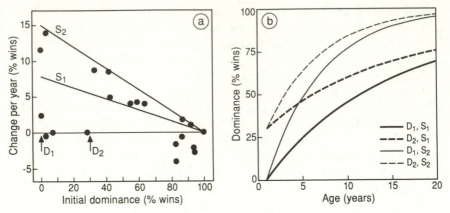

Fig. 4.15 Changes in the dominance of adult Oystercatchers against age. (a) Average annual change in dominance (*ACD*) plotted against the initial dominance (*DOM*) of the bird in the first year it was studied. Each dot refers to a marked individual on which reliable dominance scores were obtained in at least two years (Caldow and Goss-Custard 1996). D_1 and D_2 are different initial dominances, 0% and 30% respectively. S_1 is the regression through all the data; $ACD = 8 - 0.08\ DOM$. S_2 is the regression through the maximum rates of change observed; $ACD = 15 - 0.15\ DOM$. (b) Social careers, or dominance against age, expected for different combinations of initial dominance, calculated from the regression equations given in (a). For each initial dominance, the curves S_1 and S_2 allow the dominance in the next year to be calculated. The curve D_1, S_2, for example, shows the maximum rate of increase in dominance that would be expected of an Oystercatcher that began with a dominance of 0.

specifics immediately below them in the rank order also imply that they know the status and so 'threat' that a particular individual represents (D. P. Whitfield, personal communication). Accepting a low initial status may be better than risking severe injury from frequent escalated fights. Second, contests over dominance always apply to small areas, not to a general relationship between two individuals. Being dominant in a particular area is more important for the individual with the smaller home range. Such differences in resource value may override differences in fighting ability in determining the outcome of a contest. Third, the only way to assess fighting ability is via escalated fighting. For individuals that know each other, it is best to use previous experience as a cue to settle contests so that 'tradition' overrides any changes in fighting ability that may have occurred (Thouless and Guinness 1986).

Though much remains to be learned, there is evidence that local dominance hierarchies should be viewed as competitive queues for the position of top-dominant. Following Wiley (1981), we have applied life history thinking to the problem of social organization in the Oystercatcher. The newly arriving juvenile should not only survive its first winter, it should survive as many of the following winters as possible. Starting from an investigation of the minute by minute interactions among individuals, we have ended up speculating on the lifetime career decisions of individuals. It makes us realize that the models of Sutherland and Parker (1985) and Parker and Sutherland (1986) fail to take into account the true dynamics of real life.

Summary

On mussel beds, locally dominant Oystercatchers steal mussels from subdominant individuals. At increasing densities of conspecifics, caused by tidal changes in the amount of feeding area exposed, subdominants lose an increasing number of mussels to the dominants. They also find fewer mussels, which may be due to the increasing need to avoid potential aggressors. The costs and benefits of kleptoparasitic and anti-kleptoparasitic behaviour provide a sufficient explanation for these relationships and so explain the interference that occurs. However, the present data do not allow rejection of the hypothesis that subdominants find fewer mussels because they are displaced from good feeding spots. Much less is known from other feeding habitats, such as cockle beds. Here, interference may be reduced or absent, due to smaller prey and lower feeding densities. On the Exe, mussel beds are the preferred habitat. Many young birds require one or more years, before their feeding skills and dominance status have improved to a level that they can acquire sufficient food on the mussel bed throughout the winter. Surprisingly, juveniles are initially rather successful in stealing mussels. Social dominance increases with age and is very probably site dependent. The lack of knowledge on other correlates of dominance in Oystercatchers is remarkable. Indirect but weak evidence suggests that dominants have a reduced risk of dying from starvation, due to priority of access to the food supply in their pseudo-territory. As Oystercatchers have high survival rates, measurement of daily time and energy budgets for dominant and subdominant individuals in habitats of different quality may yield the most direct test of a causal link between dominance and survival. Current theories that seek to explain the distribution of unequal competitors assume fixed competitive abilities and no cost to moving. If there is a strong site-dependent element in social dominance, both assumptions will fail. It may be profitable to start thinking of Oystercatchers forming, and competing in, local queues for positions of high dominance.

Acknowledgements

John Sherman Boates and John Goss-Custard made many helpful comments on the manuscript, while Philip Whitfield spent much effort in comparing his knowledge on the social system of wintering Turnstones to our knowledge on Oystercatchers, finding remarkably few differences. Leo Zwarts unearthed his observations on interspecific kleptoparasitism. Jan van de Kam very kindly allowed us to use his photographs.

5 Where to feed

John D. Goss-Custard, Andrew D. West and
William J. Sutherland

Introduction

The food supply of wintering Oystercatchers varies considerably from place to place, whatever the scale considered. The patchiness of many macro-benthic organisms causes it to vary at a scale of a metre or less (Goss-Custard *et al.* 1993). Large variations occur in the numerical density, body size and flesh content of prey along the major estuarine gradients of salinity, low water exposure period and sediment composition (Goss-Custard *et al.* 1993). Variation occurs between entire estuaries and different areas of large coastal flats (Wolff 1969) and, in conjunction with climate-related energy demands, between regions of the entire winter range (Piersma *et al.* 1993a). A number of other factors may also vary spatially, such as the risk of predator attack (Cresswell and Whitfield 1994) and the proximity of the high water roost (Goss-Custard *et al.* 1981; Zwarts and Drent 1981; Goss-Custard *et al.* 1992). The feeding grounds of Oystercatchers can therefore be viewed as a series of interlocking habitat gradients that range in scale from a few square metres up to hundreds of square kilometres.

This chapter investigates by modelling the distribution patterns that result from the decisions made by individual Oystercatchers as to where they should feed. Were food abundance the sole consideration, all Oystercatchers might be expected to congregate in the richest parts of an estuary where their rate of energy intake is greatest, yet this seldom happens; to varying degrees, they are usually spread out. The most likely cause of this is intraspecific competition for food, due either to interference or to prey depletion (Goss-Custard 1980). Though operating over different time-scales, both processes provide a negative feedback loop from bird density to the intake rates of individuals. This chapter examines whether the interaction between the opposing tendencies to congregate where food is abundant but to avoid other birds is a satisfactory way to model the distribution of Oystercatchers over their winter intertidal food gradients.

The ideal free distribution

If their distribution is the outcome solely of these two opposing tendencies, what patterns would we expect? For simplicity, only the interaction between food supply and interference is considered first. A suitable conceptual framework is provided by game theory as applied to the ideal free distribution (*IFD*) (Fretwell and Lucas 1969; Fretwell 1972). *IFD* assumes that animals are ideal, in that they always know the location of, and move to, the site where their intake rate is highest, and that they are free, in that they are able to move readily between sites. The approach is game theoretic because the choices made by one competing individual as to where to feed are contingent on those made by all others (Maynard Smith 1982). As explained in Fig. 5.1a, this would lead to the birds in different sites all having the same intake rate. Were the intake rate temporarily higher in one site, individuals would move into that site until the

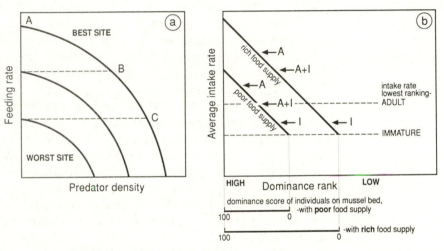

Fig. 5.1 The ideal free distribution (*IFD*). (a) With equal competitors. There are three sites that differ in prey density. At low densities of foragers (A), all individuals feed in the best site. As forager density increases, the intake rate declines due to interference until, at point B, it pays some to feed in the site of intermediate quality. If forager numbers increase still further (C), then all three sites are occupied. At any one time, the intake rate is the same in all three sites (Sutherland 1983). (b) With unequal competitors, as applied to Oystercatchers. There is a good and a poor mussel bed. Because of the greater susceptibility of subdominants to interference, the average intake rate on both beds decreases with dominance rank. The diagram shows the outcome when all birds have moved to the mussel bed where their intake rate is highest. Arrows indicate the average intake rates for adults (A) and the less dominant immatures (I). The average intake rate of adults is higher on the rich bed because the most dominant individuals are able to take advantage of the better food supply. The subdominant birds, and therefore all immatures, simply respond to the relative profitabilities of the two beds imposed on them by the dominant birds and the quality of the food supply itself. As a result, the immatures, and though not shown, the subdominant adults, distribute themselves according to the *IFD* so their intake rates are the same on both beds. The average intake rate of birds of all ages combined (A + I), however, is higher on the rich bed (Ens and Goss-Custard 1984).

increased interference removed the discrepancy. At equilibrium, no individual would be able to increase its intake rate by moving.

In the real world, many factors may prevent this outcome from being achieved. Not least, the properties of real birds may conflict with the ideal assumptions of this simple model. *IFD* only provides a starting point for developing a more realistic model, as do assumptions of random distribution and movement in other areas of biology. This section explores how far understanding can progress on the basis of such a simple model and some of its more complex derivatives.

The simplest case; equal competitors

Without interference, intake rate would depend solely on food abundance. The function relating intake rate to food abundance is known as the 'functional response'. As food density increases and the forager encounters food items increasingly often, intake rate rises until some constraint prevents it from increasing any further (Fig. 5.2a). The first formulation of this response assumed the plateau to be determined by the amount of time spent dealing with, or 'handling', food items, according to the equation:

$$N/T = a'D/(1 + a'Dh) \qquad (5.1)$$

where N is the number of food items taken, T is the length of the foraging bout, a' is the proportion of the food organisms the forager encounters that it can both detect and attack, D is the number of food items encountered in time T and h is the time taken to handle each one. According to this, intake rate at first rises rapidly as food abundance increases but then gradually slows down until all of the foraging bout is spent handling food. This function is known as the 'disc equation' because Holling (1959) first applied it to a blindfolded person 'foraging' for randomly-scattered sandpaper discs.

The functional response in Oystercatchers has been measured by relating either the numbers of prey taken per unit time (feeding rate) to the numerical prey density or the biomass ingested per unit time (intake rate) to the prey biomass density. The shapes of the functional responses in Oystercatchers feeding at low bird densities, and thus free from interference, on clams (*Scrobicularia plana*), cockles (*Cerastoderma edule*) and mussels (*Mytilus edulis*) follow the general form of the disc equation (Fig. 5.2b,c,d). But in the one case where it has been studied, the equation itself is inappropriate because the observed level of the plateau was different from that which was predicted (Wanink and Zwarts 1985). Although the equation correctly predicted the rate at which a captive bird took clams at low prey densities, they were taken faster than predicted at high densities when the bird ignored the time-consuming and less profitable closed bivalves with long handling times. The simple assumptions of the equation, that prey are identical at all densities, is

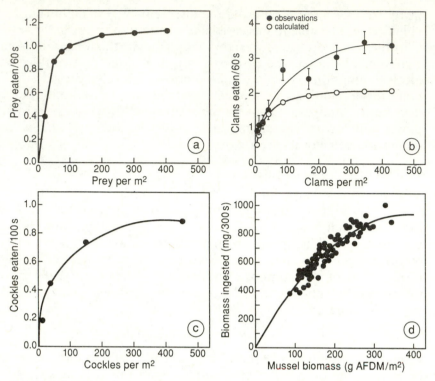

Fig. 5.2 The functional response in Oystercatchers in the absence of interference. (a) Hypothetical response for Oystercatchers feeding on one size class of mussels (Sutherland 1983). (b) Feeding rate on clams in a captive Oystercatcher in relation to the numerical density of one size class of the prey from observation and as calculated from the disc equation (Wanink and Zwarts 1985). Vertical bars show ± 1 SE. (c) Feeding rate of captive Oystercatchers in relation to the density of one size class of cockles (Hulscher 1976c). (d) Intake rate in free-living dorsal hammering Oystercatchers in relation to the biomass density of large (> 30 mm long) mussels (Goss-Custard *et al.* 1996b).

thus clearly inappropriate. This is likely to be a general finding in Oystercatchers. In mussel-eating birds, the proportion of prey that are both detectable and accessible may decrease as prey become more abundant and more mussels are hidden inside the clump (Goss-Custard *et al.* 1993). Birds may increasingly select for the least dangerous or the least parasitized prey (Hulscher 1976c; Chapter 1, this volume) or the most profitable size classes (Meire and Ervynck 1986; Goss-Custard *et al.* 1996b) as prey density increases. As Zwarts *et al.* (Chapter 2, this volume) argue, the factors that determine intake rate in Oystercatchers are unknown. A more sophisticated analysis is now required that recognizes the many competing influences that could be involved (Stephens and Krebs 1986), including the limited capacity of the gut to process food

(Kersten and Visser 1996a). It seems that the disc equation should be regarded in Oystercatchers as being adequate only in its general shape rather than in the assumptions it makes about the processes that limit intake rate at the plateau.

The strength of interference is measured as the rate of decrease in the feeding or intake rate resulting from a unit increase in forager density. In general models, both axes are transformed to logarithms on the assumption that proportionate increases in forager density lead to proportionate decreases in intake rate. The slope, m, is the interference constant (Hassell 1971). In an *IFD*, the proportion of predators in a site b_i is:

$$b_1 = cn_i^{1/m} \tag{5.2}$$

where c is a normalizing constant and n_i is the proportion of the food in each site (Sutherland 1983). In other words, the reciprocal of the exponent relating the proportions of the foragers and of their food in different places measures the average strength of interference. Were Oystercatchers to follow the *IFD*, values of m estimated from a comparison between the distribution of the birds and that of their prey should be the same as those obtained directly from field studies of foraging birds.

Using the methods detailed in Ens and Cayford (Chapter 4, this volume), m has been estimated in Oystercatchers eating cockles and mussels. Neither Goss-Custard (1977b) nor Sutherland (1982c) could detect interference in free-living cockle-eaters. However, recent studies of captive (Leopold *et al.* 1989) and free-living (P. Triplet, unpublished information) birds suggest that it may occur when bird density is sufficiently high. In contrast, interference has been widely found in mussel-eating birds. As the example in Figure 5.3 shows, plots of the untransformed data show the relationship to be strongly concave. With both axes transformed to logarithms, the relationships may be approximately linear, with slopes of < 0.5 (Sutherland and Koene 1982).

To explore the distribution consequences of interference of the strength recorded in Oystercatchers, simulations with an *IFD* game-theory model were therefore run using a range of values for m of below 0.5 (Sutherland 1992). One thousand Oystercatchers could choose between ten mussel beds, each of 1 ha, with prey density varying from 50 to 500 mussels per m². The interference-free functional response used is shown in Figure 5.2a. With little interference ($m = 0.1$), almost all birds feed in the sites with the highest prey density because there is little constraint on their congregating on the most profitable mussel bed (Fig. 5.4a). More mussel beds are occupied as interference intensifies (m increases to 0.5) so birds spread out. The resulting functional responses, which include the effect of interference on intake rate, are shown in Figure 5.4b. As would be expected of the *IFD*, intake rate is the same in all the sites in which Oystercatchers occur. Clearly, when interference

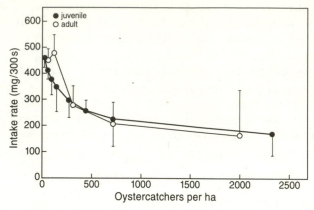

Fig. 5.3 The interference function in Oystercatchers eating mussels, as illustrated by juvenile and adult birds that open prey by stabbing (Goss-Custard and Durell 1987c). Vertical bars show ± 1 SE.

is weak so that the birds only occupy a few sites, the functional response covers only a narrow range of prey densities and is set at a higher level.

The effect of competitors being unequal

These simulations assume that all individuals are equally affected by interference. In reality, there are differences between individuals in the value of m according to their age, dominance, feeding method and thus bill morphology (Ens and Cayford, Chapter 4, this volume). The possible effect of introducing such individual differences into the *IFD* model is illustrated in Fig. 5.1b. Two mussel beds of equal size but different quality are portrayed. Adults are on average dominant to immatures and are thus less susceptible to interference. For the subdominants, the relative profitability of the two beds is determined by the density of dominant birds as well as by the food supply itself. As bird density rises on the most preferred bed, the intake rates of the subdominants decrease so that it eventually becomes more profitable for them to feed on the less preferred beds where, though the food supply is less abundant, interference is less. The lowest ranking birds move between the beds until they cannot increase their intake rate any further. At this point, the intake rates of the average immature and the most subdominant adults are the same on both beds and these birds are distributed according to the *IFD*. In contrast, the intake rate of the top dominants is higher on the rich bed because, being unaffected by interference, they are free to profit from the better quality food. Thus, in contrast to the *IFD* with equal competitors,

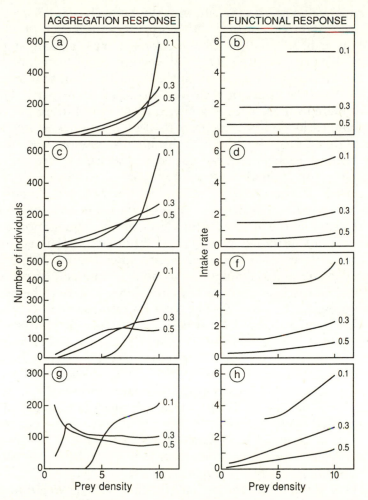

Fig. 5.4 The aggregation response and functional response—including the effect of interference—at three levels of average interference (m = 0.1, 0.3 or 0.5). In the top two diagrams, all individuals have the same value of m. In the remaining diagrams, the value of m varies between individuals, with the range of the variation increasing down the page from 0.9–1.1 (c,d) to 0.5–1.5 (Sutherland 1992).

the assumption of unequal competitors predicts a positive association between the average intake rate across all birds and food abundance.

Although this diagrammatic model conveys the main consequence of introducing inequality amongst competitors, it does make rather specific, although not clearly specified, assumptions about the relationship between intake rate and both food abundance and competitor density. The consequences of introducing inequality over a wider and specified range of assumptions were explored theoretically by Sutherland (1992). Inequality between competitors was introduced by ascribing different initial values of the interference coefficient, *m*, to different birds. For tractability, there were just ten kinds of birds (phenotypes), within each of which each bird had the same value of *m*. The interference birds actually experienced in a particular place was calculated by weighting their own value of *m* divided by the average value of *m* of all the other birds on that site. Importantly, this allowed the values of *m* to vary according to the particular competitors encountered on each feeding site; birds suffered more interference when surrounded by good competitors than by poor ones. Using a game-theory simulation procedure, the equilibrium distribution was found in which the birds in each phenotype all fed in the sites where they obtained the highest possible intake rate. The variation in competitive ability was increased or decreased by changing the between-phenotype variation around the particular average value of *m* used in any given simulation.

The simulations illustrate the effect on the aggregation and functional responses of introducing individual differences in susceptibility to interference, as measured by *m*. Birds are spread more evenly between sites than they are in the simulations without individual differences (Fig. 5.4c). The Oystercatchers in the best quality sites now have higher intake rates than those in the poorest (Fig. 5.4d), a trend which becomes even more marked as the difference between phenotypes in their susceptibility to interference is increased still further (Fig. 5.4f). This happens because more of the inferior competitors move to the poorer beds, thus reducing competitor density and so levels of interference amongst those that remain on the good quality beds. The simulations also show that, at very high levels of individual difference, the aggregation response may show unexpected patterns, such as the maximum density occurring at intermediate densities of prey. Therefore, there may be no simple and clear relationship between Oystercatcher density and prey density when individuals differ greatly in competitiveness.

These simulations show that the shape of the relationships between food abundance and both Oystercatcher density and intake rate are very sensitive to the variation in competitiveness amongst individuals and to the average level of interference. The aggregation responses also differ considerably from those expected of the simple *IFD*, implying that the average value of *m* cannot be deduced from the distribution of the Oystercatchers using equation 5.2. As the predictions of the model depend so critically upon the precise values used for just one of the parameters, empirical tests of the general approach adopted in this chapter depend upon good parameter estimation in the field and the construction of a more realistic, though initially situation-specific, model.

A model of mussel-eating Oystercatchers

Two versions of such a model have been developed for Oystercatchers foraging on the 12 main mussel beds of the Exe estuary (Goss-Custard *et al.* 1995a,b,c). In addition to including many biological details, the model differs from the formulations of Sutherland and Parker (1985) and Parker and Sutherland (1986) in being based on individuals rather than on a number of cohorts, or phenotypes, of identical animals. Each bird has its own competitive characteristics and makes its own decisions as to where to feed.

Version 1: build-up phase

This deals only with August and September when adults and juveniles from the breeding grounds join the oversummering immatures. As prey depletion is low, competition is assumed only to arise from interference. The strength of this is measured as the slope of untransformed intake rate against the logarithm of Oystercatcher density because the logarithmic transformation of both axes seemed inappropriate (Goss-Custard *et al.* 1995a). To distinguish this measure from the more widely-used quantity m, the term 'susceptibility to interference' (*STI*) was coined. The *STI* of an individual depends on its feeding method and local dominance, which varies between beds according to the quality of the competitors present. To measure local dominance in the model, each bird is given a unique value for a competitive quality which remains the same wherever it feeds; this is called 'global dominance' by Goss-Custard *et al.* (1995a) or 'fighting ability' by Ens and Cayford (Chapter 4, this volume). The

The basic idea underlying the model (diagram by S. Durell).

local dominance is calculated as the percentage of the birds present on a bed that have lower global dominances.

Individuals also vary in the rate they feed in the absence of interference; this 'interference-free intake rate' (*IFIR*) is thought to reflect a bird's foraging efficiency. Its variation across individuals has been empirically determined and shown to vary independently of local dominance (Goss-Custard and Durell 1988). The quality of a mussel bed is measured as the *IFIR* of an Oystercatcher of average foraging competence and is estimated from the biomass density of large (> 30 mm long) mussels present using an interference-free functional response derived from an empirical optimal foraging model (Goss-Custard *et al.* 1996b). From this, the *IFIR* of each individual is calculated from its relative foraging efficiency.

An individual's actual intake rate on any one bed at any one time is calculated by subtracting from its *IFIR* the reduction in intake rate due to interference, given its *STI* and the density of competitors present. The intake rate (*IR*) of an individual on any one bed thus depends on the food supply, its own feeding method, foraging efficiency and local dominance and on the density of Oystercatchers. In particular:

$$IR = IFIR - STI(\log \text{Oystercatcher density}) \qquad (5.3)$$

where *STI* is calculated from an individual's local dominance, as detailed in Clarke and Goss-Custard (Appendix 1, this volume). In each iteration, each individual is selected in random order to choose, within the 3% limits of its empirically-determined ability to discriminate (Goss-Custard *et al.* 1995a), the mussel bed where it currently achieves the highest gross intake rate.

Version 2: autumn and winter

In this version, a time-base and other modifications are added to version 1 in order to enable the impact of the birds on the mussels to be calculated and the changing distribution of Oystercatchers over the winter to be modelled. This version tracks the feeding location, intake rate and body condition of each bird on each day of its presence on the Exe from the end of the build-up phase in September to mid-March. The food supply on each bed declines as it is depleted by Oystercatchers and as the individual mussels lose condition (Dare and Edwards 1975; Bayne and Worrall 1980; Zwarts and Wanink 1993) and disappear, at empirically determined rates (McGrorty *et al.* 1990), from storms and other mortality agents. The effect of the fortnightly neap–spring cycle on the proportion of each bed that is accessible to Oystercatchers is also included. This is important as Oystercatcher densities, and so interference competition, can be very high in the reduced areas of poor quality mussels that remain exposed on neap tides (Goss-Custard and Durell 1987). In both the real world and in the model, many Oystercatchers change their feeding site through the neap–spring cycle and through the winter as the relative quality of the mussel beds change and competitors also change their foraging location.

Each bird is given an initial level of fat reserves in September drawn at random, and currently—and perhaps unrealistically (Gosler 1987; Witter and Cuthill 1993)—without regard to its dominance or foraging efficiency, from the distribution of weights measured on the Exe in autumn. High dominance is not assumed to carry additional energetic costs, an assumption that appears reasonable, at least in Dippers (*Cinclus cinclus*) (Bryant and Newton 1994). Subsequently, birds either store or metabolize fat according to how well they feed each day. Birds can feed for 12 hours in every 24 and, following Hulscher (Chapter 1, this volume), at the same rate at night as during the day. The limited capacity of the Oystercatcher gut to process mussel flesh (Zwarts *et al.*, Chapter 2, this volume), along with the known efficiency with which it assimilates mussel flesh (Kersten and Piersma 1987), limits the rate at which each bird in the model can consume food. The temperature-related energy requirements for each day are estimated from the studies of Kersten and Piersma (1987). Once an individual has obtained its daily energy requirements, it lays down any surplus as fat, deposited with an empirically measured efficiency (Kersten and Piersma 1987), up to a maximum rate of 5% of current body weight per day (Zwarts *et al.* 1990b). Each bird attempts during autumn and winter to accumulate fat at the mean rate observed by its age-class on the Exe over 15 years. The reserves are used to maintain the bird on days when it fails through foraging to meet its current requirements. As well as being individuals-based, this version of the model is thus also physiologically-structured. The mathematical definition of the model is detailed by Clarke and Goss-Custard (Appendix 1, this volume).

Model tests

As the model predictions for the build-up phase have been tested against observed trends for a number of aspects of the birds' distribution and behaviour in Goss-Custard *et al.* (1995c), they are not detailed here. Only two tests are shown in Fig. 5.5 and a number of other comparisons between observation and predictions are given below. Version 2 predicts quite well the average density of birds on 10 of the 12 mussel beds over the winter, but over-predicts densities on 2 of the most preferred ones (Fig. 5.5a). Version 1 successfully predicts the greater proportion of immatures on the lower compared with the higher ranked beds, but under-predicts by how much (Fig. 5.5b). This illustrates the general finding from both versions that the model predicts the observed qualitative trends quite well but often fails to precisely predict their magnitude.

This suggests that some parameter values need to be refined or some new functions added. For example, food quality is assumed to be uniform over a mussel bed, whereas variation occurs within beds, and to

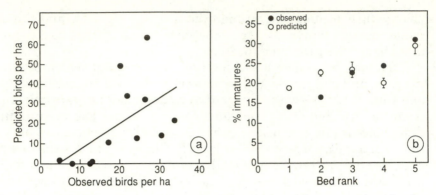

Fig. 5.5 Model predictions of the Exe estuary model compared with observed. (a) Average density of Oystercatchers on the mussel beds from October to January. (b) The proportions of the birds feeding on adjacently-ranked pairs of mussel beds that were immature. Note that the beds ranked 5 were the lowest four ranked beds combined. The observed data came from samples of individually colour-ringed birds of known age (Goss-Custard *et al.* 1995c).

different degrees in different beds, and the present values do not allow for the possibility that some beds may be more suitable for one feeding method than for another. Oystercatchers may increase their local dominance or foraging efficiency, or both, through familiarity with one area and establishing a resident advantage (Ens and Cayford, Chapter 4, this volume). This possibility has yet to be included, as has the fact that some birds also gain dominance during their adult life (Caldow and Goss-Custard 1996), and not just in their immaturity as the model currently assumes. An important untested assumption is that the amount by which the intake rate of a subdominant is depressed is unaffected by how much above or below it in the local dominance hierachy are the Oystercatchers with which it competes. This is contrary to the assumption in the models of Sutherland and Parker (1985) and Parker and Sutherland (1986) in which each individual integrates the competitive abilities of all the competitors present relative to its own. Future simulations can explore whether these alternative assumptions for estimating *STI* are important and further field studies can test which, if either, assumption lies closer to reality. But despite these reservations, the model's reasonable predictive success suggests that it provides a satisfactory starting point for exploring the spatial relationships between Oystercatchers and their main shellfish prey.

Patterns of aggregation and depletion

Build-up phase

The model predicts that, as the numbers of Oystercatchers increase through autumn, birds very quickly occupy the poorer quality beds

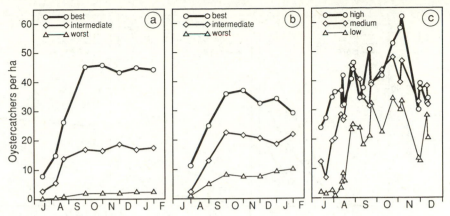

Fig. 5.6 The changes in the densities of Oystercatchers during autumn and winter on mussel beds of different quality. (a) As predicted by the Exe estuary model for groups of four mussel beds ranked according to their quality in September as the best, intermediate or worst. (b) As observed from 1976–1983 on the the same three categories of mussel beds of the Exe (Goss-Custard *at al.* 1982b). (c) As observed in the Delta region of The Netherlands on three mussel beds with high, medium and low biomass densities of prey (Meire 1991, 1993).

(Fig. 5.6a). Essentially the same pattern was observed on the Exe, although the predicted densities were higher on the most preferred beds, and thus lower on the remainder (Fig. 5.6b). The same pattern was also found amongst mussel-eating Oystercatchers in the Delta region of The Netherlands (Fig. 5.6c). In all cases, densities level off in winter because no more birds arrive after October.

The model predicts that the shape of the aggregation response changes only slightly as the population increases through autumn. Although many birds are predicted to aggregate in the richest sites at low population sizes, some also feed on the poorer beds (Fig. 5.7a). As numbers increase and the birds spread out even more, densities everywhere rise but the aggregation response remains noticeably concave. Essentially the same pattern was observed on the Exe, except that the response was linear at high population sizes (Fig. 5.7b). This difference with observed patterns again reflects the model tendency to over-predict Oystercatcher densities on the two mussel beds with high prey biomasses. Data from a small sample of sites in the Dutch Delta region suggest a similar change in the aggregation response as the total population size increases, although the sample sizes are very small (Fig. 5.7c).

The model also gives outputs for cohorts of individuals for which, unfortunately, there are no comparable data from the field. To save space, only the outputs for hammerers of all ages are shown in Figure 5.8. As would also be predicted by Sutherland's (1992) model (Fig. 5.4), the mean intake rate of all the birds on a bed increases across beds with prey

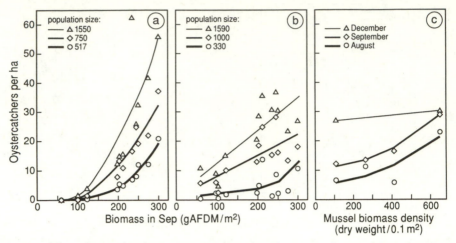

Fig. 5.7 The relationship between Oystercatcher density and mussel biomass density at different population sizes. (a) As predicted by the Exe estuary model for 12 mussel beds at three population sizes. (b) As observed at three similar population sizes on the mussel beds of the Exe (Goss-Custard *at al.* 1982b). (c) As observed at increasing population sizes in August, September and December on four mussel beds in the Dutch Delta region (Meire and Kuyken 1984).

biomass but around a lower average value as the population size, and so bird density, increases (Fig. 5.8a). The birds feeding where prey biomass is low tend to be those with the lowest global dominance (Fig. 5.8b) and the least efficient (Fig. 5.8c). As a result, intake rate correlates across mussel beds with both the foraging efficiency of individuals (Fig. 5.8d) and with their global dominance (Fig. 5.8e). In contrast to the models of both Ens and Goss-Custard (1984) and Sutherland (1992), the intake rate within a cohort of birds of approximately the same global dominance rank is not the same on all beds (Fig. 5.8f); rather, it too increases with prey biomass.

This happens because it is the more inefficient individuals within a dominance cohort that tend to feed on poor quality mussel beds. Intake rate in the model decreases arithmetically with the logarithm of bird density. For the inefficient, the reduction in intake rate due to interference on the best quality beds is greater than the reduction due to the low quality of the poorer beds themselves. To illustrate this, consider the following example of two birds of identical dominance but differing efficiencies, these being 1.25 and 0.8 times that of the average. A bird of average efficiency would achieve an *IFIR* of 500 mg per 5 minutes on the good bed and 400 mg on the poor one. The efficient individual would therefore have *IFIRs* of 625 and 500 mg on the two beds respectively while those of the inefficient individual would be 400 and 320 mg. There is no interference on the poor bed as bird density is so low; the two birds therefore actually achieve intake rates of 500 and 320 mg. But on the

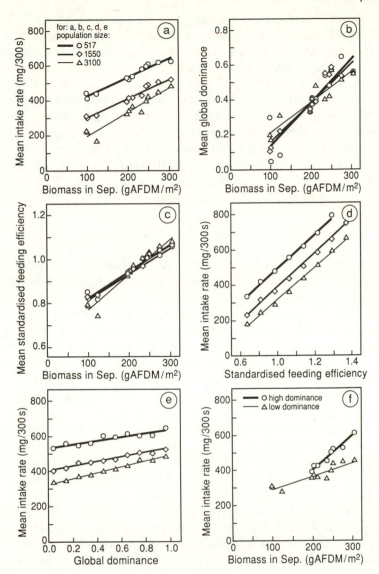

Fig. 5.8 The aggregation and functional responses as predicted by the Exe estuary model in September on spring tides at three population sizes. (a) The functional response, irrespective of global dominance and foraging efficiency. (b) The average global dominance of the birds on beds of different prey biomass. (c) The average foraging efficiency of the birds on beds of different prey biomass. (d) Intake rate as a function of a bird's foraging efficiency, irrespective of its global dominance. (e) Intake rate as a function of global dominance, irrespective of foraging efficiency. (f) The intake rates of high and low dominance birds on mussel beds of different biomass at a population size of 3100.

good bed, interference reduces the intake rate of each individual by 100 mg. Thus, the efficient individual would actually realize an intake rate of 525 mg on the good bed, 25 mg above its rate on the poor bed. In contrast, the inefficient bird would achieve an intake rate of only 300 mg on the good bed compared with 320 mg on the poor. The poorer beds, where competition is less intense, are therefore more profitable for the inefficient birds as well as for the least dominant. These findings suggest that functional responses measured in the field in circumstances in which interference affects bird distribution may thus reflect the differing efficiencies of the birds feeding in the different places as well as their susceptibility to interference and the interference-free functional response itself.

Most field studies of the functional response in Oystercatchers have been of cockle-eating birds which are thought to experience less interference than mussel-eaters. The Exe estuary model was adapted to investigate the effects of varying the average level of interference on the functional and aggregation responses, using the procedure detailed in Goss-Custard *et al.* (1995b), but leaving the food supply the same. The predicted responses at three levels of interference are shown in Figs 5.9 and 5.10, along with results from field studies. Without interference, and contrary to the predictions of simple *IFD* models, the average intake rate (Fig. 5.9a) is slightly different on the only two beds occupied (Fig. 5.10a). The model birds are only able to discriminate a difference in intake rate that is greater than 3%, so many birds feed in the site with the slightly poorer food supply because they cannot detect the penalty incurred. As interference levels increase, the functional response moves lower and extends over an increasing range of prey biomass (Fig. 5.9a) as the birds spread out (Fig. 5.10a). As also predicted by Sutherland's (1992) model, functional responses observed in the field will reflect the average level of interference as well as differences between the individuals feeding in the different places studied.

So far, it has not yet proved possible to establish the contribution that the general level of interference, along with the dominance and foraging efficiency of individual birds, make to the precise form of the functional responses measured in the field in Oystercatchers. The only noticeable pattern is that feeding rates seem to quickly reach a maximum as numerical prey density increases (Fig. 5.9b,d,e), whereas intake rates do so more slowly as prey biomass increases (Fig. 5.9c). Presumably, increasingly large prey are taken as the average size of the prey available or their density, or both, increase (Zwarts *et al.*, Chapter 2, this volume), enabling intake rate to rise even though the feeding rate has levelled off (Goss-Custard 1977b). Numerical prey density can thus be a very poor predictor of intake rate, which currently seems to be the best measure of the quality of a feeding area (Goss-Custard 1977b; Sutherland 1982c). This explains why the aggregation responses plotted against numerical prey density can show bird densities peaking at low prey densities (Fig.

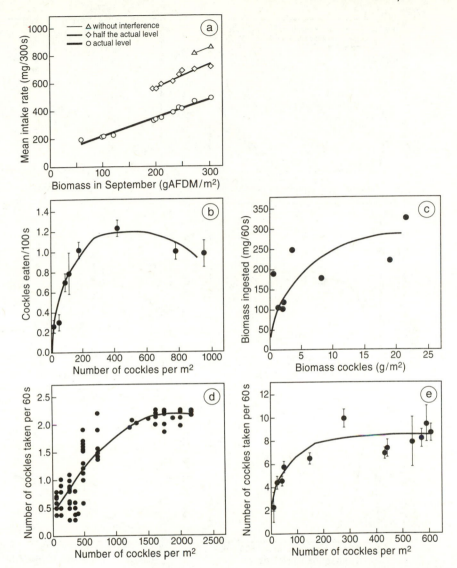

Fig. 5.9 The functional response including the influence of Oystercatcher density. (a) The predictions of the Exe estuary model at three levels of interference; actual level amongst mussel-feeders on the Exe, half the actual level and without interference. (b) The feeding rate and (c) intake rate in relation to the numerical and biomass density of cockles respectively in free-living Oystercatchers (Goss-Custard 1977b). (d) Feeding rate in relation to the numerical density of second-winter cockles (Triplet 1989a). (e) The feeding rate in relation to the numerical density of cockles (Sutherland 1982c). Vertical bars show ± 1 SE.

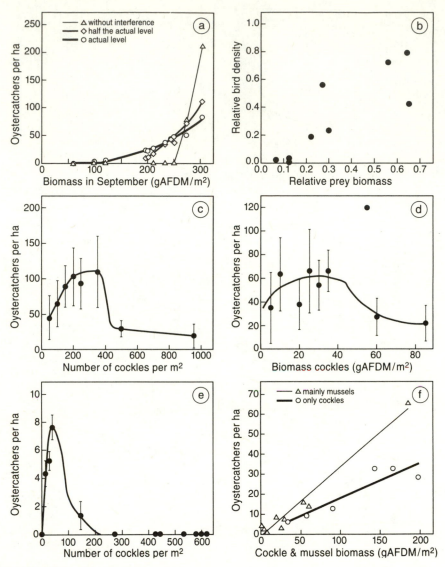

Fig. 5.10 The aggregation response in Oystercatchers. (a) The predictions of the Exe estuary model at three levels of interference; actual level amongst mussel-feeders on the Exe, half the actual level and without interference. (b) The response to the relativized biomass density of cockles (Goss-Custard 1977b). (c) Response to the numerical and biomass (d) densities of cockles (Triplet 1989a). (e) Response to the numerical density of cockles (Sutherland 1982c). (f) Response to the biomass density of mainly mussels and only cockles in December 1984 (Meire 1991, 1993). Vertical bars show ± 1 SE.

5.10c,e) whereas, with one exception (Fig. 5.10d), Oystercatchers may normally aggregate where prey biomass is highest (Fig. 5.10b,f).

At the levels of interference and differences between individuals assumed in these simulations, the Exe estuary model does not predict the same range of responses shown by Sutherland's (1992) model (Fig. 5.4). Rather, it predicts only concave or linear aggregation responses, with the relationship becoming less steep as the average level of interference increases and the birds spread out (Fig. 5.10a). Interestingly, field studies show that Oystercatchers spread out when they feed on cockles as well as on mussels (Fig 5.10). Despite the failure to detect interference directly in cockle-feeding birds, many feed in places with low cockle abundance. As the functional responses are generally so steep, it is difficult to believe that this arises because the birds cannot detect a difference in intake rates. The food distribution is also stable for periods of weeks so the birds would certainly have time to find the richer sites. The wide dispersion seems to imply that interference must occur more than has so far been directly detected in cockle-eating birds.

Autumn and winter

The model predicts a spatially density-dependent redistribution of birds (Fig. 5.11a) and depletion of the mussels (Fig. 5.11b) over the winter. Birds move increasingly to the beds that were initially less preferred as the higher rates of depletion on the preferred beds gradually equalize bed qualities. The predicted redistribution matches the observed trend qualitatively, but only poorly predicts the magnitude. Partly this arises because some birds on the Exe leave the mussel beds altogether over the winter (see below) and this option was not included in these model simulations. The predicted density-dependent mortality of the prey was not observed on the mussel beds (Fig. 5.11b). However, immigration and emigration occur in mussels of the Exe (McGrorty *et al.* 1990), and this could prevent any underlying density dependence from being detected. To provide a better test, the overwinter changes in mussel density in six areas (on three mussel beds) from which Oystercatchers were excluded were compared with the changes in six adjacent control areas where Oystercatchers were allowed to feed. The difference between the percentage rates of disappearance in the exclosure and control areas—the net depletion—was density-dependent, as predicted by the model (Fig. 5.11c).

The average level of interference influences the overwinter density dependence. The redistribution of birds was density-dependent at all levels of interference but Oystercatcher density decreased by the largest amount in the initially most preferred area when there was no interference (Fig. 5.12a). The density did not increase by the same amount on the second bed partly because of its much larger size but mainly because many birds moved to other mussel beds over the winter; since no birds used these beds in autumn, the percentage increase there could not be

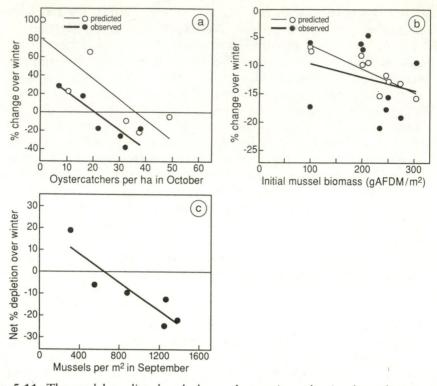

Fig. 5.11 The model predicted and observed overwinter density dependence on the mussel beds of the Exe estuary. (a) The density-dependent redistribution of Oystercatchers over the mussel beds from September to March. Beds with high densities of Oystercatchers in September suffered a net loss of birds while the initially low-density beds gained birds. Beds were combined in pairs because birds often moved prior to counting between adjacent mussels beds when disturbed. (b) The mortality of mussels from September to March across ten mussel beds; results are not shown from two beds where mussels declined rapidly from 1976 to 1983. (c) Mussel depletion in small areas in relation to the density of mussels in autumn; $r = 0.87$, $P = 0.012$, one-tailed test. Depletion is measured as the net depletion, the difference between the percentage change in mussel numbers inside an exclosure and in the adjacent control area.

calculated. This substantial overwinter redistribution of birds occurred because, without interference, most birds fed initially on the most preferred bed so that the rates of prey depletion over the winter were high (Fig. 5.12b). As the interference levels increased, the rates of redistribution and depletion decreased, but still remained density-dependent.

The total number of Oystercatchers on the Exe mussel beds is low compared with the abundance of shellfish so that the rates of depletion are rather small. Indeed, as a comparison of the two vertical axes in Figure 5.12b show, most of the overwinter loss in prey biomass was due

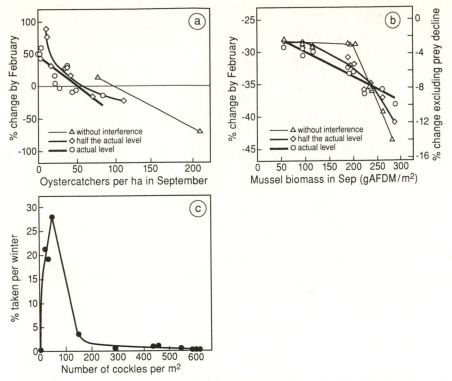

Fig. 5.12 Overwinter density dependence in Oystercatchers. (a) The density-dependent redistribution of birds as predicted by the Exe estuary model at three levels of interference; actual level amongst mussel-feeders on the Exe, half the actual level and without interference. (b) The density-dependent mortality of the prey as predicted by the model at the three levels of interference. (c) The pattern of mortality inflicted by Oystercatchers in relation to the numerical density of cockles (Sutherland 1982c).

to the loss of condition in the mussels themselves. Clearly, with many more Oystercatchers, depletion would be much greater and the over-winter responses accordingly much larger. Higher rates of prey depletion have been observed in other studies. Almost 30% of the cockles available for Oystercatchers in Traeth Melynog, Anglesey, were depleted over one winter, although the spatial pattern of loss was not density-dependent when related to the numerical density of the cockles (Fig. 5.12c). The birds actually congregated where the mussels were large and sparse rather than small and dense and so inflicted a disproportionately high mortality at quite low numerical prey densities.

The effect of Oystercatchers on their shellfish prey therefore varies from place to place but can clearly be considerable. The percentages of the standing crop of mussels or cockles removed by Oystercatchers alone between autumn and spring in the main feeding areas have been esti-mated at 40% (Zwarts and Drent 1981), 36% (Brown and O'Connor

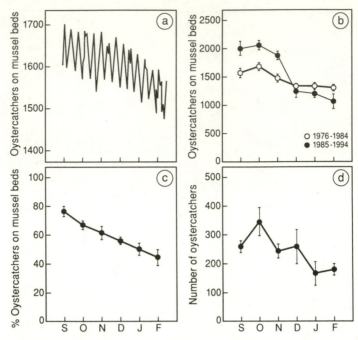

Fig. 5.13 The decrease in the numbers of Oystercatchers on shellfish beds over one winter. (a) The model prediction for the Exe estuary mussel beds, assuming a seasonally constant alternative supply is available. The fluctuations coincide with the changes in the mussel food supply associated with the neap–spring cycle. (b) The actual reduction recorded on the Exe over the winters 1976–1984. Subsequently (1985–1994), more birds wintered on the mussel beds and the overwinter reduction was greater. (c) The decline in numbers over three winters on the mussel beds of the Ythan estuary (Heppleston 1971b). (d) The decline in numbers feeding on cockles within Greyabbey Bay, Strangford Lough, Ireland (O'Connor and Brown 1977). Vertical bars show ± 1 SE.

1974), 31% (Horwood and Goss-Custard 1977), 20–25% (Sutherland 1982c), 22% (Drinnan 1957), 9–20% (Meire *et al.* 1994b), and 5–25% (Goss-Custard 1977b). Although the effects that such reductions in shellfish abundance have on intake rate have yet to be measured, in combination with the loss of prey condition, they seem to be enough for many Oystercatchers to move from shellfish to other foods as the winter progresses. With an alternative food supply that remains constant throughout the winter made available, the Exe model predicts that some 5–10% of 1650 birds settling on the mussel beds in autumn would leave over the winter (Fig. 5.13a). Encouragingly, this is similar to the actual rate of decrease observed during winters when about this number of Oystercatchers occupied the beds in autumn (Fig. 5.13b). A similar overwinter departure of birds was found on the mussel beds of the Ythan estuary (Fig. 5.13c) and on the cockle beds of Strangford Lough (Fig. 5.13d). Clearly, Oystercatchers redistribute themselves both amongst shellfish beds and between shellfish and other prey over the winter as depletion and the decline in prey quality change the relative profitabilities of the different sources of food.

Between winters

When Oystercatchers descend on shellfish stocks in autumn in particularly large numbers, they would be expected to inflict a disproportionate loss on their prey and more birds than usual would move to other prey or away from the estuary altogether. This is certainly predicted by the Exe estuary model in simulations in which the mussel food supply varied between years and an alternative food supply, that was constant both within and between winters, was supplied (Fig. 5.14a). However, this pattern was not observed on the Exe itself (Fig. 5.14b), probably because the abundance of the alternative mussels that were not included in the

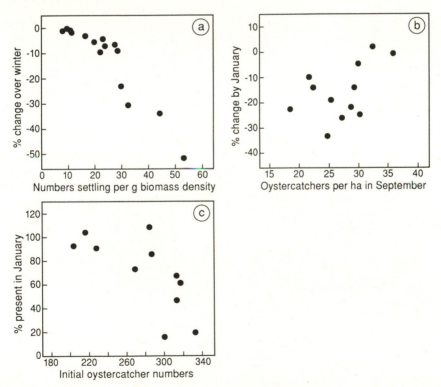

Fig. 5.14 The overwinter percentage change in Oystercatcher numbers on shellfish beds according to the initial population size in autumn. (a) The predictions of the Exe estuary model, assuming a constant mussel food supply with an alternative available. (b) The changes recorded on the mussel beds of the Exe estuary in a sample of winters since 1976 (J. D. Goss-Custard, unpublished information). (c) Cockle beds in Greyabbey Bay and Boretree Islands in Strangford Lough, Ireland (O'Connor and Brown 1977).

study itself declined over the study (J. D. Goss-Custard, unpublished information). The predicted pattern was found amongst the cockle-eating Oystercatchers of Strangford Lough (Fig. 5.14c). Since the majority of birds that disappeared from the Lough did not die, there was clearly a density-dependent redistribution of birds between winters, with a greater proportion leaving in winters with high intial numbers of Oystercatchers.

The food supply also varies between winters and, unless food abundance changes simultaneously everywhere, more Oystercatchers might winter on an estuary in years of high food stocks. This happened in the Ribble estuary over a period in which cockles were unusually abundant (Fig. 5.15a). Many of the immigrants were juveniles, probably prospecting for an estuary to spend the winter (Sutherland 1982d). In contrast, the numbers wintering on the Exe mussel beds (Fig. 5.15b) and on the Burry Inlet cockle beds (Fig. 5.15c) were unrelated to shellfish abundance,

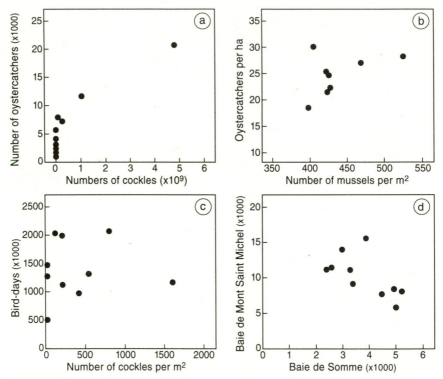

Fig. 5.15 The numbers of Oystercatchers wintering on shellfish beds in relation to the shellfish stocks present. (a) Numbers on the cockle beds of the Ribble estuary in relation to total number of cockles present (Sutherland 1982d). (b) Bird density in September on the Exe estuary mussel beds in relation to mussel biomass present in September (J.D. Goss-Custard, unpublished information). (c) Numbers on the cockle beds of the Burry Inlet in relation to the cockle stocks (Horwood and Goss-Custard 1977). (d) The numbers of birds in the Baie de Somme and Baie de Mont Saint Michel; $r = -0.62$, $P = 0.055$ (Triplet and Etienne 1991).

perhaps because in years of low stocks they were able to switch to other prey. In the Ribble, very few birds were present before the cockle spatfall arrived, so birds moved into the estuary in large numbers for the first time. The origin of the many adult immigrants involved is unclear, however. That conditions in one estuary might affect the numbers settling in another is suggested by the tendency for more birds to settle in the Baie de Mont Saint Michel when high numbers winter in the Baie de Somme (Fig. 5.15d). To explain the numbers of Oystercatchers feeding in one estuary, it seems necessary not only to monitor the feeding conditions in that estuary but also those in others.

Other factors affecting Oystercatcher distribution

Factors other than food abundance, and thus intake rate, affect bird density. The proportion of prey that can be both detected and caught varies both spatially and over time. For example, shellfish may be more available to stabbing Oystercatchers when they are submerged in shallow water so that their valves are slightly apart (Hulscher, Chapter 1, this volume); with prey density the same, stabbers may feed faster at times of the exposure period, and in places, when the mussels are submerged. Again, with the effect of other known factors taken into account, mussel-feeding Oystercatchers on the Exe estuary occur at low densities on soft and muddy mussel beds (Goss-Custard *et al.* 1981, 1992). The reasons are not clear. The suggestion that such beds are energetically expensive for foraging birds or the mussels contain high infestations of parasites (Goss-Custard *et al.* 1981, 1984) have been ruled out by Speakman (1984) and Goss-Custard *et al.* (1993) respectively. Oystercatchers also feed at lower densities on beds lying distant from the high tide roost, presumably because of the extra energy cost of travel incurred (Goss-Custard *et al.* 1981, 1992; Zwarts and Drent 1981). Whether risk of attack by raptors affects Oystercatcher distribution has not been tested, although Cresswell and Whitfield (1994) provide excellent evidence for another shorebird, the Redshank (*Tringa totanus*). Finally, the use made of mussel beds varies with their level on the shore. Upshore mussel beds that expose first as the tide recedes attract many birds for a short while before the more low-lying beds expose (Meire 1991, 1993). Although an Exe estuary model based on food abundance alone goes some way towards predicting where Oystercatchers feed, other factors will have to be included if the model is to be applied with more predictive precision to particular systems.

Discussion

The basic idea of the ideal free distribution model and its more complex derivatives seem to apply well to foraging Oystercatchers. Although other factors affect their choice, the tendency of Oystercatchers to aggregate where prey biomass and intake rate are high does seem to be countered by the need to avoid interference. Oystercatchers are also responsive to within-winter and between-winter changes in the qualities of the alternative food supplies available. As the initially most dense areas of shellfish beds become depleted, birds redistribute to previously less profitable areas. Indeed, they move to alternative food supplies altogether, especially in winters when high initial numbers of birds may cause shellfish food supplies to be particularly heavily depleted. Although most studies of their responses to these spatial, seasonal and annual differences in the main food supplies have been made within one estuary, there is some evidence that redistribution can also take place between estuaries and especially involve young birds. If so, the numbers of Oystercatchers in a particular estuary, and thus on particular parts of that estuary, may depend as much on the feeding conditions in other estuaries as on those in the estuary itself.

These small- and large-scale redistributions are the result, it seems, of individual Oystercatchers selecting at each stage of the winter the most profitable area in which to feed. The models of increasing complexity which attempt to capture this process have been based on the assumption that each bird attempts to maximize its current intake rate. While Oystercatchers may not always feed at the fastest possible rate (Zwarts *et al.*, Chapter 2, this volume), they do seem generally to choose the sites where intake rates are highest. The rate each bird achieves in each alternative site available depends on the food supply present but also on its own foraging efficiency and ability to resist interference from competing birds. These attributes appear to vary independently of each other and are also affected by the feeding method, age and, probably, site familiarity of the bird itself (Ens and Cayford, Chapter 4, this volume). These individual differences mean that the best solution at any one time varies between birds.

The individuals-based model of the Exe estuary uses a game theoretic approach to discover how hundreds or thousands of competitively different birds become distributed if each one makes its own independent decisions as to where to feed. Each model Oystercatcher selects each day the mussel bed where its current intake is highest, within its ability to discriminate. For most birds, the choices made are constrained by those made by their competitors. The model predicts the observed distribution of birds quite well and also some of the behaviour of particular cohorts. But, as a description of this particular system, it is clearly deficient. More functions need to be added and some parameter estimates need to be improved. Nonetheless, the broad correspondence between prediction and observation is encouraging and suggests that the paradigm is worth pursuing. The eventual objective is to predict accurately where individuals feed, and the rate at which they do so, in the novel circumstances brought about, for example, by the removal or creation of new mussel beds (Ens *et al.* 1996a). If this can be done, confidence in the basic model derived from *IFD* theory would be high and how Oystercatcher distribution might change in particular circumstances could be explored more generally and realistically.

Summary

The intertidal feeding grounds of Oystercatchers are a series of interlocking habitat gradients that range in scale from a few square metres up to hundreds of square kilometres. This chapter investigates by modelling the distribution patterns that result from the decisions made by individual Oystercatchers as to where they should feed. It examines whether, through its influence on the intake rates of individual birds, the interaction between the opposing tendencies to aggregate where food is abundant but to avoid other birds so as to minimize interference is a satisfactory way to model Oystercatcher distribution.

The conceptual framework for doing this is provided by game theory as applied to the ideal free distribution (*IFD*). The predictions of simple *IFD* models depend critically on the precise values used for just one of the parameters. Therefore, empirical tests of the general approach depend upon good parameter estimation in the field and the construction of a more realistic, though initially situation-specific, model. Such a model has been built for Oystercatchers eating mussels on the Exe estuary. The model is individuals-based in that each bird has its own competitive characteristics and makes its own decisions as to where to feed in the light of the decisions made by all the other individuals exploiting the same food gradient. The model tests were sufficiently encouraging to believe that the model captures reality reasonably well. However, its failure to predict all quantitative trends accurately shows that some parameter values need to be refined and some new functions added.

The model predicts how the distribution of Oystercatchers changes with increasing population size, both within and between winters, and over the winter as prey are depleted at different rates in different places. Predicted trends are compared with field data from the Exe and from studies elsewhere in Europe. The model also simulates the effects of the average level of interference on the aggregation and functional responses of Oystercatchers and on prey depletion. This suggests that interference is more widespread in cockle-eating birds than is currently thought. The model successfully predicts that Oystercatchers change their feeding areas within and between winters according to changes in the abundance in their main shellfish food supplies. However, a number of other factors could affect the distribution of Oystercatchers but are not currently included in the model. Consideration needs to be given to these before the efficacy of the overall approach can be fully evaluated. The objective is accurately to predict where individuals feed, and the rate at which they do so, in the novel circumstances brought about, for example, by the creation or removal of new shellfish beds. If this can be done, confidence in the basic model derived from *IFD* theory would be increased.

6 How Oystercatchers survive the winter

John D. Goss-Custard, Sarah E. A. le V. dit Durell, Cameron P. Goater, Jan B. Hulscher, Rob H. D. Lambeck*, Peter L. Meininger and Jamil Urfi

Introduction

Over the last 25 years, much emphasis has been placed by ornithologists on the hypothesis that, in many temperate-zone birds, food shortage is an important cause of death outside the breeding season (Lack 1954; Newton 1980). This idea has underpinned most of the research carried out on Oystercatchers on their migration and wintering grounds. It is one reason why, in the preceding five chapters of this book, Oystercatchers have been expected to make the choices while foraging that increase the rate at which they obtain energy and/or nutrients (Hulscher, Chapter 1; Zwarts *et al.*, Chapter 2; Sutherland *et al.*, Chapter 3; Ens and Cayford, Chapter 4; Goss-Custard *et al.*, Chapter 5). It is also the reason why so much concern is expressed by conservationists when winter feeding areas are threatened by the many human activities that take place on the fore-shore (Lambeck *et al.*, Chapter 11, this volume). The assumption that food shortage is an important factor that prevents some Oystercatchers from surviving the non-breeding season, or from arriving on the breeding grounds in good condition (Hulscher *et al.*, Chapter 7, this volume)—or that it could easily become so were their feeding grounds to be much changed—is thus central to our thinking about many aspects of Oyster-catcher biology outside the breeding season. As such, it demands the critical appraisal that this chapter attempts to make.

Weather conditions and survival

Large numbers of Oystercatchers can be found dead during prolonged periods of severely cold weather. This has been reported in Britain (Heppleston 1971b; Goss-Custard *et al.* 1977a; Baillie 1980; Clark 1982; Davidson and Evans 1982; O'Connor and Cawthorne 1982; Davidson and Clark 1985; Whitfield *et al.* 1988; Clark 1993), in the German (Temme and Gerß 1988; Stock *et al.* 1989) and Dutch Wadden Seas (Swennen and Duiven 1983; Swennen 1984; Stock *et al.* 1987a; Hulscher 1989, 1990; Kersten and Brenninkmeijer 1995), in the Delta area (Lambeck 1991; Lambeck *et al.* 1991; Meininger *et al.* 1991), along

* This is publication number 2001 of The Netherlands Institute of Ecology

the Belgian coast (van Gompel 1987) and in France, where hunting may increase the kill (Triplet and Etienne 1991). Many have also been found dead in severe spring weather on breeding areas in Scotland (Watson 1980) and Sweden (Marcström and Mascher 1979; Hulscher *et al.*, Chapter 7, this volume).

Body weights on such occasions are some 40% below the typical weight of live birds (Swennen and Duiven 1983; Swennen 1984; Lambeck *et al.* 1991). The birds die having used almost all of their fat (Marcström and Mascher 1979; Davidson and Evans 1982; Davidson and Clark 1985). The subcutaneous deposits are virtually completely depleted and even the lipid content of a vital organ such as the liver drops by 70% (Lambeck *et al.* 1991). Like other birds, starving Oystercatchers also metabolize non-fat tissue in the liver (Marcström and Mascher 1979) and protein from muscles, particularly the pectorals. Some protein reserves remain in Oystercatchers found dead during severe weather (Davidson and Evans 1982; Zwarts *et al.* 1996b), but this does not mean they died from causes other than starvation. Protein in shorebirds seems only to be metabolized in the very last stages of starvation (Evans and Smith 1975; Davidson and Clark 1985). Often it cannot be mobilized quickly enough to maintain body temperature, partly because protein per gram yields only half the energy of fat (Whittow 1986). According to recent work on Knot (*Calidris canutus*), the amount of protein remaining at death is closely linked to the thermostatic stress the birds experience during the final phase of starvation, while the amount of fat remaining is always about the same (Piersma *et al.* 1996). Death before protein reserves are depleted can thus be expected when thermostatic stress is high, hypothermia probably being the ultimate cause of death.

Severe weather has two adverse effects on the ability of Oystercatchers to satisfy their energy requirements. First, severe weather considerably increases the amount of energy required to maintain the body temperature. At below the thermoneutral temperature of 9–10 °C, the daily food consumption of captive Oystercatchers, even in still air, almost doubles as the ambient temperature falls to −8 °C (Kersten and Piersma 1987). The dramatic effect of wind on the insulative quality of plumage raises energy requirements in high winds considerably (Wiersma and Piersma 1994). In the wild, Oystercatchers have been seen to shelter from wind at low ambient temperatures, at least temporarily, in gardens and orchards and behind dikes, vegetation, creek edges, thick sheets of ice and, of course, other roosting Oystercatchers; flocking reduces energy expenditure in Oystercatchers by 14–19% in a wind of 9 m s^{-1} (Whitlock 1979). The precise effect on Oystercatchers of various, naturally-occurring combinations of low temperatures, strong winds and low levels of radiation have yet to be worked out. However, experiments on Knot suggest that the increase in energy demands in severe winter weather will be substantial (Wiersma and Piersma 1994).

Secondly, low temperatures and strong winds may make it more difficult to collect food. Any visual cues used to detect prey may become
scarcer at low temperatures as the prey become less active and, perhaps,
move deeper into the substrate (Hulscher, Chapter 1, this volume). The
rate Oystercatchers caught earthworms Lumbricidae in fields decreased
from 7–8 worms per five minutes to 0–1 worms per 5 minutes as the soil
temperature fell from 5° to 0 °C (Goss-Custard and Durell 1987c), presumably because the worms became less active, and so more difficult to
detect, or withdrew into the unfrozen ground and out of reach of the
birds (Edwards and Lofty 1977). Similarly on mudflats, Ragworms
(*Nereis diversicolor*) move deeper during the first cold spell of winter
(Esselink and Zwarts 1989) and the surface-feeding activity of the
bivalve *Macoma balthica* is reduced at low temperatures (Hummel
1985). However, that such changes in prey behaviour necessarily affect
intake rate cannot be assumed. There was no detectable effect of low,
but not sub-zero, temperatures on intake rate in Oystercatchers feeding
on cockles (*Cerastoderma edule*) (Goss-Custard *et al.* 1977a; Sutherland
1982c; Triplet 1989a), mussels (*Mytilus edulis*) (Goss-Custard *et al.*
1996b), Ragworms (Triplet 1989b) or clams (*Scrobicularia plana*) (J. S.
Boates, personal communication). Despite the uncertain effect of low
temperatures, Oystercatchers cannot catch intertidal prey when mudflats
freeze or when ice flows cover the flats, both of which are most likely to
occur at the higher and sheltered levels of the shore on neap tides when
the sediments are continuously exposed to the air (Triplet 1989b) and
when strong winds blow offshore (Temme and Gerß 1988). In the same
way, frozen or snow-covered ground prevents Oystercatchers physically
from catching earthworms in fields.

Strong winds may also affect intake rate. Although gales may sometimes wash mussels, cockles and even the deep-burrowing bivalve *Scrobicularia plana* out of the substrate and so make them temporarily more
accessible to Oystercatchers, they are normally more likely to decrease
intake rates. The visual cues used to locate buried prey (Hulscher 1976c,
1982, Chapter 1, this volume) can be obliterated by wind-driven surface
water as observations on cockles (P. Triplet, personal communication)
and *Mya arenaria* (L. Zwarts, personal communication) have shown.
Strong winds may directly affect foraging (i) by forcing birds to walk
into the wind to retain the insulative properties of their plumage, rather
than choosing the most profitable search path; (ii) by forcing birds to
seek shelter in what may be relatively unprofitable places, such as along
the bottom of creeks; and (iii) by making it less easy for them to manipulate the prey (P. Triplet, personal communication). Feeding may become
so difficult or unprofitable that birds stop feeding altogether during
gales, especially when the feeding grounds are also ice-covered (Evans
1976; P. L. Meininger, unpublished information). Strong onshore winds
reduce the time available for feeding because they hold back the ebb tide

and speed up the flood (Hulscher 1993); with a small tidal range, such as the Wadden Sea (1–2 m), such winds can prevent the flats from being exposed at all. But even on the Exe estuary, where the tidal range is much greater (4–5 m), the exposure period may be halved even on spring tides, especially when the river is also in flood. Furthermore, the quality of the accessible high-shore feeding areas is generally poor because of low prey biomass (Dankers and Beukema 1983), the combined effect of low numerical densities (Beukema 1976; Zwarts 1988) and low individual weights. Short immersion periods reduce prey growth rates and many are small-sized (Coles 1956; Jones 1979b; Seed 1968; Goss-Custard *et al.* 1991, 1993; Wanink and Zwarts 1993) and, for their length, contain less flesh, as has been shown in both mussels (Goss-Custard *et al.* 1993) and cockles (Kristensen 1957; Goss-Custard 1977b; Sutherland 1982b). Thus strong onshore winds may decrease both the quality of the accessible food supplies and the time available for feeding. But by the same token, strong offshore winds may have the reverse effect.

The effect on Oystercatchers of winds of differing direction is detailed by Swennen and Duiven (1983). Some 3–4% of the wintering birds in the western Dutch Wadden Sea died during 11 days of sub-zero temperatures in January and February 1976. Yet very few birds died in 1979, when sub-zero temperatures lasted for 32 days. The difference was probably due to the contrasting wind patterns. First, strong easterly winds blowing during the cold spell in 1976 caused the flats to be exposed for so long that they iced over. In 1979, by contrast, many flats remained accessible because they were still swept twice daily by the tide. Second, strong westerly winds earlier in the 1975–76 winter, before the cold spell itself began, had prevented the flats from exposing for as long, and as extensively, as normal. Unlike the Oystercatchers in the eastern Wadden Sea which suffered very few casualities that winter, the birds in the western Wadden Sea had been unable to feed in fields over the extended high water periods earlier in the winter (J. B. Hulscher, unpublished information). The western Wadden Sea birds had probably already drawn upon their energy reserves by the time the cold spell arrived, and so were less able to deal with it.

Annual and regional variations in mortality rates during severe weather have not always been so readily explained, however. Oystercatchers would be expected to have most difficulty if gales occurred either at the same time as the cold period or, perhaps, immediately beforehand. More deaths were indeed recorded in a number of British estuaries in the windy and cold spells of 1978–1979 and 1984–1985 than in the less windy, and shorter, cold spells in 1981–1982 and 1985–1986 (Clark and Davidson 1986). Similarly, the severe weather in Britain in 1985–1986 was worst along the east coast, where most of the dead birds were found (Davidson and Clark 1986). On the other hand, no more than normal numbers of dead birds have been found during some prolonged periods of cold and windy weather. Few were found on the Wash during 70 days of severe cold and wind in the 1962–1963 winter (Pilcher 1964), while many were found in Morecambe Bay (Clark 1982). The sometimes pronounced annual and regional variations in British Oystercatcher mortality rates during severe weather therefore remain to be explained.

They are several reasons why, in comparable climatic conditions, both the proportion of birds that die and the proportion of the corpses that are found may vary between estuaries and years. Wind direction relative to the shore-line affects the length of time intertidal flats are exposed and so available for feeding and the proportion of corpses washed ashore as, indeed, may ice flows. More birds may die when the severe weather co-incides with neap tides (Goss-Custard and Durell 1987c) and in estuaries where the tidal range is always small; this may be why, during the severe spell of January 1987, a greater proportion of Oystercatchers died in the Wadden Sea (Hulscher 1989, 1990) than in the Delta area (Lambeck 1991). The general feeding conditions may also vary considerably between estuaries, and affect the body condition of birds both before and during the severe period. Unlike mussels, cockles are particularly vulnerable to cold and may die in large numbers in severe weather (Hancock 1971; Beukema 1974, 1979, 1982, 1989, 1993; Beukema and Essink 1986). Unless the birds are able to turn successfully to another prey species, as Oystercatchers in the Baie de Somme may have done by eating *Nereis* (Desprez *et al.* 1992; but also see Lambeck *et al.*, Chapter 11, this volume), mortality rates may be higher in estuaries where most birds eat cockles. Birds may also move away when the severe conditions arrive (Pilcher 1964; Baillie 1980; Hulscher 1989, 1990; Hulscher *et al.*, Chapter 7, this volume), thus affecting the chances that starved corpses will be found locally. Because of the number of factors involved, it is perhaps not surprising that no clear picture has yet emerged on regional and annual variations in the occurrence of large, severe weather mortalities in the British Isles.

Because of movement, it may prove difficult to relate the overwinter mortality rates to the weather conditions. Perhaps this is why Safriel *et al.* (1984) were unable to find a relation between winter weather and overwinter survival in birds breeding on Skokholm. The weather may need to be considered over a much larger geographic scale, within which any movements may have taken place. Thus, the total number of ringed Oystercatchers recovered in Britain (Fig. 6.1), as in The Netherlands (Hulscher *et al.*, Chapter 7, this volume), is higher in years with cold winters. Although any effect of wind has not yet been taken into account, more birds are clearly found dead in winters with prolonged periods of low temperatures.

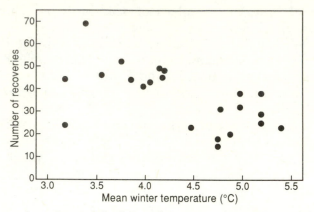

Fig. 6.1 The total numbers of dead ringed Oystercatchers recovered during a winter, and reported to the British Trust for Ornithology, in relation to the severity of the winter, as measured by the mean daily winter temperature (N. A. Clark, unpublished information).

The individuals that die

Some birds are more at risk of dying in severe weather than others. The most extensive study of this was made by Swennen and Duiven (1983) and Swennen (1984) using 380 corpses recovered in the winter 1975–76 in the Dutch Wadden Sea. First, young birds were particularly at risk, as Heppleston (1971b) had also found in North-east Scotland in 1965–66 and other workers have subsequently confirmed (Lambeck 1991; Meininger *et al.* 1991; Kersten and Brenninkmeijer 1995). Second, an unusually high proportion (19%) of those older than one year had not finished the moult of the primary flight feathers by the time the severe weather arrived in February; normally the moult of even the latest birds in the Wadden Sea finishes by mid-December (Boere 1976a). Third, the sample of dead birds contained a high (61%) proportion with anatomical abnormalities of the bill or foot. For example, in 20% of the dead birds, the bills were abnormal; the mandibles were crossed or strikingly curved upwards or downwards, or one mandible was at least 2 mm shorter than the other. Young birds and birds with deformities were therefore identified as being particularly at risk in severe weather.

Swennen (1984) stressed the significance of how well birds had fed earlier in the winter in determining which ones died. His study populations fed on two shores along which there was a gradation in the quality of both the feeding conditions and the birds. At one extreme, the flats were wide and exposed for a long time during each tidal cycle. At the other extreme, birds fed on low-level and narrow shores that only exposed for a short time. As the shores became lower and narrower, the proportion of young birds and birds with deformities increased (Fig. 6.2).

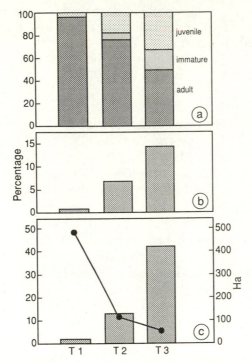

Fig. 6.2 The percentage dying and characteristics of Oystercatchers found dead during a severe spell at three roosts, T1– T3, on Texel. (a) The ages of the birds; (b) the percentage with large external abnormalities; and (c) the percentage dying and the extent of the adjacent intertidal flats (•) (Swennen 1984).

At the same time, mean weight decreased, presumably because of the assumed poorer feeding conditions, the inexperience of the younger birds (Ens and Cayford, Chapter 4, this volume) and the feeding handicap of birds with physical deformities. Some 40% of the birds in the roost with the greatest proportion of young birds and individuals with deformities died during the severe spell compared with only 2% in the roost adjacent to the widest shore (Swennen 1984). Although there is no direct evidence that deformities affect foraging, the weight of ten individuals with foot injuries (incidence 3.3% in 7350 birds examined in West Britain) was on average 6.9% (±1.7; range 0–15.0) less (sign test: $P = 0.002$) than the mean weight of samples of uninjured birds of the same sex measured at the same time of year (Dare and Mercer 1968). Swennen therefore argued that, when the birds were prevented by severe weather from feeding in February 1976, the birds using the narrow feeding areas probably had lower than average body reserves, and so were less well equipped to outlast the severe spell. The high proportion of dead birds that had retarded primary moult was consistent with this hypothesis. Additional support for the hypothesis was found in the Oosterschelde during the

1980s when the Oystercatchers that died had arrested moult and came from sections of the population that had low mean weights (Lambeck 1991). Indeed, the first birds found dead when a severe spell begins in the Wadden Sea are found within one or two days (Hulscher 1989), implying that they have very low energy reserves (Hulscher *et al.*, Chapter 7, this volume) and, prior to the onset of the bad weather, had been living close to starvation.

Not all the birds, by any means, that die in severe weather have physical deformities (Temme and Gerß 1988) and not all winter deaths occur during severe spells. Many casualities in severe winters in The Netherlands are juveniles (Swennen 1984; Kersten and Brenninkmeijer 1995). Marked differences in bill biometrics, sex- and age- composition between casualties and birds still alive after a cold spell imply that risk of death is associated with a variety of individual factors, and not just deformity (J. B. Hulscher, unpublished information). Indeed, many Oystercatchers in The Netherlands are able to feed during cold spells and survive with an almost normal body weight (Hulscher, Chapter 7, this volume); clearly birds vary in the extent to which cold weather prevents them from feeding. Birds also died during four mild winters in South-west England, and here too young birds were four to five times more at risk than adults (Goss-Custard *et al.* 1982c). So even though most deaths from starvation in severe weather may normally occur in late winter, the feeding conditions earlier in the winter and individual differences in foraging strategy need to be studied if we are to understand why certain birds are most at risk in severe weather and why mortality also occurs, especially in young birds, during generally mild winters.

Seasonal changes in feeding conditions

The food supply

For several reasons, the feeding conditions deteriorate for Oystercatchers between autumn and late winter. First, food abundance on the flats decreases. Oystercatchers usually eat the older and larger prey animals (Zwarts *et al.*, Chapter 2, this volume) and replacement of the depleted prey can only occur through the growth of animals into the size range eaten. Except in milder western and southern regions, this may not occur after late autumn. Oystercatchers can considerably reduce prey density over the winter (Goss-Custard 1980; Zwarts and Drent 1981; Zwarts and Wanink 1984; Meire *et al.* 1994a; Goss-Custard *et al.*, Chapter 5, this volume), particularly of the fraction that is actually accessible to the birds (Meire 1996a). The flesh content of many bivalves also decreases considerably between autumn and spring, thus reducing the energy value of each prey caught; typically, the mass of an animal of constant length decreases by 20–40% from August or September to March (Hughes 1970; Dare and Edwards 1975; Beukema and De Bruin 1977; Chambers and Milne 1979; Zwarts 1991; Goss-Custard *et al.* 1993; Zwarts and Wanink 1993). The decreases are smallest, however, in cold winters when the animals are metabolically less active than in mild winters (Zwarts 1991). Second, the availability of some estuarine prey decreases in winter as they burrow deeper into the substrate and become less active. This has been shown in Ragworms (Esselink and Zwarts 1989; Boates and Goss-Custard 1989; Zwarts and Wanink 1993) and in the bivalve molluscs *Macoma balthica* (Reading and McGrorty 1978; Zwarts and Wanink 1989, 1991, 1993) and *Scrobicularia plana* (Zwarts and Wanink 1989, 1991, 1993), although not in cockles or the clam *Mya arenaria* (Zwarts and Wanink 1989, 1991, 1993). In burying bivalves, the individuals in the poorest condition, and thus of least food value to Oystercatchers, may be the ones that remain closest to the substrate surface in winter because their short siphons prevent them from going deeper; these prey may also contain most parasites (Zwarts and Wanink 1991, 1993). Third, as the daylength shortens, an increasing proportion of the low water period occurs in darkness when some birds may sometimes feed at less than half the day-time rate (Hulscher, Chapter 1, this volume), at least when feeding on mussels (Zwarts and Drent 1981; Goss-Custard and Durell 1987c) or cockles (Sutherland 1982a,c). Finally, interference competition amongst Oystercatchers eating mussels intensifies through autumn and winter so that, at a given density of competitors and mussels, intake rates will be lower in midwinter than in autumn (Goss-Custard and Durell 1987a). So at the very time of the nonbreeding season when energy requirements are increasing as the ambient temperature falls and gales become more frequent, the opportunities for birds to satisfy them decrease.

Responses of the birds

The main response of the birds to these deteriorating conditions appears to be to increase the time they spend feeding (Fig. 6.3). They roost less at high water and so spend more time on the intertidal feeding grounds (Goss-Custard *et al.* 1977a). They feed more at the higher shore-levels on the receding and advancing tides, on their way out to and back from the preferred low-level feeding areas. They feed for a greater proportion of the time over the low water period (Heppleston 1971b; Goss-Custard *et al.* 1977a). Oystercatchers may also feed more at night in midwinter than at other times of year (Heppleston 1971b). In some areas, Oystercatchers feed increasingly in fields at high tide from November onwards, mainly on earthworms (Dare 1966; Heppleston 1971b; Goss-Custard and Durell 1983; Quinn and Kirby 1993), especially when it has rained recently (Quinn and Kirby 1993). However, in many places, they feed at night throughout the nonbreeding season anyway (Daan and Koene 1981; J. D. Goss-Custard, unpublished information; C. Swennen, unpublished information), apparently because the rate at which the gut can process food

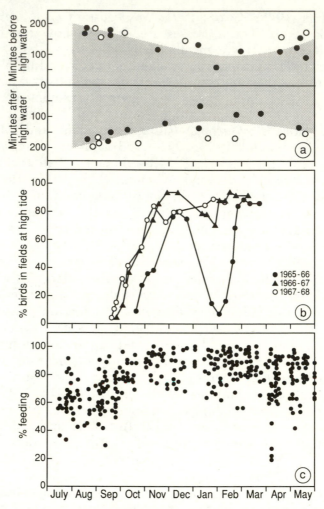

Fig. 6.3 Seasonal changes in the amount of time spent feeding by Oystercatchers each day. (a) The time (hatched area) spent not feeding at the high water roost on the Wash, as recorded on neap (•) and spring (○) tides (Goss-Custard *et al.* 1977a); (b) the proportion of birds on the Ythan estuary feeding in fields at high tide over three winters (Heppleston 1971b); and (c) the percentage of birds in flocks on the intertidal flats of the Wash that were feeding (Goss-Custard *et al.* 1977a).

limits the amount they can consume over one low water period to a quantity below their typical requirements (Kersten and Visser 1996a). Provided the food supply is sufficiently dense (Meire 1996a), Oyster-catchers may also be able to increase their intake rate, as has been demonstrated in captive birds by Swennen *et al.* (1989), although perhaps at the cost of an increased risk of damaging the bill. Thus, Oystercatchers adapt to the decline in the feeding conditions through autumn and winter by foraging more extensively and, perhaps, more intensively.

Superimposed on these seasonal changes are shorter-term fluctuations, associated with variations in weather and tide, to which the birds respond in the same way. For example, when westerly winds reduced the tidal range in the Dutch Wadden Sea, the average density of birds through the tidal cycle on a study mussel bed increased from 78 to 258 birds per hectare and the time available for feeding decreased from as much as 8 hours to as little as 2.5 (Daan and Koene 1981). Because of increased interference at the higher bird densities, the average bird spent a smaller proportion of the period for which mussels were exposed actually feeding (Koene 1978; Zwarts and Drent 1981). This, along with the shorter duration of the exposure period, reduced the mean food consumption over the exposure period from 18.2 to 6.9 g ash-free dry mass. In response, the numbers foraging inland at high tide increased from 20 to 60%. Similarly, the feeding opportunities for Oystercatchers on mussel beds of the Exe estuary are poorer on neap tides (Goss-Custard and Durell 1987c); the tide only exposes the upper and poorer parts of the mussel beds and the density of birds is much higher so that interference between foraging birds increases. In response, more birds feed in the fields at high water on neap than on spring tides (Goss-Custard and Durell 1983).

Feeding in the fields at high water may thus provide a supplementary food source when feeding conditions in the intertidal zone deteriorate during late autumn and winter (Dare 1966; Heppleston 1971b). This is confirmed by Clark (1993) and Meininger *et al.* (1994) who found that unprecedented numbers of Oystercatchers began to feed in fields at high tide, and even at low tide, besides the Wash and Oosterchelde respectively after the feeding opportunities in the intertidal zone had been considerably reduced. Similarly, unusually large numbers fed inland of the Dutch Wadden Sea when strong winds reduced the time available for feeding in the intertidal zone (Hulscher 1993). Nonetheless, Oystercatchers do not feed in fields at high water everywhere in their wintering range, either because they do not need to do so or because the nearby inland habitats are not suitable; for example, arable fields are often too dry and contain few earthworms. Supplementary feeding may also be available within the estuary itself; some birds start feeding in winter on the upper levels of the shore in the Wash and on the Exe estuary almost as soon as the tide leaves the marsh edge (J. D. Goss-Custard, unpublished information). The extent to which such supplementary sources of food are used in different seasons, weather and tidal conditions may give important clues as to the circumstances in which some birds have difficulty in satisfying their food requirements, prior to the onset of severe weather later in the winter.

Variation between individuals in supplementary feeding

The amount of feeding done at the higher shore-levels and in fields could also provide a convenient way of rapidly identifying those individuals that have most difficulty in meeting their energy requirements at low water on the estuary. Amongst a sample of 119 individually-marked mussel-feeding Oystercatchers of all ages and feeding methods studied over two winters (1989–1990, 1990–1991) on the Exe estuary, only 33% fed exclusively on the mussel beds. An additional 14% were also seen feeding on the upper flats and 40% in the fields, while the remaining 13% were seen on both the upper flats and fields. On the Exe, 34% of the 1–4 year old birds were seen in the fields at least once at high water in winter compared with 18% of adults (Goss-Custard and Durell 1983). A preponderance of young birds amongst high water field feeders has also been reported in Ireland (Quinn and Kirby 1993) and in the Dutch Delta region (P.L. Meininger, unpublished information). This is related to both the age and diet. On the Exe, more young than old birds eat clams and worms at low tide and, within one age class, birds eating Ragworms and clams occurred particularly often in the fields at high water (Goss-Custard and Durell 1983). Similarly, amongst adults, birds eating mussels and winkles (*Littorina* spp.) occurred less often in the fields at high water than birds eating clams and Ragworms, while those eating cockles were never seen there at all (Goss-Custard and Durell 1983). But, amongst mussel-feeders alone, adults have been seen in fields on 18% (SE = ±5; n = 162) of winter high tides whereas immatures have been seen on 56% (SE = ±10; n = 91), a difference that occurred in all three feeding methods. In contrast, 90% of both adults (n = 30) and immatures (n = 32) feeding on Ragworms and clams were recorded in the fields at high tide. The generally subdominant young on mussel beds therefore fed more often than adults in the fields whereas in birds feeding on mudflats, where the competition was much lower (Boates and Goss-Custard 1992), there was no difference.

Factors other than diet, however, also affect the use made of fields at high water. On the Exe, birds that stabbed mussels were many times more likely to occur there than mussel hammerers (Goss-Custard and Durell 1988), a finding recently confirmed in a larger sample (n = 36; P < 0.01) (S.E.A. le V. dit Durell, unpublished information). This difference did not arise because their blunt bill-tip prevented hammerers from taking worms; hammerers frequently feed in the fields around the Exe, although with what efficiency is not yet known. Because young birds stab mussels more frequently than adults (Goss-Custard *et al.* 1993; Sutherland *et al.*, Chapter 3, this volume), their greater occurrence amongst mussel-feeding birds in fields at high water can be understood. But the feeding method is not the whole explanation. Although no effect of sex or dominance has yet been detected (R.W.G. Caldow, S.E.A. le V. dit Durell and J.D. Goss-Custard, unpublished information), there was considerable variation between individual mussel-stabbing adults in their use of fields at high tide, with those that were least efficient at taking mussels occurring there the most often (Goss-Custard and Durell 1988).

Even amongst birds of the same age, diet and feeding method, individuals varied in the degree they supplemented their low water feeding over the high water period.

So far, these findings have not provided a convincing explanation of why young birds are most at risk during generally mild winters when the supplementary food supplies are almost continuously available. However, the varying responses of individuals to seasonal changes in the feeding conditions may eventually reveal why some birds are particularly vulnerable during severe spells. Oystercatchers that normally supplement their feeding at the upper levels of the shore and in the fields at high water cannot do so in cold weather when both food sources may freeze over. All feeding must be done during the few hours when the mid- and lower-level flats are exposed, and this will penalize the slow feeders. On mussel beds, individuals feed slowly either because they are inefficient or because they are vulnerable to interference from other Oystercatchers when bird densities are high (Goss-Custard and Durell 1987a,c, 1988; Ens and Cayford, Chapter 4, this volume). During a week of cold weather on the Exe estuary in February 1987, the many juveniles that normally feed in fields even at low water returned to the estuary with the result that some 80% of the juveniles in the population were eating mussels (J. D. Goss-Custard, unpublished information). The relatively ineffective stabbing technique they used for opening mussels, combined with their low dominance status, no doubt hindered their ability to collect the increased quantities of food required at low temperatures. In combination with their lower body reserves, this may partly explain why juveniles are most at risk during weather in which important sources of supplementary food are no longer available. However, studies on birds eating prey other than mussels are required because the intensity of competition amongst Oystercatchers (Boates 1988; Ens and Cayford, Chapter 4, this volume), and the extent to which supplementary food supplies at high tide are used (Goss-Custard and Durell 1983), vary between diets.

Seasonal and individual variations in body condition

Although the intertidal feeding conditions may decline from autumn to winter, most birds nonetheless survive (Goss-Custard *et al.* 1995d; Chapter 13, this volume) and most put on weight (Dare 1977; Hulscher 1980; Goss-Custard *et al.* 1982c; Johnson 1985; Lambeck 1991). As the example in Figure 6.4 shows, the mean weight of adults increases from August and peaks in February and March, when most depart for the breeding grounds. The autumn and early winter increase in body weight in shorebirds is usually interpreted as energy storage for later severe weather (Davidson 1981)—even though the risk of severe weather may not coincide with peak body weight. In most shorebirds, weights then decline but this does not occur in adult Oystercatchers, presumably

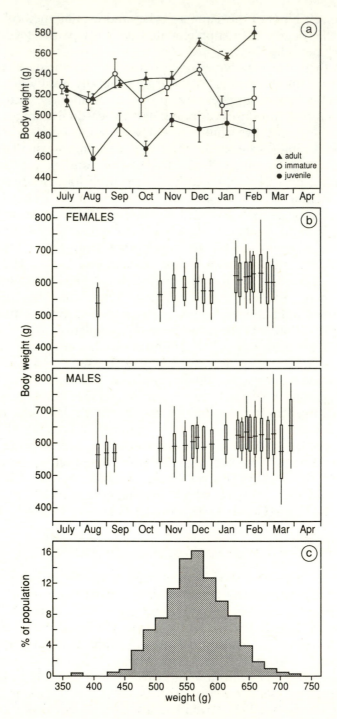

Fig. 6.4 Overwinter changes in the mean weights of Oystercatchers. (a) Three age classes on the Exe estuary—vertical bars show ± 1 SE (S.E.A. le V. dit Durell, unpublished information; (b) adult female and male mussel-eating birds from More-cambe Bay—the range is shown by the vertical lines while solid vertical rectangles show ±1 SD of the means, which are indicated by the vertical bar (Dare 1977); and (c) the weight distribution of Oystercatchers in November–February at Paesens in the Dutch Wadden Sea; sample size is 529 (Hulscher 1989).

because fat reserves are needed for their early spring migration (Dare 1977; Johnson 1985) or as an insurance against poor feeding conditions on first arrival on the breeding grounds (Hulscher *et al.*, Chapter 7, this volume). Consistent with this explanation is that the late winter decline in weight does occur in many areas in immatures (Johnson 1985) (Fig. 6.4a), which remain on the wintering grounds for the summer.

There is enormous variation in weight between individuals sampled at the same time of year. Since all birds in the sample shown in Figure 6.4b were mussel-eaters, there was clearly a considerable variation amongst individuals eating the same prey, as was also shown by Boates (1988). Hulscher (1989) found a similarly wide variation in the winter weights of Oystercatchers from the Wadden Sea (Fig. 6.4c), with some birds approaching the starvation weight of approximately 300–400 grams (Marcström and Mascher 1979; Davidson and Evans 1982; Swennen and Duiven 1983; Stock *et al.* 1987a; Hulscher 1989, 1990; Lambeck *et al.* 1991; Piersma *et al.* 1996; S.E.A. le V. dit Durell, unpublished information). These data show how large is the body weight variation amongst individuals in normal winter conditions, before severe weather occurs. Since in a sample of 79 dead Oystercatchers from the Exe, fresh body weight was as good, if not better, a predictor of protein and fat reserves as the thickness of the pectoral muscle, as measured by ultra-sound (Sears 1988) (S.E.A. le V. dit Durell, unpublished information), the variation in Figure 6.4 implies large individual differences in body condition, and thus energy reserves.

Attempts are now being made to relate these weight variations to the way in which individuals exploit their autumn and winter food resources and to their chances of surviving severe weather, but progress so far has been limited. On the Exe, the deviations in body weight in the three age classes examined were consistently below the appropriate monthly average for the whole population in stabbing birds (Table 6.1). In contrast, adults that hammered mussels ventrally were consistently heavier, although the difference was not significant in the smaller sample of immatures (Table 6.1). This might reflect the generally greater foraging efficiency of hammering birds compared with stabbers (Goss-Custard and Durell 1988), although the possible effects of dominance on con-dition has yet to be taken into account. Boates (1988) showed that adult birds eating mussels had a very much wider range in body weights than

Table 6.1. The deviation of the body weight from the monthly mean weight of the population as a whole according to age, diet and feeding method (FM) of the birds. Data from the Exe estuary, 1976-1991. (S.E.A. le V. dit Durell, unpublished information).

(a) Adults

Diet/FM	n	Mean	SE	t	p
Ventral hammerers	28	+21.57	8.5	2.54	<0.05
Dorsal hammerers	30	−9.6	11.2	−0.86	NS
Stabbers	18	−32.4	12.1	−2.68	<0.05
Worms/clams	40	+0.28	6.7	0.42	NS
Cockles	19	+27.8	13.3	2.09	NS
Winkles	12	+15.3	20.1	0.76	NS

(b) Immatures

Diet/FM	n	Mean	SE	t	p
Ventral hammerers	11	+2.27	9.31	0.24	NS
Dorsal hammerers	28	+1.4	10.6	0.13	NS
Stabbers	36	−20.36	6.95	−2.93	<0.01
Worms/clams	66	+6.21	4.63	1.34	NS
Cockles	7	+7.0	19.5	0.36	NS

(c) Juveniles

Diet/FM	n	Mean	SE	t	p
Stabbers	9	−24.2	10.7	−2.26	<0.05
Worms/clams	89	+0.65	4.22	0.15	NS

did birds eating winkles and suggested that this reflected a greater influence amongst mussel-feeders of individual variations in foraging efficiency and dominance on the rates at which birds could feed. While certainly a reasonable idea, firm evidence is lacking. It is not even clear whether any relationship between either foraging efficiency or dominance and body weight would be positive or negative. At first sight, we might expect efficient and/or dominant birds to be the heaviest. But if there is a metabolic cost to carrying reserves, or an increased risk of being caught by a predator, the most efficient and dominant birds might remain lean because they can rely on securing enough food whenever severe weather arrives; such birds might not need the, perhaps costly, insurance policy provided by extra energy reserves. There remains much variation in the body weight between individuals whose origin, and consequences for survival, still need to be established. The hypothesis that a bird's intake rate over low water, as determined by its foraging efficiency and dominance, before a period of severe weather, as well as during it, determines its chances of survival still needs critical testing.

The influence of natural enemies

A further complication is that the risk of infection by parasites (defined broadly here to include viruses, bacteria, protozoans, helminths, nematodes and ectoparasites) and of being attacked by predators may vary between individual Oystercatchers in ways that are linked to their sex, diet, feeding method, foraging efficiency and dominance. As these natural enemies may directly or indirectly influence the winter survival of Oystercatchers and their response to unfavourable winter conditions, understanding individual variation in the risk of dying requires the influence of parasites and predators also to be taken into account.

Parasites

Although both direct and indirect effects of parasites have been demonstrated for some bird populations (reviews by Loye and Zuk 1991; Hudson 1994) their overall impact on birds, and on wild animal populations in general, remains controversial. There are very few studies which have directly measured the effects of parasitism on host fitness or even on traits which may affect fitness. However, given (a) the accumulating evidence for direct effects of parasitism on host foraging, dietary intake, competitive ability and fecundity in some host-parasite systems (Minchella and Scott 1991) and (b) the knowledge that birds, and waterfowl in particular, are among the most heavily parasitized hosts yet studied (Bush *et al.* 1990), it is reasonable to consider that parasitism may play a role in Oystercatcher survival and reproduction.

Parasites of European Oystercatchers have been comparatively well studied, especially helminths. The most detailed studies have involved birds collected from the Exe estuary (Goater 1988, 1993; Goater *et al.* 1995) and the Dutch Wadden Sea (Borgsteede *et al.* 1988). On the Exe, a total of 27 species of helminth were recovered from 60 birds with an average of 2772 ± 5042 worms and 5.6 ± 1.8 (range 2–10) species per host. Three species found in Oystercatchers on the Exe and on the Wadden Sea were found in >95% of birds sampled; no others were found in >50%. Each of the common species used either cockles or mussels (or both) as intermediate hosts.

Individual Oystercatchers are extremely variable in their intensity of helminth infection. For example on the Exe, birds of similar age, sex, and diet could vary in their intensity of infection by the cestode *Micrasomacanthus rectacantha* by two orders of magnitude (Goater *et al.* 1995). Factors responsible for such variation were the focus of a field study involving Oystercatchers and their most common helminths on the Exe estuary (Goater 1988; Goater *et al.* 1995). In general, much of the variability in helminth intensity could be explained by the habitat and prey selection of individual birds. Only one species, the cecal trematode, *Notocotylus* spp., was transmitted on mussel beds. Virtually all others were transmitted through prey other than mussels, especially cockles. Thus, the diet of a bird provided the simplest explanation of the finding that most birds had a low number, and variety, of helminths. Over 80% of Oystercatchers on the Exe in winter were adult and most of these fed on mussel beds where they were exposed to very few helminth larvae. In contrast, juveniles had both the most varied diet (Goss-Custard and Durrell 1983) and had the greatest numbers, and variety, of helminths. These results were similar to those of Borgesteede *et al.* (1988) who also found higher helminth prevalence and intensity in juvenile Oystercatchers.

At a more detailed level, Goater *et al.* (1995) showed that exposure of juvenile birds to the two most common helminths was related to seasonal changes in the feeding behaviour of the birds themselves. Both species— *M. rectacantha* (Cestoda) and *Psilostomum brevicolle* (Trematoda)—had a marked seasonal pattern of infection. Juveniles arrived on the Exe from the breeding grounds with few helminths. In autumn, many juveniles fed on mussels (Goss-Custard and Durrell 1984) where they were exposed to few infective larvae. However, as winter approached, most juveniles left the mussel beds to feed on alternative prey. At this time, helminth intensities in juveniles were highest and they were at least five times greater than those in adults. This peak in worm numbers corresponded to the time when the intensity of helminth larvae in cockle intermediate hosts was also highest. In February, as adults began to leave the estuary, an increasing number of juveniles returned to the mussel beds, resulting in decreased exposure and decreased parasite intensity. In spring and summer, intensities of *M. rectacantha* and *P. brevicolle* in the juveniles dropped to levels approximating those found in adults in winter.

An individual's risk of being heavily infected also varied according to where it was feeding on the estuary. In general, intensities of helminth larval stages in cockles from muddy substrates were larger than in those collected from sandy areas (Goater 1993; Goater *et al.* 1995). Also, mussels from upshore areas had higher intensities of larval stages than those from downshore areas (Goss-Custard *et al.* 1993), as had previously been found for the infective stages of the trematode *Parvatrema affinis* in *Macoma* (Swennen and Ching 1974). Oystercatcher densities are lowest on the high shore-level mussel beds where prey are smallest (Goss-Custard *et al.* 1992) so these findings suggest that birds feeding on less-preferred intertidal habitats may be particularly exposed to infection. They also suggest that birds which supplement their winter diet by feeding on cockle beds or in fields at high tide may do so at a cost of increasing their risk of infection. Any factors which act to remove birds from preferred low-level mussel beds and onto high-level mussel beds, sand-flats, mudflats and fields are likely, therefore, to result in increased helminth infection in Oystercatchers (Goater 1993; Goater *et al.* 1995).

Unfortunately, Goater could not examine any potential pathological effects of helminth infection in Oystercatchers. Nor could he examine the role of immunity in determining the differences in helminth intensity between juveniles and adults. Most importantly, we have little idea of how interactions between the age, sex, social status and diet of the birds determine their risk of infection and, indeed, how parasites interact with environmental influences, such as competition and harsh winter conditions, in determining which birds are affected most by infection. Hulscher's (1982) observation that Oystercatchers reject heavily infected *Macoma* does suggest that there is a cost associated with ingesting helminths which the birds attempt to avoid, but its magnititude is unknown. Without experiments, the precise effect of helminth infection on the winter body condition and survival of Oystercatchers remains speculative. However, because juvenile Oystercatchers have both higher rates of helminth infection and lower rates of overwinter survival compared with adults, there is a general association between helminth intensity and overwinter survival in mild as well as cold winters. It is therefore possible that infection with parasites may at least contribute to some birds being more at risk of dying than others during the winter as a whole and during severe spells in particular. Goater's study thus provides some preliminary grounds, based on natural patterns of infection, for establishing a link between the foraging performance of the Oystercatchers, their use of sub-optimal habitat and their risk of infection with the general observation that some birds are more at risk of dying in winter than others. The next step is to test experimentally that infection levels do influence body condition and, perhaps, survival.

Predators

Oystercatchers feeding on intertidal flats are only likely to be attacked by the large Peregrine Falcon (*Falco peregrinus*). In fact, Peregrines usually take small- and medium-sized waders (Whitfield 1985b; Whitfield *et al.* 1988; Cresswell and Whitfield 1994). Even so, Peregrines—especially the larger female—regularly capture Oystercatchers in the Delta area in the South-west Netherlands, but these are generally caught at the roost (P.L. Meininger, unpublished information). Smaller shorebirds seem most at risk when they feed near to the high water mark and thus close to the cover from which raptors can launch surprise attacks (Whitfield *et al.* 1988; Cresswell and Whitfield 1994), but whether this also applies to Oystercatchers is unknown. Present evidence suggests that Oystercatchers in the intertidal zone are, in fact, rather free from attack by predators, despite reacting vigorously when hunting raptors approach.

Oystercatchers may be more vulnerable to attack at high water, given the greater opportunity there for surprise attacks and the greater number of potential predators. Amongst raptors, Peregrines and Goshawks (*Accipiter gentilis*) seem large enough frequently to attack Oystercatchers. However, of 2984 Goshawk kills in the Dutch inland province of Drente, only 9 were Oystercatchers, and all were taken during the breeding season (R. Bijlsma, personal communication). Breeding Oystercatchers are also taken occasionally by Peregrines on Schiermonnikoog (J. B. Hulscher, unpublished information), but perhaps more frequently by Goshawks and Peregrines in the White Sea breeding areas (R.H.D. Lambeck, unpublished information). There is little evidence that Oystercatchers feeding in fields in winter are taken by large raptors, but this may simply reflect lack of study (Whitfield *et al.* 1988; Cresswell and Whitfield 1994). Presumably, Oystercatchers in fields could also be attacked by mammalian predators, such as the Fox (*Vulpes vulpes*), as has been recorded once on the Exe. Partly eaten Oystercatchers have been found in the burrow of a Polecat (*Mustella putorius*), although it was not clear whether they had been scavenged after death or killed (P.L. Meininger, unpublished information). Oystercatchers in fields are certainly more exposed to other risks, such as being hit by cars, trains or overhead wires. Thus Oystercatchers that feed for longer each day, whether in the fields or on the upper flats, may indeed be at a greater combined risk from predation and accidents than birds which feed only over the low water period in the intertidal zone, but present evidence does not allow the size of that added risk to be measured.

Discussion

To summarize, most birds in most winters obtain enough to eat and few birds starve. Death from predation and accidents seems to be rare and the direct and indirect effects of parasites remain to be determined. Extreme weather conditions can cause heavy mortality because energy requirements increase while important sources of food decrease. Even though birds prepare for such conditions by storing fat earlier in the winter, many may die of starvation if the conditions are severe and persist for long periods. However, this is not the whole story as birds also die in milder winters without such severe spells, but the cause of death is unknown. The risk is not distributed equally throughout the population either. Young birds are particularly vulnerable in all winters, as are those in poor condition—for a variety of possible reasons—when severe weather occurs. The need now is to identify the factors that predispose a bird to be at risk.

The main hypothesis is that the birds most at risk are those that feed poorly on the intertidal flats at low tide. Such birds may have few reserves when hard weather arrives and also be most severely affected by the reduction in supplementary feeding during the severe spell itself. Indeed, their greater use of supplementary or marginal habitats may incur increased risks from predation and accidents in mild winters. These birds may also be at most risk of becoming parasitized and any sub-lethal but debilitating effects thereby arising could further reduce their chances of surviving in mild conditions as well as in severe.

At present, there is little direct evidence to test this hypothesis. However, studies on the intertidal feeding grounds suggest that a number of, sometimes inter-related, factors that vary between individual birds, and that also change with age, may affect the rate at which they feed. Apart from age (Goss-Custard and Durell 1987a,b,c), these include sex (Durell *et al.* 1993), diet (Boates and Goss-Custard 1989, 1992), feeding method (Goss-Custard and Durell 1988), dominance rank (Ens and Goss-Custard 1984; Goss-Custard *et al.* 1984; Goss-Custard and Durell 1987c, 1988; Leopold *et al.* 1989), foraging skill (Goss-Custard and Durell 1988) and, probably, injuries to the feet and mandibles. Clearly, the variation between birds is considerable, and the opportunity for the risk to vary amongst them is certainly present. But important information is still required to test the hypothesis, including (i) whether a low intake rate on the intertidal feeding areas is associated with increased use of supplementary feeding areas as the tide ebbs and flows and at high water, and (ii) whether this, in turn, is associated with a poorer body condition, and perhaps higher levels of parasitic infection, and by increased chances of dying.

There are some overall associations that are consistent with the hypothesis. Juveniles have generally low rates of intake and dominance on the estuary at low tide, use the fields most at high tide, have the lowest average body weights, have the highest intensity of parasites and are also most at risk of dying in severe weather. The difficulty of interpreting this is that a bird may not actually be forced to feed at a slow rate at low water. Variation between birds may simply mean that individuals adopt different strategies to maximize their fitness, and that different strategies give a different priority to foraging rapidly at low water. Animals have to make trade-offs between competing selection pressures, the one most studied being that between foraging and the risk of predation (Houston *et al.* 1993). In Oystercatchers, for example, any requirement to limit infection by helminths (Hulscher 1982) or to avoid damaging the bill while breaking into bivalves (Hulscher 1988, Chapter 1, this volume; Swennen *et al.* 1989) may depress the intake rate below the maximum. Stabbing birds may therefore feed more frequently at high tide in fields because, at low tide, they forage carefully and slowly to avoid the sometimes fatal risk of getting the bill caught between the valves of a bivalve (Hulscher 1988). The need now is for studies that evaluate further the costs and benefits of the various strategies adopted by foraging Oystercatchers and for tests of any associations between an individual's foraging strategy, its body condition and chances of survival in both normal and extreme weather conditions.

Summary

The assumption that food shortage is an important factor that prevents some Oystercatchers from surviving the nonbreeding season, or from arriving on the breeding grounds in good condition, is central to our thinking about many aspects of Oystercatcher biology in the nonbreeding season, and is examined in this chapter. Particularly large numbers of Oystercatchers are found dead during prolonged periods of severe weather when their demand for energy is increased while their opportunities to forage are reduced; both low temperatures and strong winds can hinder feeding in Oystercatchers. Birds with low energy reserves, or feeding strategies that are particularly disadvantaged in severe weather, seem most likely to succumb in these conditions, so the feeding conditions prior to the arrival of severe weather, along with individual variations in foraging strategy, determine which birds are most at risk. For a variety of reasons, the feeding conditions decline over the winter. The birds respond by feeding more intensively over the low water exposure period. But many birds also feed more on upshore food supplies, when the more profitable areas downshore are covered by the receding and advancing tide, or in fields over high water; these provide important supplementary sources of food but they are particularly likely to freeze over and become unavailable in cold weather.

The varying responses of individuals to seasonal changes in the feeding conditions should help to explain why some birds are more vulnerable than others during severe weather and in milder conditions. There is an enormous variation in body weight, and thus energy reserves, between individuals sampled at the same time of year, the origin of which, and the consequences for survival, still need to be established; age, sex, diet, feeding method, dominance and physical deformity resulting from accidents are often inter-related factors that might contribute to the variation seen. The role played by accidents and natural enemies, particularly of parasites but also of predators, in affecting body condition and survival chances have still to be elucidated, as have any reciprocal relationship each has with the feeding conditions. While there are preliminary grounds for suspecting a link between the foraging performance of the Oystercatchers on their low water feeding grounds, their use of sub-optimal and supplementary food supplies, their risk of parasitic infection, their body condition and of being killed by predators or accidents and of dying in severe winter weather, the hypothesis still needs critical testing.

Acknowledgements

We are grateful to J. S. Boates for providing unpublished information.

7 Why do Oystercatchers migrate?

Jan B. Hulscher, Klaus-Michael Exo and Nigel A. Clark

Introduction

Oystercatcher habitat requirements vary between times of the year. In the breeding season, they require the food supplies to be close to their nesting sites (Safriel *et al.*, Chapter 9, this volume), whereas in winter they can successfully forage on exposed intertidal areas that usually lie several kilometres from their land-based high tide roosting places. In northern, cold temperate regions, habitats that are suitable for Oystercatchers throughout the year are not common, so the birds must migrate, although often over relatively short distances. Oystercatchers that move north in summer, for instance to the breeding areas around the White Sea, may find highly productive habitats in which to breed, but at the cost of having then to migrate several thousands kilometres south to escape the severe cold of winter. This means not only that they must undertake long, hazardous flights, but also that they have to find suitable wintering areas which have not already been fully occupied by Oystercatchers that breed further to the south, and so closer. Alternative strategies have evolved in populations of Oystercatchers breeding in different parts of the east Atlantic flyway to resolve the conflict engendered by the contrasting habitat requirements of summer and winter. This chapter details these strategies and investigates how they relate to the environmental conditions in different parts of the Oystercatcher's range and to other important events in the annual cycle.

Methods

Ring recoveries and counts are the main tools for exploring these migration strategies. A total of 14 349 recoveries, stored in the databank of the European Union for Bird Ringing (EURING), together with 46 records of birds ringed in Iceland and recovered elsewhere (A. Petersen, personal communication), were available for analysis. Of these, 930 could not be used as full details of the date and place of ringing and recovery were not given. It was not possible either to obtain ringing data from Norway or Russia so information on these populations is incomplete. For the analysis, the ringing and recovery sites were split on the basis of region (Fig. 7.1). Ring recoveries are made either from 'controlled' birds that are still alive, as when they are re-caught and released or individual colour rings are

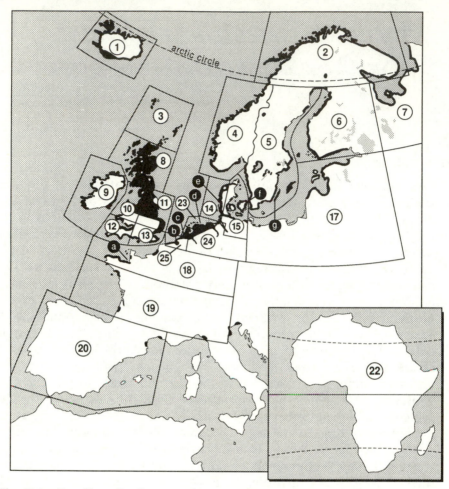

Fig. 7.1 Breeding distribution of nominate Oystercatchers (black) (Cramp and Simmons 1983; Gibbons *et al.* 1993) and the regions used for the analyses of ringing recovery data. The total number (n) of recoveries analysed were: 1, Iceland (40); 2, Lapland (0); 3, Faeroes, Shetland (324); 4, Norway (0); 5, Sweden (63); 6, Finland (68); 7, Russia (0); 8, Scotland (872); 9, Ireland (33); 10, Irish Sea (1851); 11, E England (1442); 12, SW England (744); 13, SE England (142); 14, W Denmark (428); 15, E Denmark (690); 17, Baltic (2); 18, N France (5); 19, S France (74); 20, Iberia (0); 22, Africa (2); 23, Wadden Sea islands (2170); 24, Netherlands/Germany: inland (2883); 25, Delta area (2562). The small letters show places mentioned in the text: a, Baie de Mont St Michel; b, Oosterschelde; c, Den Haag (The Hague); d, Schiermonnikoog; e, Blåvands Huk (Skallingen); f, Amager; g, Ottenby.

identified, or from birds that are shot or from birds that are simply found dead, the cause of death not necessarily being recorded and so perhaps including some birds killed by hunting. Such data must be analysed and interpreted with great caution (Perdeck 1977; Busse 1986; North 1987). For example, the likelihood of rings being recovered may vary

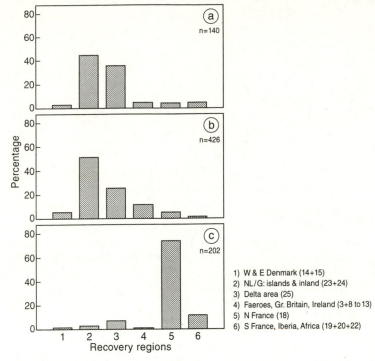

Fig. 7.2 Percentages of winter (November–January) recoveries of Continental-ringed Oystercatchers (regions 2, 4, 5, 6, 14, 15, 23, 24 and 25 in Fig. 7.1) which were: (a) controls—birds caught and released and sightings of colour-banded birds; (b) found dead; and (c) shot. Recoveries of all age classes are pooled. n = total number of recoveries. NL/G refers to The Netherlands and Germany combined.

greatly between areas in relation to the level of observation, ringing and hunting activity. Furthermore, the kind of recoveries used can influence the outcome of the analysis. The apparent winter distribution of Continental-ringed Oystercatchers, for example, depends markedly on the circumstances in which the ring was recovered (Fig. 7.2). In this case, the real distribution is probably best represented by the category 'found dead' because the number of such recoveries is much more closely correlated with the number of living birds counted in mid-winter in the same area than are the numbers of either controls or shot birds (Fig. 7.3). Therefore, most of the analyses are based on the recoveries of birds found dead, with other sources of information being used only in a supplementary capacity.

Midwinter distribution: differences between populations

The regular wintering area of Palaearctic nominate Oystercatchers ranges from 64°N in South-west Iceland (Wilson 1982) and South-west

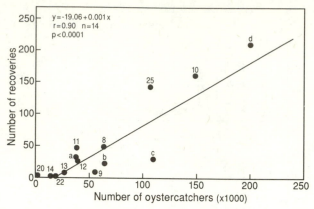

Fig. 7.3 Relation between the number of recoveries of birds found dead during winter (November–January) and the number of Oystercatchers counted in mid-January in the same regions. Numbers refer to the regions shown in Fig. 7.1, with the following exceptions: (a) France (regions 18 + 19); (b) Schleswig-Holstein; (c) Lower Saxony; and (d) The Netherlands. Regions 8–13 according to BTO data (estuarine totals, BoEE data: January average 1986–1991, non-estuarine totals from the winter shorebird counts: January 1984). All other count data according to Smit and Piersma (1989).

Norway (Thingstad 1978) down to about 12°N in Guinea–Bissau, West Africa (Smit and Piersma 1989). However, a few individuals do occasionally reach South Africa (Curry-Lindahl 1981; Hockey and Cooper 1982). Inland breeding haunts are completely deserted in winter over the whole breeding range (Dare 1970) as are the coastal breeding areas in Norway, along the White and Barents Seas, the Bothnian Gulf and the Baltic Sea (Fig. 7.1). The main wintering areas are situated in the zone characterized by mean winter temperatures above 0°C and few ice-days (Fig. 7.4). About 98% of the *circa* 875 000 nominate Oystercatchers winter in Europe, with the highest numbers occurring in the Wadden Sea (*circa* 400 000), the Delta area of the south-west Netherlands (*circa* 100 000) and around the Irish Sea (*circa* 150 000) (Smit and Piersma 1989). In these areas, the Oystercatchers congregate in large flocks, particularly in sandy estuaries where they feed on a small selection of staple foods (Hulscher, Chapter 1, this volume).

Atlantic and Continental breeding populations

Breeding Oystercatchers in North-west Europe are divided into two parts by the North Sea: the Atlantic populations, which comprise birds in Iceland, Faeroes, Great Britain and Ireland, and the Continental populations (Dare 1970). The North Sea seems to function as a barrier because an exchange of birds between the two areas is generally uncommon, especially of Continental-ringed birds (Fig. 7.5). Some 17.7% (616 in

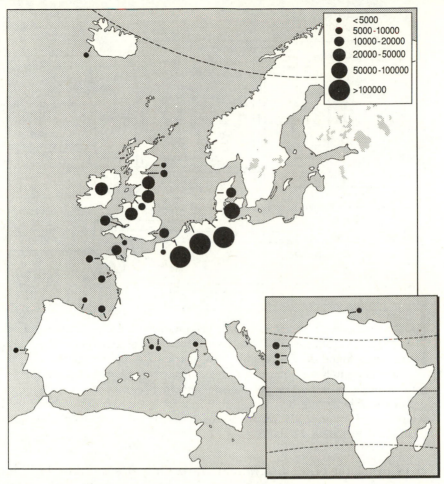

Fig. 7.4 Midwinter distribution (mid-January) of nominate Oystercatchers (after Smit and Piersma (1989); Great Britain: BoEE counts 1986–1987 to 1990–1991).

3473 across all age classes) of Oystercatchers ringed in the Atlantic region have later been found on the Continent. However, nearly 60% of these were Norwegian breeders, many of which winter in the Wash on the British east coast and along the Irish Sea. Large numbers of fully grown Oystercatchers have been ringed in the Wash and half (555) of the 1102 recoveries were Norwegian born (Anderson and Minton 1978; Branson 1989). Clearly, the unusually intensive ringing programme on the Wash has led to a disproportionately high percentage of reported North Sea crossings, leading to an over-estimation of the frequency with which birds move from the Atlantic to the Continental areas as a whole. When recoveries of birds ringed in Norway—107 of which have been in Britain alone (Mead and Clark 1993)—are excluded, North Sea

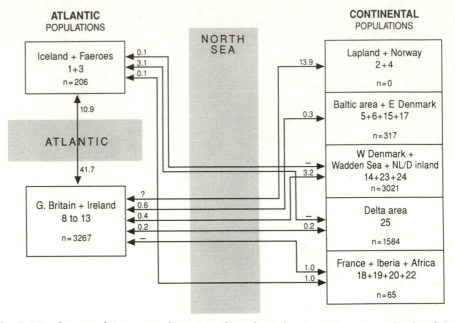

Fig. 7.5 Exchange of Oystercatchers ringed on the Atlantic or Continental side of the North Sea. Sea crossings are indicated by arrows, as percentages of the total number of recoveries found dead throughout the year (all age classes combined) per ringing region. Figures in boxes refer to the ringing/recovery regions, as given in Fig. 7.1. n = total number of recoveries per ringing region. For example, 10.9% of the 3267 recoveries of birds ringed in Britain and Ireland occurred in Iceland and the Faeroes while 41.7% of the 206 recoveries of birds ringed in Iceland and the Faeroes occurred in Britain and Ireland.

crossings in the other direction are rare. Only 0.4% (22 of 4987 birds) of the Continental-ringed birds have subsequently been found in the Atlantic region. This suggests that, with the possible exception of Norwegian breeders, few birds cross between the Atlantic and Continental areas.

This possibility can be further explored with data from birds ringed as chicks. Birds first caught and ringed as full grown outside the breeding season cannot be allocated with certainty to a particular breeding population. But as Oystercatchers usually breed within some tens of kilometres of their birthplace (Hulscher 1975b), the breeding areas of birds ringed as chicks can be determined with some confidence. The percentage of North Sea crossings in such birds is lower than in the sample as a whole: 3.9% (24 in 609) of Atlantic born chicks have been reported from the Continent (mainly from France and Spain) and 0.5% (9 in 1716) of Continental chicks have been recovered in the Atlantic area. Furthermore, in an earlier analysis of Norwegian-ringed chicks, Dare (1970) found that 20% (7 in 35) had been recovered from the British Isles and the other 80% from the Continent, including 14 recoveries (mostly shot birds) from France and Iberia. More recent information from the Norwegian ringing reports published between 1976 and 1983 shows that 52% of 41 foreign recoveries of birds ringed as chicks came from Britain with a further 22% coming from birds shot in France (Holgersen 1980, 1982; Runde 1984, 1985, 1987). This suggests that Oystercatchers regularly cross the North Sea from Norway, but not from the much more populated breeding areas further to the south. With the exception of Norway, the sub-division of breeding Oystercatchers into Atlantic and Continental populations with separate wintering areas appears to hold.

Wintering in relation to latitudinal location of breeding area

Oystercatchers breeding at higher latitudes are less likely to be able to winter successfully close to their own breeding areas than birds that breed at lower latitudes, because of the more severe winter weather (Goss-Custard et al., Chapter 6, this volume). We would therefore expect the proportion of year-round residents to increase, and the median distance between wintering and breeding areas to decrease, from north to south. As Figure 7.6 shows, this is borne out in both the Atlantic and Continental populations.

Wintering of coastal and inland breeders

Oystercatchers breeding at coastal sites that are also suitable for wintering, as along the Wadden Sea, can either winter near the breeding area or migrate. Inland breeders, on the other hand, have no choice and are obliged to leave the breeding area as soon as the young are independent. Do such birds migrate to the nearest wintering area or to more distant places? Winter recoveries of dead birds that were ringed as chicks on the Wadden Sea Islands and in the nearby Dutch and German inland areas can be used to answer this question (Fig. 7.7). Eighty per cent of the coastal breeders remain in the Wadden Sea, either on the islands or nearby on the mainland coast, whereas 20% migrate, mainly to the Delta area. In contrast, only half of the inland breeders winter in the Wadden Sea, the remainder migrating to destinations farther away (χ^2 test, $P < 0.0001$).

One explanation for this difference is that residents have an advantage over arriving migrants because they are familiar with the local feeding conditions. In particular, the dominance ranks of birds may be site-related and dependent on prior residency (Ens and Cayford, Chapter 4, this volume), and thus familiarity with the area. Immigrants to the Wadden Sea may be forced to go to less favourable feeding areas, or to more southerly wintering areas, by the presence of more dominant resident birds. In support of this idea is the observation that inland breeders that winter in the Wadden Sea are distributed between two sub-areas differently to coastal breeders. The majority of inland breeders remain on the

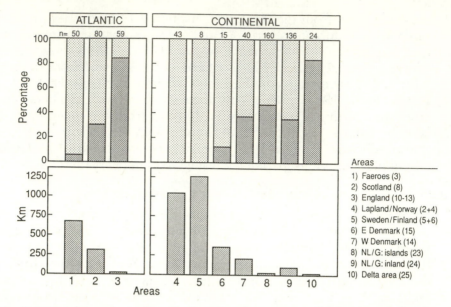

Fig. 7.6 North–south gradient in the migration pattern of Atlantic and Continental breeding populations. Indicated are: the percentages of recoveries of birds ringed during the breeding season as chicks or as adults and found dead during the winter (November–January) in the ringing area, so that the birds were residents (dark shading), and elsewhere, so that the birds were migrants (light shading); the median distances between ringing and recovery sites. Recoveries in all winters combined. Figures at the top of the bars refer to the number of recoveries. NL/G refers to The Netherlands and Germany combined. The numbers between brackets after the areas refer to the region of Fig. 7.1.

mainland coast, whereas most of the coastal breeders winter on the islands (χ^2 test, $P < 0.0001$) (Fig. 7.7). While the short distance to the mainland coast for most of the inland breeders may explain this, a difference in dominance between competing birds in the two populations could also be responsible.

Marked differences in the autumn and winter distribution of two inland populations that breed very close together in the north of The Netherlands (Koopman 1987a) illustrate, however, that other processes may be at work. Thirty-four per cent of the adult population near Drachten (16 in 47) and 80% of that near Rotsterhaule (28 in 35) respectively were recovered outside the Wadden Sea (Fig. 7.8). Such a big difference (χ^2 test, $P < 0.0001$) in wintering behaviour between birds breeding only 23 km apart is not easy to interpret in terms of dominance. Rather, the birds may stick to the routes they used when invading the inland breeding areas in the first place. Thus, the Drachten birds probably colonized their breeding area from the Wadden Sea to the north. In contrast, the Rotsterhaule birds probably came from the former Zuiderzee to the southwest, an area that was turned into a fresh water

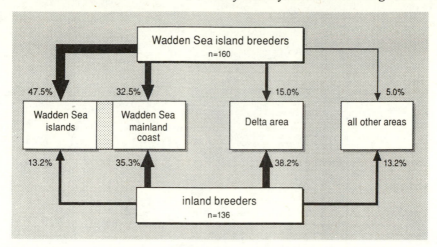

Fig. 7.7 Winter distribution (November–January) of Wadden Sea island (region 23 in Fig. 7.1) and Dutch/German inland (region 24) breeders, based on dead recoveries of birds ringed as chicks; all winters combined. n = total number of recoveries from each breeding area.

lake by the construction of the Afsluitdijk in 1932. As adult Oystercatchers, like most waders, leave the breeding grounds before the juveniles, such traditional migration routes must be passed on from adults to juveniles genetically, as in other birds (Berthold 1990). If so, the current distribution of inland and coastal breeding birds in winter in the Wadden Sea may also reflect conditions in the past rather than current levels of dominance between individuals originating from different nesting areas.

Wintering and summering sites at different ages

Winter

It is still entirely unknown how an individual juvenile Oystercatcher, migrating independently of its parents and for the first time, finds and decides to remain on a particular site over its first winter. A distinct possibility is that each bird sets off on migration along an inherited direction and tries to settle on the first suitable feeding area it encounters, unless it is forced onwards by competitor pressure. But however the choice is made, as in other shorebirds (Pienkowski and Evans 1985), many Oystercatchers remain faithful to the wintering area they used during their first winter (Goss-Custard *et al.* 1982c). Most young Oystercatchers tend generally to winter at approximately the same distance from their natal area in their first as in their subsequent winters (Table 7.1), although this is by no means true for all individuals (Goss-Custard *et al.* 1982c). In birds generally, the young of many migrating species migrate further to winter than older ones, and particularly those that migrate over short or medium distances (Berthold 1990), but there seems to be rather little evidence of this in the Oystercatcher.

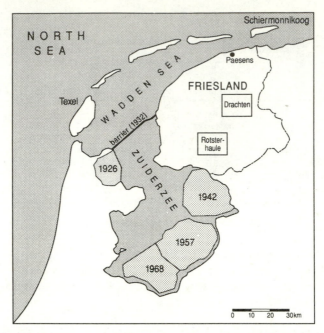

Fig. 7.8 Location of two inland breeding populations near Drachten and Rotster-haule in the province of Friesland, The Netherlands in relation to the position of the Wadden Sea and the former Zuiderzee; stippled areas reclaimed.

Once they become adult, individuals are generally very site faithful. On all wintering sites where there is regular ringing activity, large numbers of adult birds are re-trapped in subsequent winters, and usually close to the original site of ringing (Rehfisch *et al.* 1993). Furthermore, there are no records of movements between wintering sites in the EUR-ING databank. By using site-specific colour rings in five large British wintering sites, Dare (1970) was able to show that most Oystercatchers are site faithful in winter, a finding later confirmed by the high fidelity of individually colour-marked Oystercatchers to the Exe estuary (Goss-Custard *et al.* 1982c). In that study, no less than 89% of 469 adults present in one winter returned the next. Only three birds disappeared for at least one winter and were subsequently seen back on the Exe. A total of 163 birds disappeared permanently during the study but these probably died on passage or in the breeding areas, as none were seen or recovered elsewhere. Although some of these may have moved to other wintering sites but not have been seen, the evidence is overwhelming that adults generally return year after year to the same area to winter.

Summer

Once they have begun breeding, Oystercatchers may be even more faithful to their breeding sites than they are to their wintering sites (Safriel *et al.* 1984; Ens 1992). Oystercatchers do not reach sexual maturity until their third year, and most do not breed until much later (Safriel *et al.*

Table 7.1 Median distance (km) between the ringing and recovery sites of Oyster-catchers ringed as chicks and recovered (found dead) during the winter (November–January).

Ringing region (No.)	First winter (km) (n)		Second winter (km) (n)		Third winter (km) (n)		Later[1] winter (km) (n)	
Faeroes (3)	712	(10)	493	(2)			706	(38)
Scotland (8)*	254	(18)	172	(6)	24	(4)	347	(52)
England (10 to 13)	33	(17)	55	(6)	26	(5)	27	(31)
W Denmark (14)	131	(14)	314	(1)	116	(1)	275	(24)
E Denmark (15)	404	(7)	368	(1)			296	(7)
Wadden Sea islands (23)	13	(69)	108	(7)	245	(8)	46	(76)
Inland/Delta (24 + 25)	68	(73)	99	(9)	85	(8)	101	(70)

No. = number of region (Fig. 7.1); (n) = number of recoveries. Significance of the U-test* is only for the difference between the first and later winters, P = 0.026.
[1] Including birds of unknown age at ringing and recovered four or more years later.

1984; Ens 1992; Ens *et al.*, Chapter 8, this volume). Young birds of many species with such deferred maturity often linger in, or close to, their wintering area during their second or even third summer (Berthold 1990). Do immature Oystercatchers do the same?

The extent to which immature birds summer away from their natal area was analysed from recoveries between 1 May and 15 July, the period during which most birds have finished spring migration but have not yet started autumn migration. Many Oystercatchers spend their second summer far from their future breeding area, but they return to the breeding grounds from their fourth summer onwards (Table 7.2). The timing of the return varies. Half of 56 birds hatched on Schiermon-nikoog in the Dutch Wadden Sea were seen in the natal area for the first time during their second summer, 42.6% during the third and 7.4% during the fourth (J. B. Hulscher, unpublished information). By contrast, most juveniles on the Exe estuary remain there during their second and third summers and the period during which fourth summer birds are absent is still shorter (about 13 weeks) than in adults (21 weeks) (Goss-Custard *et al.* 1982c). This difference in the behaviour of the summering immatures in the two areas may be linked to the location of their future breeding areas. Thus many immatures in the Exe area originate from northern breeding grounds in Scotland and Norway (Goss-Custard *et al.* 1982c), for which migration and summering may be energetically more demanding than on Schiermonnikoog.

Further indirect evidence of juveniles summering at some distance from their eventual breeding area has been obtained by counting the numbers of young birds—birds with white neck collars—in areas where the species does not, or only rarely, breeds; for example, in the estuaries of North-west France (Triplet *et al.* 1987). Moreover, in some coastal

Table 7.2 Median distance (km) between the ringing and recovery sites of Oyster-catchers ringed as chicks and recovered (found dead) during the summer (May–mid-July).

Ringing region (No.)	Second summer[1] km	(n)	Third summer km	(n)	Later summers km	(n)	2nd-later summers U-test	3rd-later summers U-test
Faeroes (3)	297	(4)	7	(1)	0	(20)		
Scotland (8)	150	(9)			8	(17)		
England (10 to 13)	149	(6)	163	(1)	0	(30)	0.0319	
W Denmark (14)	0	(3)	0	(4)	0	(34)		
E Denmark (15)	111	(1)	824	(2)	3	(15)		0.0340
Wadden Sea islands (23)	2	(17)	19	(6)	6	(71)		
Inland/Delta (24 + 25)	58	(10)	15	(15)	7	(97)	0.0001	

No. = number of areas where studies were made; (n) = number of recoveries.
[1] First summer is the year of birth.

areas, such as the Wadden Sea, collar-bearing birds often outnumber those that might be expected to belong to the local breeding population, again indicating that many originate from breeding areas further away (Hulscher 1980). In revealing contrast, collar-bearing Oystercatchers are only occasionally observed on the inland breeding haunts (Hulscher 1976a). Clearly, although immature birds return to their breeding grounds at different ages in different populations, there is a general tendency for them to spend their first summers in areas well away from their eventual breeding sites.

The annual cycle of moult and migration: when to leave and when to return to the breeding area

Once one clutch of young has become independent and before the next is started, Oystercatchers must allocate time to moulting their flight and body feathers and, in most populations, to migrating to and from their wintering areas. What determines the timing of these activities which, as Figs 7.9 and 7.10 show, are clearly closely linked?

Location and timing of the primary moult

In order to maintain their effectiveness for flight and insulation, shore-birds renew their primaries and other wing and body feathers simultaneously once a year. Oystercatchers have a number of options for timing this moult. Like most birds, they do not start to moult before their young are independent (Harris 1967; Hulscher 1977a; Wilson and Morrison 1981); presumably, they could not mobilize either the time or energy to pursue two such demanding activities simultaneously. But they could, perhaps, complete their moult before they leave the breeding grounds. However, it takes an Oystercatcher at least 100 days to replace old primaries (Dare and Mercer 1974b; Boere 1976a; Wilson and Morrison

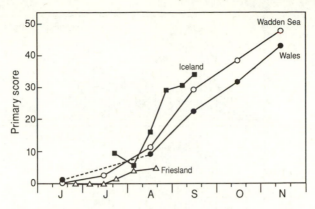

Fig. 7.9 Progress of primary moult (mean scores) of adult Oystercatchers at four different localities: SW Iceland, Faxafloi Bay, local breeding population (Wilson and Morisson 1981); Dutch Wadden Sea, Paesens, wintering area (P. M. Zegers, L. Zwarts and J. B. Hulscher, unpublished information); Wales, Burry Inlet and Conway Bay, wintering area (Dare and Mercer 1974b); and Friesland, Drachten (see Fig. 7.8) local inland breeding population (Koopman 1992).

1981) so there may not be sufficient time to complete the moult before the winter arrives, especially in the more northerly breeding areas. Furthermore, moulting before migrating would incur the risk of the birds having too short a time after their arrival on the wintering grounds to lay down energy reserves as an insurance against severe winter weather (see below). We might therefore expect Oystercatchers either to delay their moult until they have reached the wintering grounds or to begin moulting on the breeding grounds and migrate while it is in progress.

In birds generally, it is thought disadvantageous to migrate with an incomplete set of flight feathers (Berthold 1990). To do so would incur too high a cost in energy and nutrients, and thus require still more body reserves, which are themselves costly to transport, to counter the reduced flight efficiency of moulting primaries. In Oystercatchers, it would probably be particularly disadvantageous to migrate towards the end of the moulting period when the outer primaries are being replaced. On the other hand, most Oystercatchers do not have to migrate very far so the costs of migrating in moult may, in practice, be quite small compared with long distance migrants. Most Oystercatchers in West Europe are short or medium distance migrants, moving 1600 km, at most, between breeding and wintering areas (Fig. 7.6). The birds from Norway, Faeroes and Iceland make non-stop flights of only 300–800 km across the North Sea or North Atlantic in order to winter in Great Britain. The few birds breeding at high latitudes, and wintering up to 5000 km away in Africa, are the only true long-distance migrants. Most Oystercatchers therefore have only to fly relatively short distances to winter food supplies that have been replenished over the summer and are still very accessible to the

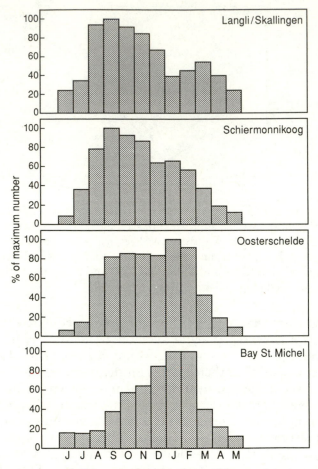

Fig. 7.10 Occurrence of Oystercatchers at four different Continental staging and wintering areas. Data per month are expressed as a percentage of the month during which the peak numbers (*n*) during the winter occurs: Langli/Skallingen (N Danish Wadden Sea), 1983–1987, September *n* = 3571 (Jakobson and Rasmussen 1989); Schiermonnikoog (E Dutch Wadden Sea), 1970–1991, September *n* = 25073 (P.M. Zegers and L. Zwarts, unpublished information); Oosterschelde (Delta area), 1978–1982, January *n* = 97460 (P.L. Meininger, personal communication); and Baie de Mont St Michel (NW France), July 1980 to May 1982, January *n* = 10000 (P. Boret, unpublished information).

the birds (Zwarts *et al.*, Chapter 2, this volume; Goss-Custard *et al.*, Chapter 6, this volume); it should therefore be possible quickly to recoup the energy and nutrient costs of a short migration. Many Oystercatchers may indeed be able to migrate while the moult is in progress.

Does this in fact happen? As it is much easier for research workers to score the progress of the moult in the primaries than in the body feathers, they have focused on the primary moult. To date, there have been four detailed studies of moult in Oystercatchers. Two concern the local breeding populations of Friesland and Iceland, with catches containing mainly post-breeders and an unknown number of nonbreeders. In the other two studies, Oystercatchers were caught in the Wadden Sea or in Wales and were of mixed origin, containing both local and foreign birds.

Because the feeding conditions are probably better on the coast, most Frisian Oystercatchers leave the inland breeding areas in July (Hulscher 1976a, 1977a), before the onset of the primary moult. Of the minority of birds that leave during the first ten days of August, only 20% have not yet begun to moult; the moult scores of the 80% which have started range from 1–19 (Koopman 1992). To reach the feeding areas in the Wadden Sea or Delta area, these birds have only to fly 40–200 km and, over such a short distance without a sea crossing, incomplete flight feathers are probably not a serious handicap. The Oystercatchers which moult in Iceland belong to the sedentary part of the population that spends the whole winter in the south-west of the island, where adequate mussel food supplies are available throughout. Birds breeding elsewhere in Iceland, where there is a shortage of winter food, set off as soon as possible for Britain, from where the first recoveries are made in August. These birds cover the 500–800 km non-stop flight over the North Atlantic with all their old wing feathers intact; primary moult has not begun (Dare and Mercer 1974b; Wilson and Morrison 1981). It could be argued that this rapid departure from the breeding grounds has less to do with avoiding being in moult while on migration than with arriving on the wintering grounds at the time of maximum biomass in cockles and mussels (Dare and Mercer 1974b). However, the primary moult in the wintering population in Wales, including many Icelandic breeders, also lags behind that of the residents in Iceland (Fig. 7.9), suggesting that migrants do indeed delay their moult until after they have arrived in the wintering areas. An alternative explanation is that early moulting by residents in Iceland may be favoured by the long period of daylight in which to feed—at least until 23 September, by which time most migratory Oystercatchers have left (Wilson and Morrison 1981). But as Oystercatchers may feed at night equally as well as they do during the day (Hulscher, Chapter 1, this volume), and as the resident birds in Iceland anyway feed mainly at night during the almost continuous darkness of winter, it seems more likely that migrants do indeed delay their moult. This is supported by the finding that Oystercatchers in the Wadden Sea also moult earlier than the birds in Wales, and this is again associated with the presence of many resident birds. It is not known whether the Continental populations that breed in the far north start to moult before they migrate south. However, their early arrival in the Wadden Sea in July and August suggests that there is little, if any, wing moult beforehand (Dare and Mercer 1974b). Thus, the evidence strongly suggests that when Oystercatchers migrate, especially over the sea, most birds delay the onset of moult until they have arrived at their destination. Given the short distances that are often involved, this gives an indication of the probable high costs of making sustained flight without a complete set of primary feathers, whether old or new.

Moulting is not necessarily restricted to breeding and wintering areas because some populations moult at staging sites before moving on to their winter quarters. For instance, counts indicate that many Oystercatchers leave the northern part of the Danish and German Wadden Sea (Busche 1980) and the eastern Dutch Wadden Sea in October (Fig. 7.10). Large numbers also leave some British estuaries in late autumn; for example, the Solway Firth, the Wash (Anderson and Minton 1978) and Carmarthen Bay (Fig. 7.11). Most of these birds have probably finished their moult before moving on to the wintering grounds. In the Wadden Sea, the first adults have finished primary moult by the beginning of

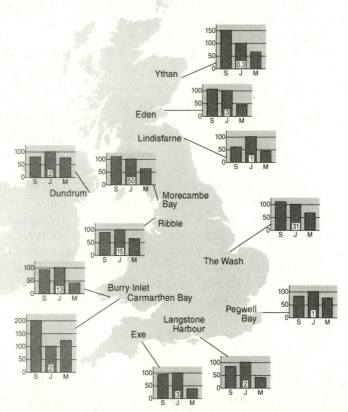

Fig. 7.11 Mean number of Oystercatchers in 12 British estuaries in September (S), January (J) and March (M), expressed as a percentage of the January population (January population = 100%; numbers in thousands are shown) (BoEE counts 1987–1988 to 1991–1992).

September (Boere 1976a) while, in Wales, this happens during the second half of that month (Dare and Mercer 1974b). It seems, therefore, that Oystercatchers start wing moult as soon as possible after breeding and go to the nearest area where the process can be completed quickly, before the onset of difficult winter conditions (Goss-Custard *et al.*, Chapter 6, this volume). Thereafter, we may speculate that, as food supplies begin to dwindle, some Oystercatchers are forced away by competition from more dominant birds (Ens and Cayford, Chapter 4, this volume), with those birds which have completed moult being able to migrate with their feathers in good condition.

The timing of autumn migration

Although they can be used to track migration routes, ringing recoveries tell us little about when Oystercatchers move between breeding and wintering areas (and, indeed, vice versa; see below). Direct observations of migrating birds are more informative, but these are only available for Continental populations.

Northern Continental populations

Oystercatchers passing near Ottenby, in South-west Sweden, and Amager, in East Denmark, originate from breeding populations along the Baltic Sea, the Bothnian Gulf, the Barents Sea and the White Sea, east of the 30° E meridian. Although some cross South Sweden and the Skagerrak, birds from the last two areas largely take an overland route, crossing Karelia and Finland to reach the Baltic before continuing on to Denmark (Bianki 1977; Meltofte 1993; Lambeck and Wessel 1993; Lambeck *et al.* 1995). Birds from Estonia and Latvia probably also migrate past Amager. Oystercatchers passing near Blåvands Huk, in West Denmark, probably come from the west coast of Norway (Thelle 1970).

Autumn migration past these three observation sites starts at the end of June and ends early in September (Fig. 7.12). However, the peak time, when most birds pass through, varies between them: it is the third week of July in Amager, the first week of August at Ottenby and mid-August along the Danish west coast. The relatively late passage at Ottenby and Blåvand is associated with the late completion of breeding in the northern part of Norway and Russia (Meltofte 1993). The birds passing through Blåvand, Ottenby and Amager all go on to the Wadden Sea, where they mingle with the local inland and coastal breeders. As they leave the breeding areas after the adults, young birds do not arrive at Blåvand until mid-August but then comprise an increasing proportion of the Oystercatchers passing through until migration ends in mid-September (Meltofte 1993).

Dutch inland population

The pattern of autumn migration in a Dutch inland breeding population was revealed by regular counts of the numbers of Oystercatchers visiting a roost near Drachten in Friesland (Fig. 7.8). The birds first gather at night roosts from which they begin migration, leaving in small groups in darkness before midnight. Their departure is synchronized by long periods of continuous calling beforehand (Hulscher 1973, 1976a, 1977a). Migration starts slowly at the end of June and reaches a peak in the last ten days of July, all birds leaving by the end of August. Young birds comprise an increasing proportion of the birds visiting the roost from mid-July, again indicating that they leave later than their parents (Fig. 7.13). The migration of these inland Oystercatchers can also be detected at Den Haag, The Netherlands, where they pass through before the peak numbers arrive (Fig. 7.12). Probably, most Oystercatchers passing Den Haag

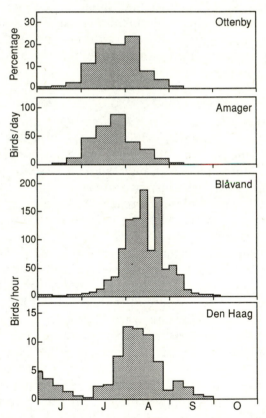

Fig. 7.12 Observations on active migrating Oystercatchers at different localities on the Continent, shown in Fig. 7.1: Ottenby on Öland (SW Sweden), per cent of 10-year total (1947–1956)/10 days (Edelstam 1972); Amager, near Copenhagen (E Denmark), mean number of birds/day over 10-day periods (1974–1978) (Meltofte 1993); Blåvand (W Denmark), mean number of birds/h over 5-day periods (1963–1971) (Meltofte *et al.* 1972); and Den Haag, North Sea coast (The Netherlands), mean number of birds/h over 7-day periods (1974–1979) (Camphuysen and van Dijk 1983).

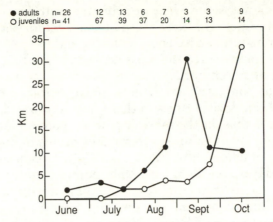

Fig. 7.13 Difference in the onset of autumn migration of adult and juvenile Oyster-catchers of the Dutch and German inland breeding population. Median distances between ringing and recovery sites per half-month periods are indicated. Recovery distances of adults and juveniles were significantly different (*U*-test) for the periods July ($P = 0.024$) and the second half of August and first half of September combined ($P = 0.033$). n = number of recoveries.

from August onwards originate from breeding populations much further to the north.

In summary, the timing of autumn migration varies between breeding populations, with juveniles always migrating later than adults. In general, the populations from further north migrate later than those that breed in the south. They will therefore arrive on potential wintering quarters later than the resident birds in areas such as the Wadden Sea and it is probable that they will be at a competitive disadvantage. It would, therefore, be advantageous for them to migrate on to wintering areas further south where there are fewer residents.

Spring migration

Continental populations

Departure from the southern Continental wintering areas starts at the end of January and beginning of February and only finishes at the end of May (Fig. 7.10). The numbers of Oystercatchers on passage further to the north, as in the Wadden Sea in Schleswig-Holstein and Denmark, increase during March. These areas function as staging sites, probably for Norwegian Oystercatchers (Busche 1980; Meltofte 1993). The Danish east coast, on the other hand, is a staging area in March and April for Oystercatchers breeding in the Baltic and White Seas (Meltofte 1993).

The southerly breeding Oystercatchers of the Dutch inland population arrive on the breeding grounds early; in mild winters, some may even be present in January (Hulscher 1975a). Usually, however, the first birds arrive during the second half of February, with most doing so during the first half of March. All experienced breeders return before the end of March whereas immatures arrive later, up to the end of May (Hulscher 1985).

The first adults reach the most northerly breeding grounds around the White Sea at the end of April, with the majority arriving during the first ten days of May; few immatures arrive before the beginning of June (Bianki 1977). Arrival times in these northern breeders is thus some two months later than in the more southerly Dutch inland breeding population. Yet egg laying in the northern sites begins in mid-May and 50–80% of the chicks hatch between 20 and 30 June (Väisänen 1977), which is only ten days later than on Schiermonnikoog (Ens *et al.* 1992). The breeding cycle is thus compressed at these northern latitudes. The poor spring weather conditions do not allow the birds to arrive any earlier and egg laying must start quickly so that breeding can end before unfavourable weather conditions return in August and September.

Atlantic populations

In the British Isles, the timing of the departure of birds for their breeding areas differs markedly between regions (Fig. 7.11). Birds which winter in estuaries along the western and south-western coasts include many early breeders from Scotland, the Faeroes and Iceland. Those wintering on the eastern and south-eastern coasts consist mainly of Norwegian birds, which start breeding later (Dare 1970).

Departure dates vary between individual birds and between years. The departure times for the Exe estuary in spring—and arrival times in autumn—of individually colour-ringed adults are very consistent and probably linked to the location of the breeding area (Goss-Custard *et al.* 1982c). The variation between years can be illustrated by birds breeding on the Westman Islands in southern-most Iceland where the dates of first arrival varied from 1952 and 1990 between 17 March and 13 April. First arrival times of British inland breeders may vary between years by as much as four to six weeks, depending on spring weather conditions (Dare 1970). Colour-ringed Scottish breeders at Finzean (Aberdeenshire), on average, returned ten days earlier in the mild spring of 1989 than in the much colder spring of 1987 (Picozzi and Catt 1989). Although the timing of spring migration in different individuals and populations of Oystercatchers reflects the date at which breeding areas become available, it seems that it also varies between years according to annual differences in the weather.

The need to return early

Competition for territories, particularly for high quality ones, is very strong amongst Oystercatchers (Harris 1967; Ens 1992; Ens *et al.*, Chapter 8, this volume). Apparently as a consequence of this, territories on Schiermonnikoog, for example, are occupied three to four months before egg laying starts. It seems highly likely that an early return to the breeding area, as soon as weather conditions permit, is very important in a species where prior residency gives a great advantage to individuals that are defending, or trying to acquire, a territory (Ens 1992; Ens *et al.*, Chapter 8, this volume). But clearly, this must be tempered by the increased risk of encountering severe weather when birds arrive early; particularly in inland breeding birds, a late spring snowfall can kill many Oystercatchers (Marcström and Mascher 1979; Watson 1980; Picozzi and Catt 1989). Indeed, one advantage of wintering as close as possible to the breeding area may be the greater opportunity it provides for monitoring weather conditions there and being able to leave for them as soon as circumstances become favourable (Alerstam and Högstedt 1982). The extreme case of this, of course, is year-round residency, when very close monitoring can be carried out. For example, Dutch birds that move to inland breeding areas in January in mild winters return to the Wadden Sea if a frosty or snowy spell arrives (Hulscher 1976b). In contrast, birds that breed far inland in Scotland do not return to the coast once they have left, even in severe weather conditions. In March 1979, for example, many such Oystercatchers starved in severe weather (Watson 1980; Picozzi and Catt 1989) as did many territory holders on Öland, in Southwest Sweden, during a cold spell in late April (Marcström and Mascher 1979). The heavy penalty incurred by these birds which occupied breeding territories too early in a late spring implies that, to offset the risk, there must be considerable benefits to an early return to the breeding grounds in clement springs. The most likely candidate for this is the increased chances of acquiring and retaining a good quality territory associated with an early arrival on the breeding grounds.

Dealing with severe weather in winter

Between their return to the wintering grounds and completion of their moult, Oystercatchers must focus on surviving and ending the winter in good enough condition to allow an early return to the breeding areas in spring. In normal circumstances, the main obstacles to their achieving these goals arise from severe weather which, not only increases their demand for energy, but also makes food more difficult, and sometimes impossible, to obtain. Severely cold winter weather prevents Oystercatchers from feeding in either inland or intertidal habitats because their prey are either inaccessible under ice or snow or buried too deeply and inactive (Goss-Custard *et al.*, Chapter 6, this volume). In the main wintering areas (Fig. 7.4), which have mean winter temperatures above 0 °C, Oystercatchers are usually able to feed throughout the winter. When exposed over the low water period, intertidal feeding areas start to freeze only below a temperature of −2 °C, the freezing point of sea water. At such temperatures, sites with a small tidal range, such as the Wadden

Sea, run a considerable risk of becoming totally frozen, especially when strong winds prevent the tide from flooding the flats. Sites with a large tidal range, such as the Delta area and most British estuaries, are much less vulnerable to frost because the ice can be dispersed over the high water period. Nonetheless, even here, feeding conditions deteriorate in cold weather because supplementary food supplies, lying upshore of the main intertidal feeding areas, are likely often to be inaccessible under ice, especially on neap tides when they are exposed over high water. Furthermore, supplementary food sources in fields adjacent to the flats are also inaccessible in these conditions. Nonetheless, Oystercatchers are well adapted to the average weather conditions in their winter quarters. They can survive a couple of days without any food, living on the body reserves accumulated since their return from the breeding areas (Hulscher 1989). But when a severe spell continues for about a week, huge numbers of Oystercatchers may leave the area in a 'hard weather movement', 'cold rush' or 'exodus'. Since these circumstances represent a severe potential threat to Oystercatchers, their adaptations to such conditions is an important aspect of their biology.

Frequency of hard weather movements

No large-scale hard weather movements have been reported for the British Isles (Clark 1982), although local movements between estuaries may occur (Pilcher 1964; Dare 1970). In contrast, they have been observed in at least 10 (11%) of 92 years in the Wadden Sea (Fig. 7.14).

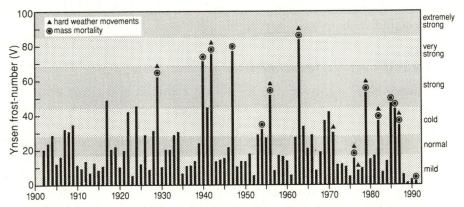

Fig. 7.14 The severity (mild to extremely strong) of the winters in The Netherlands (De Bilt) according to their Ynsen frost-number (V), after Ynsen 1988. (V = 0.000275 v^2 + 0.667 y + 1.111 z, where v = number of frost days (daily minimum temperature minus zero), y = number of ice days (daily maximum temperature minus zero), z = number of very cold days (daily maximum temperature minus 10 °C)). Shown are years with mass mortality, when many frost victims were found in the Wadden Sea and/or Delta area, and years when hard weather movements were observed in The Netherlands (J.B. Hulscher, unpublished information).

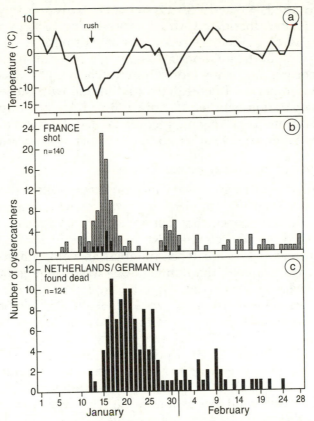

Fig. 7.15 The hard weather movement from the Wadden Sea on 14 January 1987. Indicated are: (a) the daily mean temperatures in The Netherlands (De Bilt); (b) the total numbers of Dutch-ringed Oystercatchers recovered shot in France in January and February 1987, including colour-marked birds from two breeding populations in the Wadden Sea (Schiermonnikoog and Texel, black bars); and (c) found dead in the Wadden Sea and Delta Area (regions 23, 24, 25 in Fig. 7.1).

Eight occurred during cold winters when the Ynsen frost-number (V) exceeded 39. Often, mortality also increases in severe weather (Figs 7.14 and 7.15); mass mortality occurred in The Netherlands in 14 (15%) of 92 years, with up to several thousand casualties being found (Meininger *et al.* 1991). In Great Britain, mortality in extreme weather is less severe. For example, the numbers of corpses found over four years in the whole country varied between only some tens and 780 in the severe weather of 1963 (Dobinson and Richards 1964; Clark 1982).

Analysis of the recoveries during cold winters of birds ringed in the Wadden Sea reveals not only where Oystercatchers go during a hard weather exodus but also the causes of death. During the 57 mild ($V < 28.4$) and 23 cold ($V \geq 28.5$) winters that occurred in The Netherlands from

1910 to 1989, 61 and 180 Oystercatchers respectively were recovered shot in France, Iberia and Africa, representing 83% of all recoveries. On average, seven times more recoveries of shot birds were made in the cold winters than in the mild ones. Assuming a constant hunting pressure and ring reporting rate, many more Oystercatchers visit France in a cold winter than in a mild one. Interestingly, and in line with our earlier conclusion on the separation between Continental and Atlantic populations, there is no indication from ringing recoveries that Oystercatchers leaving the Wadden Sea in cold weather go to the British Isles; or, indeed, vice versa.

Direct evidence of hard weather movements

Since the rate of ring recovery in France may reflect the local conditions there more than the numbers of Oystercatchers immigrating from elsewhere, direct evidence that hard weather movements do take place is also needed. In fact, counts have been made in North-west France that show increased Oystercatcher numbers in cold winters; in the Somme estuary, for example, the numbers of Oystercatchers far exceed normal winter levels in some cold winters (Triplet 1989a). Furthermore, hard weather movements from the Wadden Sea itself have been confirmed by direct observation on two occasions. Between sunrise and sunset on 30 December 1976, 40 000 Oystercatchers were seen passing the Dutch Wadden Sea island of Schiermonnikoog. Snowfall on the 28th and 29th and a drop in temperature from above zero on the 26th to $-7\,^{\circ}$C during the night of the 29–30th preceded the exodus, during which a strong (8–9 m s^{-1}) SSE wind was also blowing (van Eerden 1977).

The other well-documented exodus occurred on 14 January 1987 when 60 000 Oystercatchers were counted flying along the Dutch North Sea coast near The Hague (Keijl and Mostert 1988). It began after the daily mean temperature in The Netherlands had already been below zero for seven days (Fig. 7.15). The birds were on a heading for North-west France, where many were shot one or two days later. When the temperature dropped to below zero again at the end of January, recoveries of shot birds in France increased once more. The relatively high numbers of recoveries of shot birds in February that year suggests that many Oystercatchers remained there until the beginning of March. The origin of the Oystercatchers shot in France that winter differed markedly from that of the birds that had been recovered there during the preceding eight winters of 1978–1979 to 1985–1986 inclusive. The exceptionally high proportion of 27.3% (62 out of 227 recoveries) German-ringed Oystercatchers recovered that winter, compared to the 3.9% German of 180 recoveries in the preceding eight (P. Triplet, personal communication), strongly suggests that the exodus in January 1987 originated in the northern and eastern parts of the Wadden Sea. In contrast, whereas no British rings were recovered in France in 1986–1987, 16.1% of the much smaller number of rings recovered over the preceding eight years had been British. These two occasions thus confirm that very large numbers of birds are involved in these hard weather movements and that the high rates of recovery of rings from France in the winters in which the movements occur accurately reflect the large scale of the exodus that takes place.

Options for Oystercatchers in periods of severe weather

When a cold spell starts, an Oystercatcher is probably unable to predict its length yet it must decide whether to stay and live on its reserves or to move to areas further to the west and south, where the weather may or may not be less severe. Leaving immediately is generally a poor option as cold weather often lasts for only a couple of days, which most birds can survive on their reserves. Furthermore, Oystercatchers are not all equally vulnerable to severe weather. For example, long-billed birds specialize in feeding on deeply burrowing prey, and are particularly numerous at the higher levels of the shore, whereas short-billed birds mainly feed on the large and sturdy bivalve prey found at low shore-levels (Hulscher and Ens 1992; Hulscher, Chapter 1, this volume; Sutherland *et al.*, Chapter 3, this volume). Since high-level flats are more liable to freeze than low-level ones, long-billed Oystercatchers in the Wadden Sea may be more badly affected by severe weather than are the short-billed birds. Individuals would therefore be expected to vary in their response to the onset of severe weather and three categories of Oystercatchers can, in fact, be distinguished. First, there are birds which cannot find food at all during the severe spell itself and are also in poor condition at the start; they may be diseased or have fed badly earlier in the winter due to bill damage or some other disadvantage (Hulscher, Chapter 1, this volume; Goss-Custard *et al.*, Chapter 6, this volume). Their body reserves are probably insufficient for migration and hence they are doomed to die; the first mortalities are usually found one to two days after the beginning of a cold spell, and probably consist of such birds. The second category consists of birds which can no longer find enough food, but wait a few days before leaving the area with the exodus. Finally, there are birds which, by virtue of their speciality, can still find sufficient food and hence stay in the area and remain in good condition, as surviving birds are known to do (Boere 1976b; Koopman 1987b; Hulscher 1989).

Even after the exodus of 14 January 1987, counts showed that about half the normal winter numbers were still present in the Wadden Sea as a whole (Hulscher 1989). Clearly, many birds do opt to remain when a severe spell arrives. How long a bird can survive then depends on its food intake rate, body reserves and the prevailing weather conditions. The survival times of birds which do not feed and live solely on their reserves during a severe spell can be estimated from data drawn from a number of sources. The fat and protein contents of healthy adults in winter and of birds that had starved during cold spells has been measured

(L. Zwarts and T. Piersma, personal communication). As is the case in most shorebirds found dead after severe weather (Davidson and Clark 1985), the dead birds had used up all their metabolizable reserves. By comparing starved and healthy birds, the normal fat and protein reserves of adults could be calculated by subtraction and then converted into energy. Energy expenditure of Oystercatchers at different temperatures in the laboratory has been measured by Kersten and Piersma (1987). Assuming the birds die when all their reserves have been used up, the data allow the putative survival times of a non-feeding average adult Oystercatcher weighing 580 ± 48.2 (SD) g (Hulscher 1989) to be calculated at different temperatures under standard laboratory conditions (1 × Standard Metabolic Rate (SMR)). At 1 times SMR, an Oystercatcher having that weight at the start of the cold spell in January 1987, and unable to feed for the next 4.5 days, would have reduced its body reserves by 46% at the temperature prevailing at the time. But in the field, of course, energy expenditure is much higher because birds are more active and they are frequently exposed to the chilling effect of wind. Assuming free-living Oystercatchers use energy at roughly 2 times SMR (Wiersma et al. 1993), the reserves of an adult Oystercatcher of 580 g will last, for instance, at an ambient temperature of −5°C for only five to ten days. In reality, many birds are able to continue feeding, although probably at a reduced—but currently unknown—rate (Goss-Custard et al., Chapter 6, this volume). Therefore, unless rates of energy expenditure have been seriously under-estimated, birds healthy at the start would be expected to survive quite long severe spells

Birds that are unable to feed at an adequate rate during the severe spell, and so are beginning to starve, then have to consider the option of moving away. At an ambient temperature of −5°C, and assuming a strong tailwind, an average bird would consume another 24% of its initial reserves during the 7-hour flight of 700 km to France (T. Piersma, personal communication). The remaining reserves on arrival in France would last, at the most, for only another one to two days of full fasting at −5°C. A starving bird in the Wadden Sea thus has a difficult choice to make between staying and leaving. In the particular case of January 1987, leaving the Wadden Sea would seem, at first sight, to have been the better choice for our hypothetical bird; the extreme cold lasted for another seven days after the exodus, which surely would have caused its death had it remained. But, in actual fact, feeding conditions were hardly better in France in 1987 because many estuaries there were also largely covered with ice. In the Somme, for example, large numbers of Oystercatchers were seen trying to feed in a few small ice-free areas bordering the deeper channels (P. Triplet, personal communication) and, in such competitive circumstances, immigrants might be at a disadvantage. Clearly, for a starving Oystercatcher in the Wadden Sea, either staying or leaving is risky because neither the length of the severe spell in The

Netherlands nor the conditions on arrival in France can presumably be predicted with certainty.

Certainly, many birds do die in severe weather in The Netherlands. Mortality in the Oystercatchers which stay in the Wadden Sea or Delta area rises up to 5%, being two to three times higher than the normal winter rate (Hulscher 1989; Meininger *et al.* 1991). Mortality rates amongst the birds that move, and which are not shot on arrival in France, are more difficult to assess. During the spell in 1987, nearly all Oystercatchers from Schiermonnikoog left the area. At least 19% of the breeding population did not return the next spring and presumably had died; this compares with a mortality rate of 1–5% during normal winters (Ens 1992; Goss-Custard *et al.*, Chapter 13, this volume). Similarly, 16% of breeders on Texel did not return that winter (C. Smit, personal communication). Clearly, the birds that left paid a high toll.

As argued above, Oystercatchers can survive in freezing temperatures without food for ten days at the most. Severe spells of this length or longer are quite common in The Netherlands and an increase in mortality has been observed in at least 15% of winters in the Wadden Sea (Fig. 7.14). Nonetheless, with an average life span of 19 years (Schnakenwinkel 1970), an Oystercatcher in the Wadden Sea is likely to encounter severe weather on average three times during its lifetime. Since the mortality rate on such occasions only increases by a few per cent, severe weather does not seem to be the main threat in winter during the lifetime of an average Oystercatcher in The Netherlands, despite the large numbers of birds that may starve. That the mortality is at least as high, and perhaps higher, in the milder British Isles than on the Continent (Goss-Custard *et al.* 1995d, Chapter 13, this volume) supports this conclusion. A variety of factors that operate in most winters, including intra-specific competition for food, may be more important than occasional severe spells in determining the lifetime risk of dying in winter for most Oystercatchers.

Migration as a survival strategy in Oystercatchers

Food shortage due to low winter temperatures and intra-specific competition for food are probably the main factors shaping the autumn migration patterns of European Oystercatchers. We suggest that birds breeding in areas unsuitable for wintering, such as inland habitats and coastal habitats at northern latitudes, encounter in the nearest favourable wintering areas resident Oystercatchers which, by virtue of their prior residency, are generally dominant to them. Consequently, many of them are forced to continue migration to areas where competition is less intense. Thus, en route to their southerly wintering areas, northerly breeders pass over the Oystercatchers that breed in the centre of the sub-species range, birds which themselves migrate only a short distance or are even resident throughout the year, unless forced to make temporary hard-weather movements. This was called 'leapfrog migration' by

Table 7.3 Leapfrog migration of Oystercatchers. The percentages of the recovery totals (all categories) reported during the winter from France to Africa are indicated for the different geographical breeding populations. Recovery percentages are higher when breeding far away (at high latitudes) from the main wintering areas, or inland compared to when breeding close to these areas.

Ringing region	No. of birds recovered	Birds found in France, Iberia, Africa	
		(*n*)	(%)
Sweden (5)	63	7	11.1
Finland (6)	68	7	10.3
E & W Denmark (14, 15)	1118	75	6.7
Wadden Sea islands (23)	2170	186	8.6
NL/D[2]: inland (24)	2883	438	15.2
Delta area (25)	2562	113	4.4
Iceland (1)	40[1]	6	15.0
Faeroes (3)	324	23	7.1
Br. Isles (8–13) & Ireland (9)	5084	227	4.5

[1] Exclusive of recoveries from Iceland itself, but few Oystercatchers winter there.
[2] Netherlands and Germany.

Salomonsen (1955). Its possible role in competition has been discussed by Pienkowski and Evans (1985) and Holmgren and Lundberg (1993). On the basis of an analysis of Fenno–Scandinavian ringing recoveries, Salomonsen (1955) concluded that Oystercatchers are leapfrog migrants. Thus, more birds from the Swedish and Norwegian populations than from the Danish population were recovered in France during the winter, a trend which still continues (Table 7.3). Recent work confirms that Oystercatchers are leapfrog migrants. Oystercatchers which have to evacuate the breeding areas completely every year (the Norwegian, Finnish, Swedish, Dutch and German inland birds and most of the Icelandic and Faeroes birds) migrate further to the south than those which can winter close to the breeding areas (the Danish, Wadden Sea and British birds). The high proportion (80.6%) of adults among several thousand Oystercatchers wintering on the Banc d'Arguin in Africa, and their late departure between mid-March and the beginning of May (peak in last week of March), also suggests that they too originate from breeding areas in the far north (Swennen 1990). Leapfrogging probably also occurred during the exodus of 14 January 1987 from the Wadden Sea when Oystercatchers passed over the Delta area without stopping—because there were no recoveries from there—and continued on to France. At that time, more than 90 000 Oystercatchers were already in the Delta feeding on an area substantially reduced in size by ice coverage (Lambeck 1991), so birds moving overhead probably could not settle because of competition from local birds (Hulscher 1989).

In common with most bird species, Oystercatchers leave for the wintering areas long before feeding conditions on the breeding grounds have actually deteriorated seriously. This is true for both the northern coastal Scandinavian and Russian populations as well as the inland breeding populations in The Netherlands and Germany. In The Netherlands, for example, huge flocks of Lapwings (*Vanellus vanellus*), Black-headed Gulls *(Larus ridibundus)*, Common Gulls *(L. canus)* and Starlings *(Sturnus vulgaris)* feed from August until October in the grassland areas that breeding Oystercatchers have just vacated and largely eat the same prey, mainly earthworms (*Lumbricidae*). Furthermore, the young Oystercatchers of the year feed in the breeding areas for some time after their parents have left. Those born inland, for example, roam the vicinity of the nesting place for a couple of weeks after they become independent and before they start their migration to the wintering areas (Fig. 7.13). As adult Oystercatchers could apparently have remained inland for longer, there must be an advantage to reaching the wintering areas as soon as possible. The estuarine food stocks are certainly high at that time of year but, as the main winter food supplies of Oystercatchers deplete quite slowly over the autumn (Goss-Custard *et al.*, Chapter 5, this volume), there would seem to be no great urgency to return to them. Based on studies showing that competition can be intense on the wintering grounds and that an individual's success may increase with the period of its residency (Ens and Cayford, Chapter 4, this volume), we suggest that, after breeding, Oystercatchers return to the wintering areas as soon as possible to establish themselves before competition from other returning Oystercatchers intensifies.

Young Oystercatchers probably leave the breeding grounds later than their parents because they do not moult flight feathers during their first autumn. We do not know whether young birds reach the first distant wintering area in a single non-stop flight or in several steps. The young of some other shorebirds frequent a wide range of areas and habitats before settling in their final wintering area (Pienkowski and Evans 1985). Sutherland (1982d) suggests that the dispersion of young Oystercatchers when they first arrive is determined by their search for potential feeding areas; for example, a rich but short-lasting cockle spatfall on the Ribble Estuary in North-west England was mainly exploited by young immigrant Oystercatchers. The role that competition from more dominant and already established adults plays in governing any movements they make during their first autumn and where they eventually establish themselves for their first and subsequent winters remains to be established, yet may be the key to understanding their distribution and migration patterns. Unfortunately, this is very difficult to study because the dominance of juveniles relative to adults changes rapidly over this period (Goss-Custard and Durell 1987a,b,c), because the precise time when young birds decide where to stay is unknown and because the spatial scale over which studies are required is so large. Yet this may be a critically important phase in the annual and life cycle because it affects the chances for all Oyster-catchers of surviving subsequent winters. In the case of Continental Oyster-catchers, which are confronted with long spells of severe winter weather three times during their lifetime, this period could also provide crucial information of the location and quality of alternative wintering areas; indeed, this may be the reason why some young birds from southerly breeding coastal populations migrate during their first summer, whereas their parents do not. Finding out where young birds go during their first few months on the wintering grounds and investigating the decisions they make, the basis upon which they are made and the consequences they have for their lifetime fitness, is one of the most challenging aspects of the annual cycle and biology of Oystercatchers still to be tackled.

Summary

This chapter discusses the alternative migration strategies that have evolved in Oystercatchers and investigates how they relate to the environmental conditions in different parts of the Oystercatcher's range and to other important events in the annual cycle. In northern temperate regions, habitats that are suitable for Oystercatchers throughout the year are uncommon, so the birds must migrate. However, Oystercatchers breeding towards the centre of the range migrate over rather short distances, as do inland-breeding birds in Germany and The Netherlands. In fact, many coastal-breeding Oystercatchers in these regions are normally resident throughout the year. By contrast, Oystercatchers that breed in the north migrate several thousand kilometres south to escape the severe cold of winter and, in so doing, overfly less migratory populations towards the centre of the range. It is argued that this leapfrog migration arises because resident birds have a competitive advantage on these wintering grounds over immigrants, which therefore move further south to find areas where there is less competition for food. Even though the migration distances are often rather short, few Oystercatchers begin their moult until they have arrived on the wintering grounds, or in some cases, staging areas; this implies the energetic or other costs of flying with some primaries missing may be considerable. Birds that migrate begin their moult later than those which winter on or near to their breeding areas. In the Continental Oystercatcher population, severe winters occur on average three times during the lifetime of an Oystercatcher and many individuals die, even though they can survive on their stored energy reserves without feeding for a number of days, depending on the severity of the conditions. Many birds move away, but this too is risky as feeding conditions elsewhere are unpredictable and the migrants may again be at a competitive disadvantage with the local birds. In comparison, in the largely distinct Atlantic populations which winter in milder climates, hard weather movements occur only on a very small scale. In spring, competition for territories on the breeding grounds again favours an early return but at the risk, in some places, of birds being killed by severe weather in late springs. It is concluded that food shortage due to low temperatures and intra-specific competition for food in winter and competition for breeding territories in spring are probably the main factors shaping the distribution and the migration patterns of European Oystercatchers. Since individuals become site faithful as they mature, the critical processes probably occur in young birds during their first autumn and winter. These processes are difficult to study but they represent one of the most challenging aspects of the annual cycle and biology of Oystercatchers still to be tackled.

Acknowledgements

We acknowledge all those people who have caught and ringed Oystercatchers, or have participated in one of the many wader counting schemes in operation. Their efforts provided the necessary recoveries and local patterns of changing numbers of Oystercatchers on which the migration pattern of the species could be based. We particularly wish to thank EURING databank in Heteren (The Netherlands), where Rinse Wassenaar made the recoveries available to us. Dr A. Petersen of the Icelandic Museum of Natural History did the same for the Icelandic ringed Oystercatchers. Hans Meltofte allowed us a pre-publication reading of his book on wader migration through Denmark (Meltofte 1993).

8 Life history decisions during the breeding season

Bruno J. Ens, Kevin B. Briggs, Uriel N. Safriel and Cor J. Smit

Introduction

According to current evolutionary insights, animals in a stable population should behave in such a way that they maximize the number of offspring they produce during their lifetime. Yet, field studies have revealed a remarkably high variation in lifetime reproductive success between individuals, with only a minority producing most of the young (Clutton-Brock 1988; Newton 1989). This variability demands an explanation: what prevents the many unsuccessful individuals from raising more offspring? On the principle of individual optimization, each animal should make the choices during its development and later life which, in current conditions, are most likely to maximize its expected lifetime reproductive success. Instead of expecting each bird to show the same optimal behaviour (Lack 1968), it is now assumed that each individual will fine tune its behaviour to optimally exploit its own particular characteristics and circumstances under which it lives. This viewpoint arose from the growing awareness that individuals that laid more eggs probably did so because they were equipped to raise more young (Perrins and Moss 1975; Högstedt 1980; Drent and Daan 1980). Thus, in order to understand individual variation in lifetime reproductive success, we should ideally measure the conditions experienced by each individual during its development and explore its unique optimal solution to the problem of maximizing the production of young. For this, we would also need to understand how individuals become distributed between conditions. Neither issue can be addressed by studying which fitness component (survival, fecundity etc.) makes the greatest contribution to variation in lifetime reproductive success, as advocated by Clutton-Brock (1988) and Newton (1989); see also Grafen (1988). There are also considerable practical difficulties in measuring lifetime reproductive success in a species such as the Oystercatcher that may live for several decades.

Thus, this chapter focuses on the costs and benefits in given conditions of decisions that directly influence life history, i.e. the age-specific birth and death rates (Lessells 1991; Stearns 1992). We use the word decision to indicate that alternative behavioural options exist for the animals, but do not want to imply conscious choice. Following Wiley (1981), we assume that life history decisions, like on the age of first breeding, in animals with complex social organisations like the Oystercatcher, are best viewed within his conceptual framework of *'ontogenetic trajectories'* (Wiley 1981), which describe the movement of the individual through successive *social positions*. It is useful to think of social position as a 'pact' with other individuals over access to resources that are vital for survival and reproduction. This includes partners, since these can be viewed as a resource which contributes to the raising of offspring (Davies 1991). As suggested by Ens (1992), we distinguish in this context career decisions from reproductive decisions. Career decisions affect the social position of the individual, whereas reproductive decisions are concerned with such attributes as clutch size and the amount of parental care, which directly affect breeding success. Within a season, career decisions will usually precede reproductive decisions. As a consequence, reproductive decisions are made under the constraints imposed by the social position so far achieved through career decisions. In contrast, career decisions are made on the assumption that the most productive reproductive decisions will be made from the social position that will eventually be achieved.

Although quantification of the costs and benefits of these decisions lies some way ahead, a consistent qualitative framework is beginning to emerge for Oystercatchers. It appears that the two most basic aspects of the life history of the Oystercatcher are its long life span, which may restrain the birds from heavily investing in the current breeding attempt so as not to imperil future reproduction, and intense competition for breeding space of high quality. Within this perspective of the long-term development of an individual, this chapter first discusses decisions over the acquisition and use of breeding space before going on to consider the mating system and life history decisions in chronological order. The hope is that a full understanding of these decisions, in combination with a knowledge of the distribution of environmental conditions within which the decisions are taken, should allow us to predict the frequency distribution of lifetime reproductive success across individual Oystercatchers.

Habitat saturation and the social system

Territorial behaviour

Throughout the breeding season, male and, to only a slightly lesser extent, female Oystercatchers are highly and conspicuously territorial against intruders and neighbouring pairs; detailed descriptions of the various displays are given in Makkink (1942) and Cramp and Simmons (1983). Six types of territorial encounters can be distinguished:

1. **Butterfly flight**. The bird flies slowly with a very deep and deliberate wing beat over the territory, while repeatedly uttering a penetrating call. Often no obvious stimulus triggered the display.

2. **Calling.** One or both of the pair perform the piping display while standing still; this is similar to the 'solitary piping' seen on the wintering grounds (Ens and Goss-Custard 1986). Although without an opponent in the immediate vicinity, a distant target can usually be identified, such as Oystercatchers flying overhead or intruders landing in a distant part of the territory.

3. **Chasing intruder(s).** The displays used when chasing away an intruder vary from a slow approach in an ambivalent threat posture (Fig. 8.1a) to rapid running with piping and short flights (Fig. 8.1b), sometimes culminating in a prolonged chase flight.

4. **Border dispute** (Fig. 8.1c,d). These sometimes very time-consuming displays are directed against one or more neighbours along the territory boundary. Usually, the piping display is performed first, followed or interspersed by periods of bobbing. Border disputes rarely escalate to fighting.

5. **Hovering ceremony** (Fig. 8.1e). From three to 15 birds, predominantly nonbreeders, perform this most spectacular piping display while hovering.

6. **Fighting** (Fig. 8.1f). Oystercatchers stand face to face and attempt to peck each other with the bill. Attacks and retreat are often initiated on the wing and may end with a prolonged chase flight or, in case of neighbouring territory owners, a retreat beyond the territory boundary.

Nonbreeding birds do not defend a nesting territory but may defend a feeding territory on the mudflats and are usually paired when doing so (Ens *et al.* 1996e). When not feeding, nonbreeders gather in 'clubs' consisting of flocks of 20–200 birds scattered throughout the breeding area. In inland sites, nonbreeders may feed around the club.

Evidence that breeding space is limited

The Oystercatcher is one of six bird species where removal experiments have demonstrated that territory density is limited and that there exists a non-territorial surplus capable of breeding (Newton 1992). Harris (1970) removed one male and five females before egg laying and two males, five females and four pairs after it had begun. All the experimentally widowed birds found new mates of which two were known, and three were suspected, of coming from the nonbreeding 'flock'. Vacant space following pair removal was taken over by neighbouring pairs in three cases and a pair of nonbreeders in one other. The speed with which replacements took place was remarkable; four males studied acquired new mates in 1–4 days. In a similar experiment, Ens *et al.* (1996e) removed males after their mates had started egg laying. In four of the six cases where the male was only removed for 36 hours, probably nonbreeding males mated with the female and settled in the territory. When eight males were removed permanently, the female lost the territory in five cases, after one or more days of fighting, and obtained a new mate within 24 hours in three, again probably from the nonbreeding 'flock'. In another study, a female that died during egg laying was quickly replaced by a new female from the 'flock' which started to lay eggs herself within 7 days (Briggs 1985). In contrast, Vines (1979) found no replacement when she removed two breeding pairs. However, this was done rather late in the breeding season (end of June) and, in one case, breeding densities were very low. On the whole, the evidence that intense competition occurs for territories seems overwhelming.

Rapid replacement of removed breeders requires nonbreeders to be in the vicinity. Nonbreeding birds in full adult plumage have, in fact, been reported in all detailed studies, though the abundance may vary: 13% in Kandalaksha Bay (Bianki 1977); 16% along the River Lune (Briggs 1985); 30% on Skokholm (Harris 1967); 4% on the Åland Archipelago (Nordberg 1950); 9% on Rockcliffe Marsh (Rankin 1979); 31% on Schiermonnikoog (Ens 1992) and 10% in the inland polder Arkemheen (H. v. d. Jeugd, personal communication). These data suggest that the percentage of nonbreeders may be lowest in the far north, intermediate in inland sites at more southern latitudes and highest along the coast at temperate latitudes where the species has its major stronghold.

Nonbreeding birds include young that have never bred and former breeders that have been ousted from their territory; in fact, intermittent breeding is surprisingly common and provides further evidence of the intensity of the competition for space. Young birds have to compete intensively to breed, as is shown by the difference between the age when young birds are capable of breeding and the age when they actually start. Based on the recorded ages of first breeding, the earliest possible age of breeding is probably three in females and four in males. Plumage characteristics are consistent with this view because three-year-old birds in the summer of their fourth calendar year no longer carry a white neck collar and are hard to distinguish from adults (Cramp and Simmons 1983). Furthermore, whereas one- and two-year-olds invariably spend the summer on the wintering grounds, the length of the summer absence in four-year-olds is indistinguishable from that of adults (Fig. 8.2). However, data from birds marked as chicks on Mellum and Skokholm indicate that the majority of individuals do not breed until much later (Fig. 8.3). Indeed, the data may underestimate the average age of first breeding because some late starting birds may have bred only after the studies finished. Despite this, females clearly breed at an earlier age than males. As there is no evidence that breeding adult females have a higher mortality rate than males (Goss-Custard *et al*, Chapter 13, this volume), and assuming an equal sex ratio at laying, it is likely that the sex difference in the age of first breeding arises because fewer females than males survive their first few nonbreeding seasons (Durell and Goss-Custard 1996). Whether this reflects a lower dominance in young females when competing for food, as shown by Davies (1992) for severe

(a)

(b)

(c)

(d)

(e)

(f)

Fig. 8.1   The various ways in which Oystercatchers defend their territories. (a) A pair chasing an intruder in an alert posture at the far right; the left bird is in diplomatist attitude, while the middle bird is pseudo-sleeping. (b) A pair chasing an intruder while performing the piping display. (c and d) A single territory owner in a typically hunched posture pushing back (compare the position of the white marker in the two photographs) a pair of piping territory owners during a border dispute. (e) A hovering ceremony. (f) Fighting Oystercatchers (J. van de Kam).

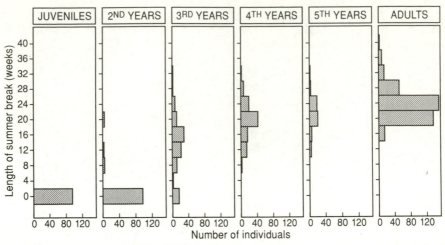

Fig. 8.2 Changes in behaviour with age. The graph shows the average number of weeks absent from the Exe estuary during summer (S.E.A. le V. dit Durell and J.D. Goss-Custard, unpublished information).

winter mortality in the Dunnock (*Prunella modularis*), remains to be determined in Oystercatchers (Ens and Cayford, Chapter 4, this volume; Goss-Custard *et al.*, Chapter 6, this volume).

What is high quality breeding space and why is it limited?

Oystercatchers intensively compete for high quality breeding space where many chicks can be raised at low cost, because the demand for such space is high relative to the supply. This prompts two questions: (1) what makes some space to be of high quality; and (2) why are there so many competitors for it?

Nesting in Oystercatchers is risky. Nests can be flooded during high tides, washed away by rains, trampled by cattle, or robbed by predators. In addition to these risks, chicks can also be stabbed to death if they venture into the territory of neighbouring Oystercatchers or they may be killed by disease or starvation. Although these primary causes of egg and chick loss have been established and are known also to vary between areas (Safriel *et al.*, Chapter 9, this volume; Goss-Custard *et al*, Chapter 13, this volume), the characteristics that go to make up high quality breeding space are, surprisingly, largely unexplored. Perhaps we are closest to being able to define territory quality on Schiermonnikoog, to which much of the remainder of this chapter is devoted. Here, pairs that occupy territories where they can take the chicks to the food fledge many more chicks than pairs occupying territories where the nesting and feeding area are separated, so the parents have to transport all the food to the chicks over a considerable distance (Ens *et al.* 1992; Safriel *et al.*, Chapter 9, this volume). The first type is referred to as a 'resident' territory,

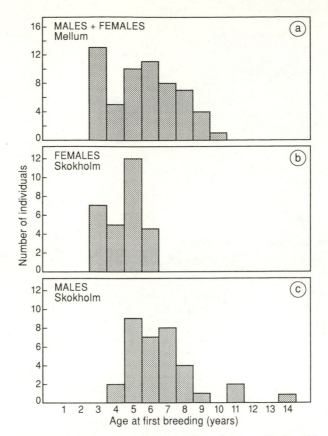

Fig. 8.3 Distribution of the age of first breeding for: (a) birds of both sexes on Mellum (from Schnakenwinkel 1970); (b) females on Skokholm (M.P. Harris and U.N. Safriel, unpublished information), and (c) males on Skokholm (M.P. Harris and U.N. Safriel, unpublished information).

the second as a 'leapfrog' territory (Fig. 8.4). Due to the intense competition, it is very likely that birds in the good quality resident territories would benefit from an enlargement of their territory, but there are no experimental data to support this suggestion. We are a long way from understanding the territory size at which residents would achieve maximal reproduction.

This brings us to the second question on the number of competitors. In estuarine environments, there is usually no shortage of good feeding areas during summer. In the Dutch Wadden Sea, for example, less than 50 000 of the 200 000 wintering Oystercatchers remain to breed (Hulscher 1983), even though the biomass of the benthic food supplies reaches its peak in summer (Beukema 1974). Although there are also vast saltmarshes available for nesting, there are relatively few areas where birds can both nest and collect food for the young within the same

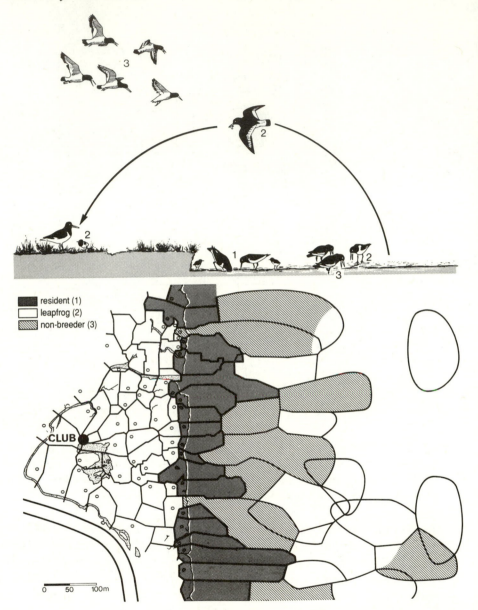

Fig. 8.4 The location on Schiermonnikoog of resident territories and the nesting and mudflat feeding territories of leapfrog Oystercatchers, spring 1986. Also shown are mudflat territories defended by pairs of nonbreeding birds. The top diagram shows how leapfrogs have to transport all food while residents can take their chicks to the feeding grounds. Furthermore, nonbreeders engage in border disputes and hovering ceremonies (Ens *et al.* 1995).

territory. The habit of feeding the precocial young thus reduces the area of 'usable' mudflats to a small fringe bordering the saltmarsh. The Oystercatcher therefore seems an example of a 'survival species', where the amount of habitat where individuals can survive but not reproduce much exceeds the amount where individuals can reproduce (Alerstam and Högstedt 1982). Oystercatchers that breed inland must winter along the coast. If the number of Oystercatchers breeding inland continues to increase (Goss-Custard *et al*, Chapter 13, this volume), the Oystercatcher should become less of a survival species, so that the age of first breeding may decline. In the meantime, many aspects of the mating system can be understood as an adaptation to the intense competition for breeding space.

Monogamy and cooperation in territorial defence

Oystercatchers are nearly always monogamous. Why are they not polygamous? In fact, polygamy, in the form of polygyny, does sometimes occur. Polygyny in birds usually arises by males monopolizing females indirectly, through the control of patchily distributed scarce resources such as food or nest sites (Davies 1991). These conditions seem to apply in many Oystercatcher breeding areas yet probably less than 1% of males in any population is mated polygynously (Cramp and Simmons 1983; B.J. Ens, unpublished information). Certainly, there are no obvious reasons why males should not benefit from polygyny. Indeed, a polygynous male from the River Lune fledged four chicks in 1979 and three in 1980, while successful monogamous males fledged only 1.5 and 1.9 chicks in the same years (Briggs 1984a). Clearly, the interests of the female must also be considered. With the influence of territory quality taken into account, birds that worked harder on Schiermonnikoog raised more chicks, because they transported more food to the chicks (Ens *et al.* 1992). Thus, with polygyny, a second female would probably reduce the parental care given by the male to the first brood and so reduce the reproductive success of the first female. Though based on circumstantial evidence, this argument does explain the generally high aggressiveness of females towards other females and the several cases in which early associations between one male and two females ended before breeding (Ens *et al.* 1993b). Thus, Oystercatchers may be monogamous, because female aggression prevents polygyny.

Females not only chase females from the territory, they also chase males. Why? If a territory provides exclusive access to resources, females would be expected to chase from the territory individuals that do not assist in raising the brood. On the other hand, neighbouring males should only assist if they have some chance of paternity. Thus, Davies (1992) showed that the amount of effort a second male Dunnock is willing to invest depends on how frequently he has copulated with the female before she produced the clutch. However, such copulations by the female are against the interests of her own male, as well as against the interests of the neighbouring female. There is also evidence for the

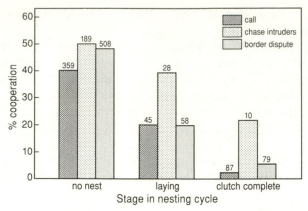

Fig. 8.5 The probabilities that the members of a pair cooperate in defending their territory when both are present in relation to the stage of the nesting cycle (no nest; laying eggs; clutch complete) and type of territory defence employed (calling; chasing intruders; border disputes). The number of observations is shown (Ens 1992).

Oystercatcher that birds that regularly copulate with non-mates risk losing their mates (see below). The open habitat frustrates sneaky sexual contacts (see later), so there is little chance that the female can acquire some help from a neighbouring male without risking losing her mate, and she may not even try. Under these circumstances, the neighbouring male is a rival competitor for territorial resources throughout the breeding season. The scarcity of high quality breeding space may thus select for females that assist their mate in territorial defence.

In fact, assistance is more than simply chasing intruders irrespective of their sex. As both members of a pair perform the same type of aggressive display towards the same target (Fig. 8.1), Oystercatchers may be said truly to cooperate. Surprisingly, Ens (1992) could not show that cooperation increases the effectiveness of territory defence, although cooperation may predominantly occur when the threat to the territory is high. Indeed, male and female do not always cooperate and cooperation is necessarily abandoned when the birds start incubating the eggs (Fig. 8.5). Nonetheless, cooperative defence is clearly of high significance because an incubating parent may still on occasions leave the nest to assist its mate, thereby exposing the eggs to a high risk of being taken by predators (D. Heg, personal communication). Probably, not assisting the mate in such circumstances might lead to a partial or complete loss of the territory which, as will become clear, is a genuine risk in Oystercatchers.

Territory acquisition

The life history problem of deferred maturity

This section explores the relationship between competition for breeding space and the age at which Oystercatchers first breed. Each year not devoted to reproduction is a wasted year, unless postponement increases future success (Lessells 1991; Stearns 1992), so it is likely that there is a high cost to breeding at an unusually early age. It has been proposed for long-lived seabirds that early breeding would be disproportionately costly because of the low feeding skills of young birds (Lack 1968; Orians 1969; Ashmole 1971). However, whether to breed at a particular age is not the result of a single decision, as implied by this feeding skills hypothesis. At least two consecutive decisions are involved: (1) at what age to return from the wintering to the breeding grounds; and (2) at what age to try to acquire a territory.

The first career decision facing a young Oystercatcher that has survived the winter is whether to return to the breeding grounds in summer or to stay in the wintering area. Most remain, and only a few start visiting the breeding grounds when three years old (Fig. 8.2). Although one year old and older immatures may be less efficient feeders than adults (Norton-Griffiths 1968; Goss-Custard and Durell 1983, 1987a), it is not clear whether this affects their decision as to when to return to the breeding areas. Nor is it known whether feeding skills are of decisive importance during territory establishment, the second major decision facing a young Oystercatcher. On the other hand, it is very likely that the delay between returning and first breeding is due to intense competition for territories. In the following, the issues associated with this social competition hypothesis are elaborated.

Why settle in a territory of poor quality?

On Schiermonnikoog, potential recruits should attempt to acquire a resident territory because the reproductive success there is so much greater than in leapfrog territories (Ens *et al.* 1992). In fact, the majority do not. Why? Five hypotheses erected by Ens *et al.* (1995) are discussed in turn.

1. **The poor discrimination hypothesis** proposes that Oystercatchers cannot discriminate among territories on the basis of their quality, and is very unlikely. Nonbreeders intrude frequently into territories before they first breed and so have many opportunities to observe the differences in quality, which are highly consistent over years. There is also ample evidence that Oystercatchers can detect differences in quality. The intense competition suggested by the lateral compression of the feeding territories of residents (Fig. 8.4) and the small size of their nesting territories (Fig. 8.6), and those of the leapfrog Oystercatchers located close to the edge (Fig. 8.6), implies the birds can and do detect differences in quality.

2. **The low cost–low benefit hypothesis** proposes that the low annual reproductive success in the poor territories is offset by a high annual survival because the poor quality territories have lower defence costs. As there is no indication that leapfrogs had a higher survival on

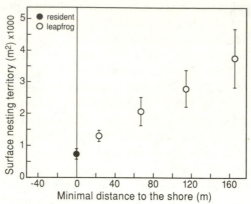

Fig. 8.6 The size of the nesting territory on Schiermonnikoog in relation to its distance from the edge of the saltmarsh (Ens *et al*. 1995). The vertical bars show 1 SE.

Schiermonnikoog between 1985 and 1992 (Fig. 8.7), this hypothesis can be firmly rejected.

3. **The breeding skills hypothesis** proposes that acquisition of breeding skills, even in a poor territory, augments the chances of acquiring a high quality territory later in life. Although some 25% of leapfrog Oystercatchers do eventually succeed in obtaining a resident territory, the average leapfrog has a much lower annual probability of recruiting into a resident territory than the average nonbreeder (Fig. 8.7). Clearly, nonbreeders seeking a resident territory should not first settle into a poor territory, because they have little chance of moving once settled.

4. **The inferior phenotype hypothesis** proposes that birds in poor territories are poor fighters that simply cannot beat the good fighters occupying the high quality territories; settling in a poor territory is their only option. While it is highly unlikely that there are no differences in fighting ability, it has so far proved more or less impossible to measure fighting ability independently of site-related social dominance, and the hypothesis remains largely untested. The only test, which did not favour the hypothesis, assumes that large birds are better fighters: despite considerable individual differences, residents were not larger, nor weighed more than leapfrogs.

5. **The queue hypothesis** proposes that nonbreeding Oystercatchers form queues differing in length for territories differing in quality. The fundamental assumption here is that nonbreeders are not free to move between queues within a year. If this is true, the reproductive advantage of settling in a high quality territory is offset by many years of waiting in the queue and the associated risk of dying before acquiring one. Under this hypothesis, potential breeders must decide whether to queue with the *'hopeful leapfrogs'*—and trade-off early and more certain breeding against low reproductive output in any one year—or the *'hopeful resi-*

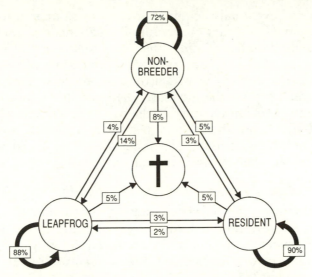

Fig. 8.7 The annual probabilities of Oystercatchers changing status on Schiermonnikoog (Ens *et al.* 1995).

dents', where the reverse trade-off applies. On the assumption that the number of poor and good territories is fixed, and that competitors do not differ in fighting ability, Ens *et al.* (1995) calculated that, at evolutionary equilibrium, the average hopeful resident has to wait 10 years before establishing a territory, with an even chance that it will die before doing so. Clearly, this is a simplistic calculation, but it does indicate how long birds should be prepared to wait to obtain a resident territory. A more realistic model would include the possibility, for example, that the better fighters might be able to improve their position in the queue—in effect, combining the queue and inferior phenotype hypotheses. However, as individual differences in fighting ability have yet to be measured, a more realistic appraisal of the reproductive consequences of obtaining a leapfrog or resident territory cannot yet be made.

Evidence for the queue hypothesis

Although its quantitative predictions cannot yet be tested, several lines of evidence provide a qualitative test of the hypothesis. On Schiermonnikoog, fledged chicks have a 50% chance of returning to the natal breeding area (Kersten and Brenninkmeijer 1995). Of 46 returned chicks, only nine have so far recruited into the breeding population, and all have taken leapfrog territories (Ens *et al.* 1995). This is consistent with the prediction that it takes fewer years to acquire a leapfrog territory. There is also some evidence that, as predicted, it takes many years to obtain a resident territory. On the mudflats, pairs of nonbreeding hopeful residents defend feeding territories adjacent to those of residents (Fig. 8.4). Several of these pairs have gradually extended their feeding territory upshore and eventually acquired an area of saltmarsh for nesting. Based on the number of years marked individuals have occupied such territories, and the number of times they have succeeded in gaining a nesting territory, we can derive an average waiting time of nine years. To this must be added the age at which the nonbreeders obtained their hopeful resident territory on the mudflat. The only clue here is a marked chick that obtained such a territory at the age of seven. If 16 years is typically required by a hopeful resident to acquire a nesting territory, they are, as expected, considerably older when they first breed than the average Oystercatcher when it first breeds at the age of six (Fig. 8.3).

Site-dependent dominance and escalated fighting

Two questions arise when examining the queue hypothesis in more detail. First, why does a bird wait in a queue rather than challenge a territory owner and engage in escalated fighting until it has either acquired a territory or died? Second, what prevents an individual from switching queue?

In fact, the two questions are related. On the basis of a simple model, Ens *et al.* (1995) calculated that on Schiermonnikoog a nonbreeder should fight for a resident territory if the probability of success exceeded the probability of death. This is not a trivial result and strongly depended on the assumption that the estimated future lifetime production of fledglings by the average resident is twice that of the average leapfrog. At first sight, this condition might seem easily fulfilled. Equal opponents stand an equal chance of winning, so that a nonbreeder has only to challenge an owner that is slightly weaker than itself to satisfy the condition. However, this reasoning assumes that one escalated fight is sufficient to gain the territory and that beating the owner also establishes dominance over the other queuing nonbreeders. But this is most unlikely, because dominance is site dependent.

Site-dependent dominance means that the outcome of contests depends on the site where the individuals meet (Brown 1963). Clearly, dominance relationships among breeding Oystercatchers are site dependent: each individual wins over all others inside its territory but loses when outside. Here, the difference in importance to each of the contestants of winning the contest apparently determines the outcome, instead of any difference in fighting ability (cf. Parker 1974). Being dominant over another Oystercatcher must become increasingly less valuable the further the bird is situated from the centre of its territory, its area of greatest commitment. This is presumably why territory owners respond more slowly to intruding conspecifics the further the intruders are away from the nest (Vines 1979), since it is most unlikely that the fighting ability of a bird declines away from the centre of its territory. Observations on the club on Schiermonnikoog, where territory owners and nonbreeders meet at high water, confirm that the further the owner is away from its territory, and thus

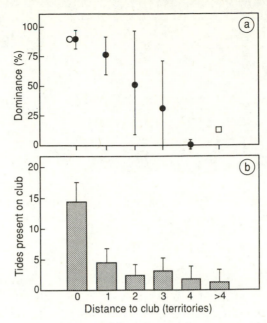

Fig. 8.8 (a) The mean dominance of breeders in the communal club, measured as the percentage of encounters won, in relation to the distance, measured in numbers of intervening territories, of their nesting territory to the club (•; bars show 1 SE). Also indicated are the average dominance scores for nonbreeders attempting to establish a territory on the club (○) and for hopeful residents (□); only in the latter case was the distance from the territory to the club measured. (b) The number of tidal cycles birds were present at the club in relation to the distance of their territories from the club. The vertical bars show 1 SE (Ens *et al.* 1996e).

the less important the encounter, the less often the bird visits the club and the lower is its dominance score (Fig. 8.8). Importantly, the same applies to nonbreeders attempting to obtain a nesting territory; the nonbreeders that are highly aggressive on the club are apparently the ones attempting to establish a territory in the vicinity. Again, hopeful residents have a very low dominance on the club, reflecting the large distance between their area of commitment and the club. We thus suggest that, over the course of several seasons, nonbreeders become gradually committed to a smaller and smaller area and, in parallel, the value of being dominant in that area increases, so the birds fight more vigorously. When a breeder disappears, only the locally dominant of the nonbreeders or breeders can replace it. Indeed on Schiermonnikoog, only leapfrogs with nesting territories directly adjoining the resident territories have succeeded in taking one over. For each territory, there is a certain number of potential recruits with an associated dominance ranking. Were a nonbreeder to move somewhere else, it would lose its historically acquired social position, which explains why nonbreeders do not switch

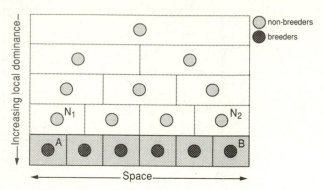

Fig. 8.9 Scheme depicting the social organization of breeders and nonbreeders. The territory size is represented by the size of the box. Birds with smaller ranges dominate birds with larger ranges.

queues. The longer a nonbreeder has waited and increased its local dominance, the greater its chances that it will finally obtain a breeding territory when an existing owner disappears. But clearly, at a global level a random element remains. In Fig. 8.9, if territory holder A disappears, nonbreeder N1 is in the best position to take the territory. But if B disappears, nonbreeder N2 is most likely to succeed.

It seems that, before birds commit themselves to joining one queue or the other, they sample widely over the breeding area. As this is completely occupied by territorial pairs, newly arrived young birds are not free to move about and they spend much time in the clubs, visiting several during the breeding season. From these, they make sallies into the surrounding territories, fleeing instantly when the territory owner approaches. They are often attracted by unusual events, and gatherings of up to 50 intruders may quickly accumulate. Overall, young nonbreeders appear to be 'curious' (K.B. Briggs, B.J. Ens and U.N. Safriel, unpublished information). Given the large differences in territory quality, mate quality, intensity of competition for breeding space and the apparently limited opportunities for change once settled, it is clearly advantageous for young birds to spend time sampling before deciding which queue to join.

Vacancies and the actual process of territory establishment

The simplest way in which the stable numbers of territories observed on Schiermonnikoog might be achieved is for all vacancies to be taken up, but only by nonbreeders. In fact, the process is more complicated. Primary vacancies occur through the death of one or both territory owners. As most mortality occurs outside the breeding season and rarely exceeds 10% (Safriel *et al.* 1984; Hulscher 1989; Goss-Custard *et al.*, Chapter 13, this volume), the probability that both members of a pair will die during the same winter is less than 1%. Territories are thus seldom

Table 8.1 Fate of experimentally widowed birds defending a breeding territory, separated for males and females, and the timing of the removal experiment; a distinction is made between early removals (before egg laying) and late removals (during incubation of the eggs). Data from Skokholm (Harris 1970) and Schiermonnikoog (Ens 1992) are lumped.

Fate experimental widow	Female removed		Male removed	
	Early	Late	Early	Late
Marry widowed bird	2	—	—	—
Marry divorced bird	2	1	—	2
Marry nonbreeder	1	4	—	7
Ousted from territory	—	1	1	5

vacated completely and then simply re-occupied by another pair. Usually, a returning breeder simply finds itself without its mate, and so acquires a new one. Occasionally, widowed birds fail to attract a mate, and sometimes are even ousted from the territory. Although some nonbreeders take up vacancies arising from the death of a pair member, widowed birds are just as likely to pair with another breeder (Ens *et al.* 1993b). Apparently, nonbreeders are more likely to fill a vacancy when it occurs later in the season: experimentally widowed breeders were more likely to pair with a nonbreeder when the mate was removed during egg laying than before (Table 8.1). Furthermore, five out of eight breeding males that were deserted by their mate during the breeding season attracted a nonbreeding female, while the other males lost the territory (Ens *et al.* 1996e).

Instead of accounting vacancies, we can tabulate how nonbreeders recruit. On Schiermonnikoog, 26 nonbreeders became breeders by filling a vacancy while 13 chased away one of the territory owners and mated with the partner (Ens 1992). But 52 nonbreeders, and so the majority, attained breeding status by conquering space as a pair from pairs holding territories. It was striking that this often happened in May, when the breeders were distracted from territory defence by incubation (Fig. 8.5). Contrary to what might be implied to some nationalities by the word 'queue', nonbreeding Oystercatchers are not simply waiting to fill space vacated by dead birds; rather, they are continually pushing and being pushed. Oystercatchers must regularly perform aggressive displays to maintain their position in the queue so that they are well positioned to take over a territory when an opportunity, for whatever reason, arises. This necessarily restricts the individual to a small geographical locality, making it impossible to maintain high status elsewhere and favouring the formation of either short or long queues of competitors seeking an opportunity to breed.

Fundamental trade-offs

The queue hypothesis proposes a trade-off between waiting a long time for a high quality territory or settling immediately in a low quality territory. From our description of the social organization and recruitment of nonbreeders we suggest that the more fundamental underlying trade-off is between ranging over a large area, to maximize the chance of spotting a vacancy, and defending a position of high local dominance, to maximize the chances of securing a vacancy there (see also Zack and Stutchbury 1992). The fundamental constraint seems to be that it is impossible to range widely as well as be regularly present in a small locality. The queue hypothesis does not specifically address the pair bond. However, the presence of a partner may be a handicap when prospecting for a suitable single-sex vacancy early in the season, but may be vital in conquering territorial space later in the season.

Studies that follow the behaviour of individual nonbreeders and their social relationships in much greater detail will be necessary to quantify these trade-offs. Ideally, such studies should include experiments that manipulate the social environment of the particular potential recruit, a tall but inescapable order.

Divorce and cuckoldry

Our description of recruitment made clear that in Oystercatchers pair formation and territory establishment are hard to separate. Oystercatcher pairs may breed many years together, yet new pairs can form within hours when one mate disappears. In this section, the focus shifts from the territory to the pair bond.

Divorce

Having survived the winter, most breeding birds return to their previous territory and old mate. However, birds sometimes change mate even though their previous partner has also returned; on Skokholm, this happened in 8% of the 560 cases where both birds returned and bred in successive years (Harris *et al.* 1987). What causes some pairs to divorce, while the majority of birds remained faithful? As in many bird species, young birds on Skokholm were more likely to divorce than old birds (Harris *et al.* 1987). Unsuccessful pairs were also more likely than successful pairs to divorce; 15% of pairs that did not rear chicks divorced compared with 7% of pairs that raised at least one. Such observations have often been taken as evidence for the incompatibility hypothesis which proposes that incompatible partners divorce while compatible ones reunite (Coulson 1972). However, this hypothesis provides only an incomplete explanation because, of 148 pairs that did not even hatch eggs, 76% reunited the next season. Furthermore, the correlation between low success and high divorce rate may be spurious since young birds are both divorce prone (see above) and often unsuccessful (Fig. 8.17). In Kittiwakes (*Rissa tridactyla*), divorce rate strongly declined with female age, but for females of a given age, there was very little effect of reproductive success on subsequent probability of divorce (Coulson and Thomas 1985).

As an alternative explanation for divorce, Ens *et al.* (1993b) proposed the better option hypothesis. This states that individuals only divorce when there is a better pair bond option available. The basic assumption is that the quality of potential mates varies according to the quality of their territory or to the abilities of the individual as a parent or as a defender of space. Each individual is then assumed to attempt to pair with the mate of highest quality. As the number of high quality mates will be limited, most individuals necessarily take a mate of submaximal quality. That such individuals rarely change mates may reflect the high price generally involved in fighting a rival, especially when it is a better fighter. An individual would only be expected to divorce when a mate of better quality is available at low cost, as when the rival dies in winter. When one individual *chooses* to divorce, its old mate is simply *the victim of that choice*; only choosers are therefore expected to benefit from divorce.

What is the evidence for this hypothesis in Oystercatchers? Choosers and victims have only been identified after prolonged observation of pairs on Schiermonnikoog (Fig. 8.10). Seven females, but only one male, deserted their mate and moved to a neighbouring territory; chooser and

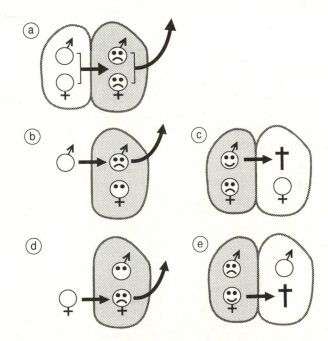

Fig. 8.10 Scheme showing the ways in which divorce occurs. (a) A pair loses its territory to neighbouring pairs. (b) A male is chased out by a male usurper. (c) A male deserts his female when another becomes available on the death of her previous mate. (d) A female chases out a female usurper. (e) A female deserts her male when another becomes available on the death of his previous mate (Ens *et al.* 1993b).

victim were clearly identifiable in these cases (Fig. 8.10). Divorces also arose when one mate was ousted from the territory by a usurper; there were ten female usurpers and seven male. The ousted individual was clearly the victim, although the identity of the bystander was less clear (Fig. 8.10); it remains to be seen if usurping attempts can be successful if the partners of an established breeding pair continue to cooperate in territorial defence. Perhaps, the erstwhile mate decided to withhold cooperation and let the two rivals fight it out, or it may have encouraged the usurper by copulating with it; if so, it was clearly a chooser. Finally, some nesting territories are usurped by neighbouring pairs in early spring. The previous owners often remained paired for a while as non-breeders and defended a feeding territory, but invariably divorced in the end. Divorce can thus happen in several ways.

Did the choosers gain from divorce and the victims lose, as would be expected from the better option hypothesis? The greater tendency on Schiermonnikoog for females to move when changing mates was also apparent on Skokholm, and also applied to widowed birds (Fig. 8.11). Thus, females are the moving sex and may also be the choosing partner in divorce. If so, females should, on average, benefit more from divorce than males. Comparing fledgling production in the post-divorce year to fledgling production in the pre-divorce year, Harris *et al.* (1987) recorded an improvement, which was especially marked in females, but which was not significant in either sex. This might have been due to the inclusion of too many victims in the comparison, although this is not very likely as most victims on Schiermonnikoog did not breed in the post-divorce year (Ens *et al.* 1993b). An alternative possibility is that reproduction with the new mate is initially less efficient so that there is an efficiency cost of mate change. This idea was tested by comparing the change in success of divorced birds with the change in success of

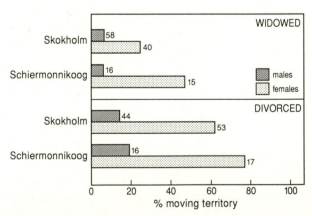

Fig. 8.11 The probability of changing territory in males and females following the death of a mate or divorce on Skokholm and Schiermonnikoog (Ens *et al.* 1993b).

widowed birds. These birds were not expected to benefit from changing mates but would bear any efficiency cost of mate change. In both males and females, widowed birds suffered a decline in reproductive success, but only in females was this decline significantly different from the improvement in divorced birds (Ens *et al.* 1993b). If there is an initial inefficiency with the new mate, we would also expect reproductive success to increase with the number of years that members of the pair have been together. In pairs where females were first-time breeders or divorced, fledgling production increased in later years with the new mate, but this was not so for widowed females. Since young birds are especially likely to divorce, widowed birds will generally be older than divorced birds. Thus, the effect of pair bond duration on reproductive success may just have been due to the females in the sample becoming older. The evidence for an initial inefficiency with the new mate is thus equivocal, so the evidence that females benefit from divorce remains equivocal too.

Although the better option hypothesis is rather more in accord with the data on divorce in Oystercatchers than the incompatibility hypothesis, firm evidence is lacking. More observational data are required so that increased sample sizes will allow all the potentially confounding factors to be taken into account. But studying the effect of the experimental removal of males and females on the pair bond of neighbouring pairs is likely to be the most profitable research strategy. When Harris (1970) removed ten females during the breeding season, neighbouring females deserted their mate in three cases, showing that the proposed experiments are feasible. However, to be interpreted, a methodology must be developed to measure the quality of a mate before it reproduces. While there are many suggestions as to what constitutes a high quality mate (foraging ability, fighting ability, resistance to parasites etc.), there are no empirical data.

Cuckoldry

If Oystercatchers initiate divorce when better options become available, they must be able to recognize them as such. Following Heg *et al.* (1993), we argue here that copulations with a bird other than the mate, or extra-pair copulations (*EPCs*), serve this purpose for both male and female Oystercatchers. Males are generally believed to perfom *EPCs* to increase their reproductive success without the cost of parental care (Trivers 1972). Females were first thought to be disadvantaged by *EPCs*, but subsequently a variety of benefits has been proposed to account for the observation that females often actively solicit them. However, all the possible benefits for both sexes of *EPCs* listed by Birkhead and Møller (1992) refer only to the current reproductive attempt and Heg *et al.* (1993) find no support in the Oystercatcher for any of them. Instead, their observations suggested that *EPCs* enable Oystercatchers to sample

(a)

(b)

(c)

(d)

Fig. 8.12 A pair of Oystercatchers (a) resting together; and (b–d) copulating (J. van de Kam).

the opportunities for mate change, implying that the benefits relate more to future reproduction.

Before egg laying, Oystercatcher pairs often rest (Fig. 8.12a) and feed very close together. Both male and female often make pre-flight calls, as if to reach agreement on the decision to depart. Clearly, both partners adopt measures to remain together. The mates also cooperate in defending the territory and regularly copulate (Fig. 8.12b,c,d), with both sexes initiating copulation. On Schiermonnikoog, Oystercatchers begin cop-

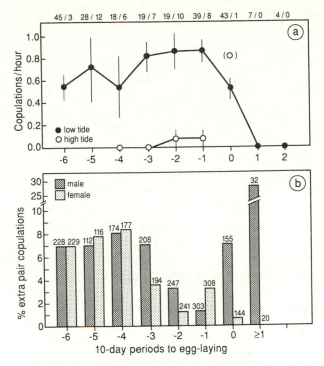

Fig. 8.13 (a) The frequency of copulation at high and low tide and (b) the probability that males and females occupying a breeding territory were engaged in extra-pair copulations, as a percentage of all copulations during that period, in relation to the timing of egg laying (Heg *et al.* 1993). In (a), the vertical bars show 1 SE. The number of observations are shown in both graphs; in (a), the left-hand value refers to low tide and the right-hand value to high tide.

ulating more than two months before the first eggs are laid and continue to do so at a rate of almost one copulation every two hours during low water until the clutch is complete (Fig. 8.13), yielding about 700 copulations before the first clutch is produced (Heg *et al.* 1993). Very few bird species have a higher copulation frequency (Birkhead and Møller 1992). Why do Oystercatchers copulate so often? As high copulation frequencies are especially common in species where the mates are regularly separated, they may be the way in which males guard their paternity (Birkhead and Møller 1992). However, Oystercatcher pairs are seldom separated, yet they copulate particularly frequently. Furthermore, as Oystercatchers have small reproductive organs until one month before egg laying (Hulscher 1977b), many of the frequent copulations early in the season could not lead to sperm transfer and fertilization. It seems more likely that, in Oystercatchers, copulations contribute to forming and maintaining the pair bond. Since the piping display is aggressive, copulatory displays are the only ones that can be unambiguously identified as courtship.

Both males and females engage in *EPCs*, although they are not common and are most often observed early in the season (Fig. 8.13b). Females rarely engage in *EPCs* during egg laying, and have never been observed doing so after the clutch was complete. Males also rarely copulate once the clutch is laid, probably because of the demands of incubation, but when they do, it is often an *EPC*. If *EPCs* are an attempt to change or sample other potential mates, they would be expected to occur early in the season when they would cause the least interference with the current breeding attempt. Six further lines of evidence are consistent with the mate change hypothesis (Heg *et al.* 1993):

1. Of two females studied over several years, one switched to a neighbouring male after two years of regular *EPC* with him (Fig. 8.14). The other has not yet switched, but has copulated increasingly often with one particular neighbouring male. With both females, the territory of their old mate had become very small, suggesting a gradual deterioration in his fighting ability. It was the females that took the initiative, intruding on the territory of the neighbouring male and inviting copulations.

2. Males generally engage in *EPCs* on their own territory, while females more often move to another. This corresponds neatly with the observation that males that change mate usually stay in the territory, while females often move (Fig. 8.11).

3. Most *EPCs* by males were with nonbreeding females, which will almost certainly leave no offspring; they were therefore not increasing their current reproductive success, as would be predicted by the hypothesis of Trivers (1972). However, these females could become alternative mates.

4. DNA-fingerprinting showed that of 65 chicks in 26 clutches, only one was not fathered by the male partner, but by a neighbouring male which, in any case, took over the female the following year (Fig. 8.14).

5. Male breeders whose mate was absent sometimes evicted soliciting female intruders instantly. Contrary to the Trivers' (1972) hypothesis, this suggests that *EPCs* are not necessarily beneficial in the short run, even when there was no apparent risk of a penalty being incurred by the mate.

6. Birds that stand to gain most from a change of mate should put most effort into sampling potential mates. As expected, members of pairs occupying high quality territories rarely engaged in *EPCs*. At the other extreme, nonbreeders without territorial commitment copulated with many different partners in a single season.

Several observations suggest that birds attempt to conceal their sexual contacts with potential new mates (Heg *et al.* 1993). For instance, neighbours sometimes flew some distance to copulate out of sight of their partners. If one of the partners nonetheless joined the flight, *EPCs* were less likely to occur. Males that had engaged in *EPCs* within their territory chased the female away when their mate returned. A possible risk of

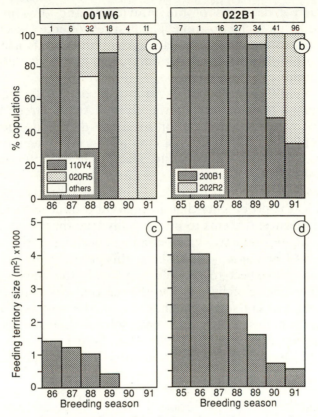

Fig. 8.14 The case histories of two female breeders who engaged in extra-pair copulations. (a) The percentage of all copulations by the female 001W6 that were with her old male 110Y4, with whom she bred from 1986 until 1989, the male 020R5 to whom she switched in 1990 and two other males. (b) The percentage of copulations by the female 022B1 that were with her mate 200B1, with which she bred from 1984 to 1991, and the neighbouring male 202R2. (c and d) The sizes of the feeding territories of males 110 7 and 200B1, respectively. In (a) and (b), numbers above the bars show sample sizes (Heg *et al.* 1993).

being 'found out' is that the cuckolded partner might also start to seek *EPCs* and so encourage potential usurpers. Clearly, many ramifications arise from the idea that *EPCs* in Oystercatchers are an attempt to find and acquire a new mate and extremely detailed observations will be necessary to test it further.

Reproductive decisions

Introduction

So far, we have concentrated on career decisions associated with competition for breeding space and mates. The next step is to see how the reproductive decisions of the individual, and thus its fitness, are influenced by its social position. The first reproductive decisions taken each breeding season concern the laying date and clutch size. Later decisions relate to parental care, and are discussed by Safriel *et al.* in Chapter 9 (this volume).

Of the early reproductive decisions, the date of laying the first egg has been studied in most detail in the Oystercatcher. Drent and Daan (1980) proposed that natural variation in laying date and clutch size is due to individuals in good nutritional circumstances laying earlier and producing larger clutches. Subsequently, Daan *et al.* (1989) concluded that two seasonally changing variables are important in the annual organization of reproduction: food availability and the reproductive value of eggs, defined as the probability that an egg will produce an individual that will itself reproduce. It seems to be generally true that, in single-brooded bird species, the reproductive value of an egg declines during the course of the season. As Daan *et al.* (1989) show, this gives the prediction that individuals that can expect to be able to provide the chicks with much food should lay large clutches early in the season, while birds under poor conditions should lay small clutches late in the season. Good quantitative evidence for this theory has been obtained from the Kestrel (*Falco tinnunculus*) (Daan *et al.* 1990), but there is some way to go before studies of Oystercatchers reach the same level of sophistication.

Seasonal decline in reproductive value of an egg

Oystercatchers are single-brooded and repeat clutches are only produced when the eggs or chicks are lost sufficiently early in the season (Fig. 8.15). In any one place and year, there may be a two month difference between the start date of the first and last clutch. Between years however, the breeding period is remarkably fixed.

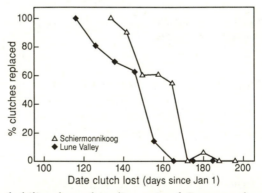

Fig. 8.15 The probability that a breeding pair of Oystercatchers will start a repeat clutch in relation to the date when the previous brood of eggs or chicks was lost on Schiermonnikoog (B.J. Ens and J.B. Hulscher, unpublished information) and in the Lune Valley (Briggs 1985).

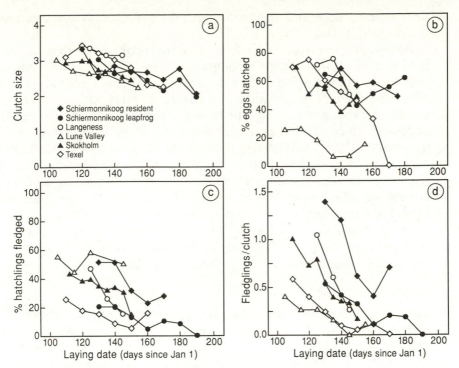

Fig. 8.16 The components of breeding performance plotted against laying date: (a) clutch size; (b) the percentage of eggs that hatched; (c) the percentage of hatchlings that fledged; (d) the number of fledglings produced per clutch. Data are from Texel over the years 1983–1993 (C.J. Smit, unpublished information); Langeness for the year 1986 (Becker 1987); Lune Valley for the years 1978–1980 (Briggs 1985); Skokholm for the years 1963–1977 (Harris 1967, 1969, and M.P. Harris and U.N. Safriel, unpublished information).

There is a small decrease in clutch size over the season (Fig. 8.16a). But does the reproductive value of an egg also decline as we would expect from the hypothesis of Daan *et al.* (1989)? In coastal sites, hatching success does not change over the season (Fig. 8.16b), whereas fledging success declines strongly (Fig. 8.16c), leading to a steep decline in reproductive success (Fig. 8.16d). In contrast, in the Lune valley, the only inland site studied, it was a decrease in hatching rather than fledging success that caused fledgling production to decline over the season (Fig. 8.16d). Texel and Skokholm are partly inland and partly coastal sites as the birds nest on meadows, but obtain a significant amount of food from the coast. They are also intermediate in the sense that both hatching success and fledging success decrease over the season. For Schiermonnikoog, a coastal site, the seasonal decline in reproductive success is not due to birds in good territories laying much earlier than birds in poor territories; early residents and early leapfrogs start laying on the same date and for both, reproductive success declines over the season (Fig. 8.16d).

Although these studies suggest that the probability that an egg will lead to a chick that fledges strongly declines over the season, much may happen to the fledgling before it finally becomes a reproducing adult. Hatching date does not influence the probability either of an Oystercatcher surviving more than one month after fledging (Harris 1969) or of a fledged chick returning to the breeding grounds (Kersten and Brenninkmeijer 1995). Admittedly, the sample sizes are small but, if anything, early born chicks seem more likely to return, thus enhancing the seasonal decline in the reproductive value of an egg. A more fundamental problem is that we relied completely on natural variation. If only 'high quality' birds lay large clutches and breed early, we need experimental manipulations of clutch size and laying date to determine for each class of birds how the reproductive value of an egg would change with clutch size and laying date (Daan et al. 1990; Tinbergen and Daan 1990).

Feeding conditions

How do the feeding conditions change over the egg laying period from late April to June? They almost certainly improve for coastal Oystercatchers as the days lengthen, the temperature rises and the biomass and availability of benthic prey species increases (Hulscher, Chapter 1, this volume; Zwarts et al., Chapter 2, this volume). Thus, Hulscher (1982) recorded a threefold increase during May in the feeding rate of a male Oystercatcher while Ens et al. (1992, 1996b) found intake rates were higher during the incubation than pre-laying period. It is less clear whether feeding conditions continue to improve as chicks are being reared from June to August. They might even decline (Ens et al. 1992, 1996b).

It is well established that inland breeders start laying eggs considerably earlier than coastal breeders (Heppleston 1972; Wilson 1978; Briggs 1984b; Beintema et al. 1985). Inland breeders feed mainly on earthworms (Lumbricidae) and leatherjackets (larvae of Tipulidae), whose seasonality almost certainly differs from that of coastal prey (cf. Zwarts and Wanink 1993). Heppleston (1972) found that the majority of inland chicks hatched slightly after the peak in available biomass of earthworms and leatherjackets. Briggs (1984b) found that earthworm numbers and biomass declined during May, as the chicks start to hatch, while the biomass of leatherjackets reached its peak. However, the numbers of other insect larvae taken from cow-pats increased throughout the season. As several prey species are taken, it is difficult without data on intake rates to gauge when feeding conditions are best for inland breeding birds. Furthermore, it is likely to be the availablity rather than the abundance of prey that has the greatest influence. One possible measure of availability is soil resistance. Thus, Hulscher (1985) showed that inland breeders sustain considerable lateral wear to the bill-tip as a result of probing into hard soil. In his study, soil resistance reached a minimum in early May, when chicks start to hatch, after which it increased as the soil dried out. Clearly, more information on seasonal changes in food abundance and availability are required if the relationship between the timing of breeding and the feeding conditions are to be understood.

The differences between individual birds may depend on a number of factors, including the quality of their feeding territory, their foraging ability and the time they can devote to feeding rather than territorial defence. Current data on individuals are of remarkably little help in distinguishing between these alternatives. On Skokholm, individuals tended to lay earlier as the number of years since they had first bred increased, especially during the first five years (Fig. 8.17). But as experienced females lay only ten days earlier than when they first bred, and as the breeding season spans more than 40 days, only a part of the variance in laying date can be due to age differences between females. Indeed, there

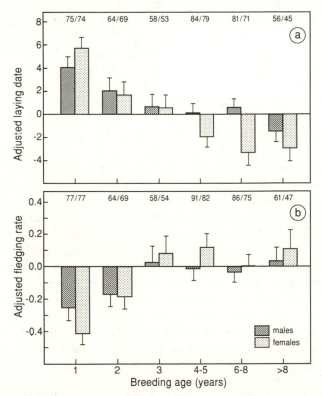

Fig. 8.17 (a) The laying date and (b) reproductive success as a function of the number of years since the individual first occupied a breeding territory. Both scores are represented as deviations from annual means (Ens *et al.* 1993b). The vertical bars show 1 SE. Numbers are sample sizes.

is evidence for consistent differences between individuals, independent of age: on both Skokholm (Ens *et al.* 1993b) and Schiermonnikoog (B. J. Ens and J. B. Hulscher, unpublished information), females that bred early in their first year tended also to breed relatively early in subsequent years. The cause of such individual variations is unknown, however. Laying date does not correlate with female size, as measured by bill and wing length (Briggs 1985). Without further detailed studies, it is not possible to disentangle the separate influence of individual variations and seasonal changes in the food conditions on the date of first laying in the Oystercatcher.

Predation during egg laying

Although it has been assumed to this point that laying dates can be measured without bias, this is almost certainly not the case. Egg predation is very much higher during egg-laying than during incubation (Ens 1991). Thus, without very frequent nest visits, the eggs may be predated before the investigator can first locate them. Thrice daily searches on Schiermonnikoog revealed that first eggs have a 40% chance of disappearing before the second egg is laid (Ens 1991). As the laying interval between the first and second egg in 175 nests was just under 38 hours (Briggs 1985), the majority of the predated first eggs would have been missed had the searches been made once a week. As the loss of eggs does not cause them to lay replacements, Oystercatchers seem to be determinate layers, even though they may move to another nest scrape after the loss of the first egg (B.J. Ens, unpublished information). Indeed, it is quite possible that all Oystercatchers lay four eggs and that differences in clutch size simply reflect the extent of egg loss during egg laying. As all the eggs are sometimes lost during laying, repeat clutches may be mistaken for the first clutch, causing the date of first laying to be estimated incorrectly. If such birds do not produce a repeat clutch, or also lose the repeat clutch very quickly, it may be wrongly concluded that the birds did not lay that season. In fact, most territorial pairs classified as 'non-laying' on Schiermonnikoog (Ens *et al.* 1992) and Skokholm (Harris *et al.* 1987) may actually have lost their clutch very quickly. The sometimes high predation rates on Oystercatcher nests can therefore cause important quantities in their reproduction to be estimated incorrectly.

According to Ens (1991), the high egg loss during laying is due to incomplete nest defence. When the clutch is complete, one of the partners incubates the eggs while the other either stands guard or goes to feed; the eggs are therefore always protected. But during egg laying, they are often left unguarded when the parents are elsewhere together. One explanation of why the parents remain so closely associated at a time of high risk to their first eggs is that the male stays near the female to prevent rival males from fertilizing eggs he later tends. As Oystercatchers continue to copulate during egg laying, the chance of being cuckolded continues and it may pay the male to accompany his mate wherever she goes during this phase of the breeding cycle.

Summary

Career decisions, which relate to gaining access to scarce resources vital for reproduction and survival, can be distinguished from reproductive decisions, which deal with the different ways in which these resources can be utilized by individuals. Costs and benefits of both career and reproductive decisions must be measured in terms of the expected contribution to the number of chicks fledged over a lifetime and subsequently able to raise young themselves, but this has not been achieved in any study. However, Ens *et al.* (1995) show how a description of Oystercatcher society as composed of different social positions, in combination with estimates of the reproductive success of each position and the transition rates between positions, allows an estimation of the expected future reproductive success of each social position.

For Oystercatchers, the major career decision concerns at what age, and in what place, they should attempt to acquire a breeding territory. Competition for space of high quality is intense and only a few are able to settle in the best territories. Prospecting nonbreeders are faced with a choice; to settle quickly in a territory of poor quality or to wait a long time for a high quality territory, with the associated loss of potential breeding years and the greater risk of dying before they are able to breed. Because dominance relationships are site dependent, waiting birds form local queues within which their dominance, and thus chances of obtaining a territory, gradually increase; in the case of the best territories, individuals may queue for many years before acquiring one. The reason that so many Oystercatchers settle in poor quality territories is probably the large number of nonbreeders queuing for territories of high quality.

Forming a pair bond is regarded here as a second career decision, although as male and female cooperate in territorial defence, the distinction between the formation of a pair bond and territory establishment may be more conceptual than real. Divorces are rare, probably due to the high cost of evicting rival partners. Extra-pair copulations, where a male or a female copulates with a bird other than the mate, are interpreted as attempts by both males and females to locate potential and available mates of a higher quality than their current mate. In Oystercatchers, there is no evidence that males engage in extra-pair copulations simply to increase their immediate reproductive success, as much theory has assumed. Rather, their purpose seems to be more long-term and to involve a career rather than a reproductive decision.

Our current knowledge is too meagre to allow us to predict for a bird of a given social position the outcome of reproductive decisions on the optimal combination of clutch size and laying date. Experimental manipulations are required to confirm the much repeated observation based on natural variation that the reproductive value of an egg declines over the season. At present, little is known about the underlying causes of this decline because the necessary studies on seasonal changes in the feeding conditions and of variations between individual Oystercatchers in their timing of breeding have not yet been made.

The general conclusion is that individual Oystercatchers can be thought of as prisoners of their historically acquired local social relationships. Although the birds are physically free to move, the costs of moving territory and/or changing partner involved in establishing new social relationships and foregoing the benefits of the current social order will usually not outweigh the benefits associated with the new social position. The next step is to derive the social order of the Oystercatcher from the decision rules of the individuals. For this, it will be necessary to know why exclusive access to space is so important and how the benefits of this exclusive access depend on territory size and quality, and intruder pressure.

Acknowledgements

We are very grateful to Dik Heg and Joost Tinbergen for commenting on the manuscript and to Jan van de Kam for allowing us to use his fine photographs.

9 Rearing to independence

Uriel N. Safriel, Bruno J. Ens and Alan Kaiser

The Oystercatcher's puzzle—why feed a precocial chick?

The Oystercatcher provides a test of how far existing theory helps to understand the behavioural and population ecology of a well-studied species (Goss-Custard, Introduction, this volume). By the same token, research on a carefully selected, and such a well-studied species, should advance general theory. Is the Oystercatcher not only relatively easy to study, but also a 'typical' shorebird whose study is applicable to many other species, not only of shorebirds, but also of birds in general?

In at least one feature, the Oystercatcher is unique. The young leave the nest upon hatching and, though fully precocial, are fed by their parents until well after fledging; they thus have precocial mobility while enjoying a fully altricial parental nourishment. In this respect, the Oystercatcher represents an additional state (parent-fed mobile chick, or 'precocial 5'; Table 9.1) along the precocial–altricial gradient proposed by Nice (1962). It occupies an intermediate position between two other more widely occurring conditions: the semi-precocial condition of partial precocial mobility and full parental nourishment, typical of gulls and terns, and the 'precocial 4' condition of full precocial mobility and partial parental nourishment, typical of grebes, rails, and some Charadriiformes, such as Stone Curlews (*Burhinus oedicnemus*) and Pratincoles (*Glareola* spp.). The Oystercatcher's is a singular condition, even within the Charadriiformes which, as an order, is more varied than any of the others in demonstrating a strong evolutionary conservatism with respect to the precocial–altricial gradient (Ricklefs 1983). The broad uniformity across taxa suggests that selection has worked strongly against

Table 9.1 Categories of precociality in Charadriiformes (modified from O'Connor (1984))

	Sedentary young	Parent feeding	Example
Precocial 2	-	-	Charadriidae
Precocial 4	-	+	Burhinidae
Precocial 5[1]	-	+	Haematopodidae
Semi-precocial	+	+	Laridae
Semi-altricial 1	+	+	Dromadidae

[1] An additional category with respect to Nice's (1962) classification.

transitional states (Ricklefs 1983). Yet the Oystercatcher seems to have been able to straddle the basic dichotomy between the precocial and altricial conditions.

How has the Oystercatcher managed to occupy such a unique position? Nice (1962) suggested that strength and skill are required to tackle the prey eaten by any precocial young, hence only anatomical maturation and a long learning process enable them to become independent. Given the Oystercatcher's diversity of food types and flexibility of feeding techniques (Hulscher, Chapter 1, this volume; Sutherland et al., Chapter 3, this volume), the 'valley' of low fitness between the two 'adaptive peaks' of the precocial and the altricial condition could have been crossed in either of two ways. One way could have been to turn to fully precocial behaviour by transporting food types to the young that they could deal with, with the young gradually acquiring the feeding specializations of adults only after fledging. Alternatively, the 'valley' could have been crossed through a fast maturity attained by more sedentary chicks fed by parents, as in Larii, and rarely in Charadrii, i.e., the Crab Plover (*Dromas ardeola*) (Cramp and Simmons 1983). Adherents of the 'adaptationist program' (*sensu* Gould and Lewontin 1979) need therefore to do two things: they should identify disadvantages for Oystercatchers in adopting either of these two extreme conditions and demonstrate the disadvantage for most other shorebird species of the intermediary condition occupied by the Oystercatcher.

A likely trade-off in the balance of advantages between the altricial and precocial conditions is that between speed of growth and exposure to predation. The longer chicks stay in the nest, the more likely they are to be predated. Parents can shorten the risky period by increasing feeding rates and thus accelerating growth. This in turn increases predation risk, since a high visitation rate to a fixed site discloses the nest location to predators. The altricial condition allows a fast growth but at the cost of high exposure to predation (e.g. Ricklefs 1983). On the other hand, the precocial condition minimizes predation risk for several reasons: (a) there are no food-delivery trips by parents; (b) though the young are at risk during foraging, the parents are free to warn and protect them; and (c) the young can defend themselves to a certain extent by running, hiding and 'freezing', whereas altricial young are helpless once discovered (Evans 1970). However, this reduced risk is likely to be achieved at the cost of slower growth of the precocial young because an adult will be more successful than an unfledged young at foraging, due to its physical maturity and experience.

The problem of explaining how the Oystercatcher has come to occupy its unique position boils down to explaining how it has apparently overcome this trade-off between predation rate and growth rate. The general hypothesis developed in this Chapter is as follows. Feeding is performed by the adult; hence, provided there is sufficient food, growth is maximized, as is thought generally to be the case in the altricial condition (Ricklefs 1969; Shkedy and Safriel 1992). Predation risk, on the other hand, is lower than is usual in altricial birds for two reasons. First, being precocial, the mobile Oystercatcher young protects itself better than altricial young. Second, though Oystercatcher parents feeding their young behave as central-place foragers, the 'central place' is 'mobile'; the Oystercatcher's chick often changes its hiding site so delivery trips to the same site are fewer than those made by an altricial parent visiting a nest. If predators do indeed learn the position of the young by watching a foraging parent, predation risk should be lower in the Oystercatcher than in altricial birds. The Oystercatcher may thus reap the main benefits, but minimize the main costs, that accrue to typical altricial and precocial birds. While obtaining the 'altricial' benefit of being fed by mature and experienced parents, they gain the precocial benefit of anti-predator mobility. It is this unusual combination that gives the Oystercatcher its interest to general theory.

Hypothesis and predictions

A number of testable predictions can be derived from this line of reasoning. As compared to fully precocial birds, predation risk will be greater in the Oystercatcher if parental feeding trips do indeed attract predators. For the Oystercatcher's 'precocial 5' condition to be advantageous over the condition of full mobility and self-feeding ('precocial 2'), the foraging efficiency of parents as compared to that of self-feeding young should, therefore, be unusually high. It needs to be high enough for the resulting acceleration in growth, and hence shortening of the risky period, to more than compensate for the increased predation risk resulting from the attraction of predators by the young-feeding parent.

The high degree of individual specialization, and associated elaborate searching skills and handling techniques, along with its robust morphological adaptations, may indeed enable the Oystercatcher to exploit a resource that is largely un-utilized by other species and so allow a relatively high foraging efficiency (Hulscher, Chapter 1, this volume). On this argument, the intermediate 'precocial 5' condition could only be advantageous for a species with such an enhanced efficiency, and would be disadvantageous in other Charadriiformes without exceptionally efficient foraging techniques. Accordingly, the following predictions can be made: (a) the daily amount of food obtained should be larger and/or the daily energetic cost of foraging per unit of obtained energy should be lower, for an Oystercatcher pair feeding their young than for the parents and brood combined in a Charadriiform species that does not feed its young; (b) provided that brood size is optimal, Oystercatcher chicks should grow faster than other Charadriiform chicks that feed themselves; and (c) predation on Oystercatcher chicks should be lower than on parent-fed semi-precocial Charadriiform chicks that remain near the nest until fledging.

The problem with such between-species predictions is that it is difficult to find species sufficiently similar to the Oystercatcher in all other respects except those to be tested. The alternative approach is to make within-species comparisons and inferences. If the required within-species variations in Oystercatcher behaviour do indeed exist, then a failure to detect the following predicted trends would weaken confidence in the hypothesis: (a) the chick's growth rate should increase as the parents' foraging efficiency increases; (b) the chick's mortality should increase as the number of parental feeding trips increases; (c) the chick's mortality should increase with an increasing length of the fledging period; and (d) the fledging period should shorten with increased growth rate. In other words, successful individuals are defined as those that fledge more young than others either through maximizing growth rate at the cost of a larger number of risky feeding trips or minimizing the number of feeding trips at the cost of a reduced growth rate.

Only some of the data required for testing these predictions are available. But the great asset as a study organism of the adult Oystercatcher is its large size and open habitats, which facilitate field measurements on feeding skills, habitat selection and foraging tactics. Data on chicks can be difficult to obtain, due to their precocity. However, their reliance on parents for food makes them easier to locate and study than the young of fully precocial species. In the following, we use available data to evaluate some of our hypotheses and predictions.

Why do Oystercatchers not have sedentary chicks?

First we explore why the Oystercatcher has not acquired the more common semi-precocial habit of the Larii. Larids nest in colonies and/or on ledges. In obligate colonial breeders in open terrain, such as sand dunes, the chicks' restricted mobility assists parents in locating offspring (Nice 1962). For chicks on cliff ledges, mobility is lethal (Evans 1970). Nesting in colonies and/or on ledges in the first place is associated with the need to provide some protection for chicks while the parents are absent foraging on distant, and often clumped and unpredictable, sources of food (Lack 1968). As chicks cannot be taken there to feed, they remain close to the nest until fledging and so the parents transport food to a rather fixed site. Although on our hypothesis this would be expected to result in a higher predation risk than in Oystercatchers, the evolution of multiple-prey loading in Larids may have reduced the risk of predation. This ability to transport several prey at once has required the development in Larii of anatomical and/or functional adaptations that allow storage and regurgitation (Orians and Pearson 1979), adaptations that are not shared by the single-prey loading Oystercatcher.

Table 9.2 Chick's mobility, food-transport strategy, the resulting growth rate and implied predation risk.

	Chick mobility	Food transport trips to:	Growth constant[1]	Predation risk
Altricial (passerines)	None (nest)	Fixed point	0.201[2]	Highest (repeated trips to some point)
Semi-precocial (Larii)	Partial (around nest)	Fixed site	0.114[3]	Medium (nests in naturally protected sites, or colonies)
'Precocial 5' (Oystercatcher)	Full	Several points (hiding sites, 'mobile nest')	0.081[4]	Low (feeding trips but not to same point)
Precocial 2 (Charadriidae)	Full	No trips	0.057[5]	Lowest

[1] Expected Gompertz growth coefficients of an Oystercatcher-sized species (asymptote of growth curve 466 g (3)), as predicted by regression equations of growth coefficients as functions of asymptotic mass.
[2] Equation from Ricklefs (1968) for 40 passerine and 12 raptor species.
[3] Equation from Drent and Klaassen (1989) for 26 species of gulls and terns.
[4] Data from Drent and Klaassen (1989).
[5] Equation from Beintema and Visser (1989) for 7 Charadriidae and 12 Scolopacidae.

Why, then, has a sedentary chick not also evolved in the Oyster-catcher? There would seem to be no phyletic constraint (Gould and Lewontin 1979) to their doing so. First, the sedentary habit of chicks has evolved elsewhere in the Charadrii, although rarely. In the Charadriid Crab Plover, the (single) young is fully sedentary at a permanent site for a long period (Cramp and Simmons 1983). However, this site is a deep burrow dug by parents, in which the young are well protected from predation, but mostly from the fierce heat prevailing in this bird's habitat. This 'semi-altricial 1' condition (Table 9.1) is even rarer than the 'precocial 5' Oystercatcher condition, and is probably associated with the very specific features of the habitat and region occupied by this bird. But this does not mean that Oystercatchers could not have evolved it, if pushed. Second, other Charadrii have evolved colonial nesting, such as the Lapwing (*Vanellus vanellus*) (Hale 1980). Finally, there would also seem to be no major constraint on the evolution of multiple-prey loading in Oystercatchers. As discussed by Zwarts *et al.* (Chapter 2, this volume), Oystercatchers already store up to 80 g wet flesh weight of prey when flying to roost and regurgitation would seem easy to evolve, given its widespread occurrence in birds. Probably, then, the 'mobile nest' (Table 9.2) of Oystercatchers is the only available solution for reducing predation and for fooling canny predators in a system in which a widespread, predictable (compared to Larii) and nearby food supply favours nesting pairs being spread out rather than being concentrated within colonies.

Why do Oystercatcher chicks not feed themselves?

A common answer to the question of why the Oystercatcher has not evolved partial self-feeding or fully self-feeding precocial chicks is that the evolution of feeding specializations in Oystercatchers precludes self-feeding until the young develops the physical ability and acquires the necessary expertise (Norton-Griffiths 1969). But the specialized feeding adaptations of the Oystercatcher have not evolved at the expense of the traditional probing methods used by other Charadrii: Oystercatchers also use probing for hunting worms (Hulscher, Chapter 1, this volume). Presumably, the young could feed themselves by probing and begin eating molluscs after fledging. One argument for why this has not evolved (Norton-Griffiths 1969) is that cockles and mussels seldom occur in the same place as worms, hence a probing chick could not be guarded by an adjacent mollusc-eating parent. Yet, parents that feed themselves when they have young mainly by probing for worms or clams still feed their young even though their young could easily feed alongside them on the mudflats. Thus, parent-feeding in the Oystercatcher is not necessarily associated with feeding specializations, and another reason is required to explain why Oystercatcher chicks do not self-feed.

Predation risk: the trade-off between parent- and self-feeding

A general feeding-strategy-threshold model

U.N. Safriel and A. Kaiser have modelled the trade-off in predation risk for Charadriiform birds with fully mobile chicks (precocial 2, 4 and 5) which, on our hypothesis, determines the threshold between self-feeding (the common condition, precocial 2) and parent-feeding (partial in precocial 4, full in precocial 5) (Table 9.1), as has evolved in Oystercatchers. The model assumes that predation risk for a self-fed chick increases asymptotically with the time spent foraging by the chick (T). It also assumes that predation risk to a parent-fed chick increases asymptotically with the number of parental delivery trips (N). Both risks can be described by either a Poissonic or a binomial probability of predator's encounter distribution and can either be an exponential asymptotic or a Michaelis–Menton function, respectively. Parent-feeding becomes advantageous when predatory risk per unit time per energy harvested by parent-feeding is less than that for self-feeding. On both sides of this inequality, terms of the risk functions cancel to yield:

$$b_{p}N < b_{s}T \tag{9.1}$$

in which b_s and b_p are the rate constants of the self-feeding and the parent-feeding risk functions, respectively. Thus, whether the rule for the predator's probability of encountering chicks is Poissonic or binomial is irrelevant to the model results. Furthermore, T and N, the time required

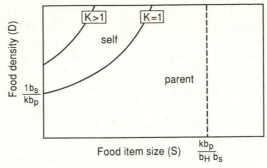

Fig. 9.1 Parent-feeding and self-feeding regions on the food density–prey item size plane, simulated by the threshold model. With large prey item sizes, to the right of the asymptote (vertical broken line), only parent-feeding occurs. For item sizes below the threshold, high food density makes it more likely that self-feeding will occur. When the prey item size taken by the parent is larger than that taken by the chick ($k > 1$, see text), the area in which parent-feeding is advantageous increases. In the intercepts, b_p and b_s are constants of the risk function and b_H is the slope of handling time function (see text).

and the number of trips needed, respectively, to meet a unit of energy demand, are a function of item size, S. Assuming a linear increase of handling time $H(S)$ with item size ($H(S) = b_H S$, b_H being the slope), the food density threshold D for parent-feeding is given by:

$$D < \frac{1}{\dfrac{b_p}{b_s} - b_H} \qquad (9.2)$$

and is plotted in Fig. 9.1. The curve is an equal-risk isocline, which divides the space between the food density and item-size axes into a region in which parent-feeding should prevail and another region in which self-feeding should occur. Evidently, when very large items are taken, only parent-feeding occurs (the area right of the asymptote, given by $b_p/b_H.b_s$ on the S axis in Fig. 9.1). When smaller items are taken and their density is high, self-feeding becomes advantageous. The model predicts that breeding Oystercatchers should use habitats with relatively large food items, otherwise their chicks could not be parent-fed. The model also predicts that when chicks move to habitats containing smaller items at very high densities, parent-feeding becomes less beneficial than self-feeding. Fledged Oystercatcher chicks do gradually begin to feed themselves and the model predicts that they should only do this on habitats rich in relatively small food items.

These model predictions are only valid on the assumption that all item sizes can be captured equally effectively by either chick or parent. However, there is a large difference in bill-length between the parent and a newly-hatched Oystercatcher chick; the mean values in samples of adults and chicks from a Dutch population are 74.2 mm and 15.9 mm respectively (B.J. Ens, unpublished information; Hulscher 1985). This yields a parent to hatchling bill-length ratio of 4.66. For a typical self-fed shorebird species with much shorter adult bill, the ratio is far smaller; for example, in the Little Ringed Plover (*Charadrius dubius*), adult bill-length is 13.2 mm and the ratio 2.03 (U.N. Safriel and A. Kaiser, unpublished information). The large difference in bill size in the Oystercatcher presumably enables adults to capture much larger items than chicks, which must feed on very small prey. By denoting the ratio between the size of prey taken by the parent and the size of prey taken by the chicks as k, the isocline becomes:

$$D < \cfrac{1}{k\,\dfrac{b_p}{b_s} - b_H S} \tag{9.3}$$

Note that when adult and young take the same item size, $k = 1$, but with $k > 1$, when adults can capture larger items than their young, the isocline is higher and the area where parent-feeding prevails becomes larger (Fig. 9.1). The model therefore predicts that, when parents can transport to their young items much larger than those that the young could have taken for themselves, parent-feeding should prevail, especially where item sizes are large.

Model implications for Oystercatchers

The prevalence of parent-feeding in Oystercatchers is thus explicable, given the large ratio in parent/hatchling bill size. When this ratio converges to 1 after the young fledge, the area over which self-feeding is advantageous increases, and a gradual increase in self-feeding should occur. The model implies that self-feeding is only advantageous in a breeding habitat in which small items occur at very high densities. This suggests that parent-feeding in Oystercatchers has evolved in environments with large food items. Though exploitation of large items may require adaptations and expertise that the Oystercatcher chick does not have, the clear message of the model is that parent-feeding in the Oystercatcher is not necessarily a handicap imposed by morphology. Rather, it may be an advantageous strategy for a basically unspecialized bird (Heppleston 1971a) that can also exploit large food items, whether these require elaborate techniques (e.g. limpets) or not (e.g. worms).

If this line of reasoning is correct, a more long-term prediction can be made. Compared to intertidal habitats, the incidence of large items in inland terrestrial habitats may be relatively low. Given the apparent evolutionary plasticity implied by the relatively new and fast spread of inland breeding in Oystercatchers (Safriel 1985), the self-feeding threshold may now have been reached in many inland areas. Self-feeding in Oyster-catcher chicks would therefore be predicted to become gradually wide-spread. Indeed, inland-breeding Oystercatcher chicks do seem to be proficient at finding food for themselves at an earlier age than coastal chicks, and become independent from the parents at an earlier age (B.J. Ens, personal observations). Finally, inland chicks have been observed self-feeding on small insect larvae in dung pats (K. Briggs, personal communication). This is a food source of high density and small item size, falling in the top-left corner of Fig. 9.1, exactly where self-feeding should prevail.

Predation risk: the trade-off between length of fledging period and number of transportation trips

Young-rearing strategy and tactics

The large size of the parent's bill relative to that of the young dictates that parent-feeding should have evolved in the coastal habitats where, until recently, all Oystercatchers bred. Given this, and the inevitable separation between onshore nesting sites and offshore feeding areas, predation can be minimized by reducing the length and number of trans-portation trips. In a sense, the trade-off in predation risk between parent-feeding and self-feeding dealt with by the threshold model determines the strategy. But since predation can also be minimized by short-ening the fledging period (Hockey 1984), the trade-off in predation risk between the length of the fledging period and the number of trips determines the tactics of the growth rate. In the following, we evaluate some of the predictions based on this reasoning by making between- and within-species comparisons with respect to food item size, growth, fledging period, and predation risk.

Between-species comparisons

Growth rate

If fast growth does shorten the period during which young are exposed to predation, and parent-feeding is more efficient than self-feeding, then parent-feeding should prevail when predation risk is high, and parent-fed chicks should grow faster than self-fed ones. As Table 9.2 shows, the high predation risk of the fully sedentary altricial chicks does indeed seem to have selected for fast growth (Ricklefs 1983). The partially-mobile gull chicks grow somewhat more slowly; though they are fed by their parents for the whole fledging period, multiple-prey loading reduces the number of visits required and chicks are protected by relatively in-accessible sites and/or by coloniality. Though Oystercatcher parents are single-loaders, they have a 'mobile' food-delivery site so that their chicks may be less endangered than would otherwise be the case with frequent visitation; the selection pressure for fast growth may thus be weaker in Oystercatchers than in gulls. Fully mobile self-fed precocial chicks should grow more slowly than Oystercatcher chicks due to the higher efficiency of parent-feeding. Furthermore, food acquisition by the Oystercatcher chick is not affected by adverse weather; the chick is then warmed by the non-feeding parent, and at the same time can be fed by the other parent. In contrast, self-fed precocial young do not forage during such times (Beintema *et al.* 1990), so further reducing the rate at which they are provisioned in comparison with Oystercatchers.

To test whether self-fed Charadrii chicks grow more slowly than Oystercatcher chicks, we calculated the growth rate Gompertz constant, K, of Oystercatcher-sized self-fed shorebirds from the regression equation for shorebirds, $K = 0.390A^{-.312}$ where A is the asymptotic mass in grams (Beintema and Visser 1989). The value of 0.057 is much lower than the observed rate constant of 0.081 in Oystercatcher chicks. This is also higher than the value of 0.054 recorded in Lapwings, with an asymptotic mass half that of Oystercatcher, and higher than the 0.051 found in Curlew (*Numenius arquata*), which has an asymptotic mass more than twice that of the Oystercatcher (Drent and Klaassen 1989). Assuming we are correct in assessing the relative risk of chick predation in Oystercatchers (Table 9.2), these between-taxa comparisons in growth rate are consistent with our hypothesis.

Food item size

In order to accelerate growth without increasing the number of feeding trips, prey size should be large. Are Oystercatcher chicks fed larger items than those eaten by the self-fed precocial young of other shorebird species? This possibility was tested in three ways. First, we examined the size of items taken by the self-fed chicks of four shorebird species foraging in Dutch pastures (Beintema *et al.* 1990). Compared against a slight and unexpected trend for the proportion of large items taken to decrease as adult bill-length increases, the Oystercatcher chicks—being fed by adults in the same pastures—clearly deviate (Fig. 9.2). The bill-length of adult and hatchling Oystercatchers (74 and 16 mm respectively) are intermediate between those of Redshank (*Tringa totanus*) (41 and 13 mm) and Black-tailed Godwit *(Limosa limosa)* (100 and 17 mm). Whereas these two species have similar prey size frequency distributions, the Oystercatcher, as expected, has the highest proportion of large items in its diet (Fig. 9.2: $\chi_2^2 = 78.1, 57.6$ for comparisons with Godwits and Redshanks, respectively; $P < 0.001$ for both).

A similar comparison can be made using data collected by Holmes and Pitelka (1968) from four Calidrine sandpipers in Alaska, eating items ranging between 5 and 19 mm long. We regressed the weighted mean length of larval and adult food items from the stomachs of the four sandpiper chicks against hatchling bill-length to predict the size of prey

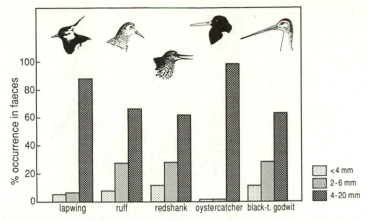

Fig. 9.2 Food item sizes in the Oystercatcher compared to those of other shorebirds feeding in the same habitats. Size categories of prey items in chicks' faeces collected in Dutch pastures, arranged in an increasing order of bill length. Data re-analysed from Beintema *et al.* (1990): size categories from Table 2 (largest class comprises earthworms, snails, spiders and 15 insect groups; smallest size comprises Nematocera species; the proportional occurrence in faeces samples were taken from Table 3).

expected to be taken by Oystercatchers. Both regression equations (n = 8, 4 species, two seasonal sections; B = hatchling bill length; mean length of larva = 2.85 + B × 0.54 ± 0.27mm (SE); mean length of adult insect = 2.47 + B × 0.56 ± 0.18mm (SE)) predict a mean length of terrestrial food item of 11.5 ± 1.1mm (95% confidence interval) for Oystercatcher chicks. In fact, on Skokholm, the length of Tipulids, moth caterpillars and earthworms brought to chicks, estimated from the bill-length of the feeding parent, was always ≥20 mm. Indeed, the means of 3 Tipulids and 3 earthworms presented to chicks and measured directly were 30.3 and 68.3 mm, respectively (Safriel 1967). In this case also, Oystercatcher chicks took larger items than would be expected in self-fed precocial young with the same length of bill.

Finally, we compared the difference in item size between chicks and adults in self-fed *Calidris* species from Alaska and parent-fed Oystercatchers from Skokholm. Birds took insect larvae in Alaska and both insect larvae and earthworms on Skokholm (Fig. 9.3). In the *Calidris* spp. with the longest bill, chicks ate items less than half the size of those eaten by their parents. In contrast, Oystercatcher chicks ate items twice as large (Fig. 9.3). Parent Oystercatchers ate smaller earthworms, Tipulid larvae and moth caterpillars than those they transported to their chicks (Fig. 9.4), although the difference in caterpillars, which were rare in the diet, was statistically not significant. The chicks ate 4.4 times more caterpillars and only 1.8 times more earthworms than adults, whereas adults ate 1.3 times more Tipulids than did the chicks (Fig. 9.4). In terms of their food value, measured as mg dry mass, kJ per gram and kJ per item

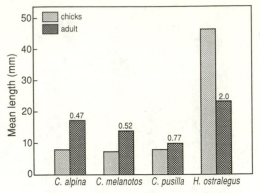

Fig. 9.3 Weighted mean prey item size (insect larvae and earthworms) in chick and adult Alaskan *Calidris* sandpipers (Holmes and Pitelka 1968) and Skokholm Oystercatchers (Safriel 1967). The numbers are the ratio of the chick/adult prey item size. *Calidris* species arranged in decreasing bill-length from left to right.

respectively, these correspond to 51, 19.2 and 0.984 for earthworms, 42, 23.0 and 0.933 for Tipulids, but 106, 28.0 and 2.972 for caterpillars (Safriel 1967). To conclude, whereas parents of self-fed chicks take larger items than their chicks, the reverse occurs in Oystercatchers. Overall, the results of the three comparisons support the prediction that the parent-fed chicks of the Oystercatcher take much larger prey than do the self-fed chicks of other Charadriid species.

Predation risk

Are transportation trips risky, as is assumed by our model? Nest visits by parents, either for feeding young, or for incubation, can attract predators. In parent-fed species, nest visitation rates are greater at the chick stage, when the young are being fed, than at the egg stage, when parents are incubating; in other words, feeding trips are much more frequent than incubation shifts. If the frequency of visits does affect the risk of predation, predation on chicks would be expected to be greater than on eggs. In self-fed species, on the other hand, the reverse would be expected, because chicks are not 'visited'. In Oystercatchers, the risk of predation on chicks relative to that on eggs would be expected to be intermediate because, although the chicks are visited, the visiting site is 'mobile'.

In Oystercatchers (Table 9.3), overall mortality (for which data are often more reliable and more readily available than for predation) and predation itself, are significantly lower for eggs than for chicks, except in the two sites with small sample sizes. When all available data are combined (Table 9.4), predation on chicks is significantly higher than on eggs. Egg predation, though, may be underestimated, since much of it happens during egg laying, and so may go undetected (Ens 1991). How-

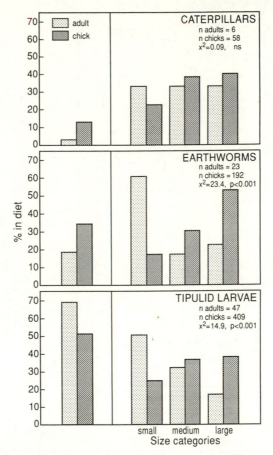

Fig. 9.4 The right-hand columns show the percentages of large, medium and small prey size categories in the terrestrial diets of adult and chick Oystercatchers in 25 families on Skokholm, Wales. Large, medium and small are > 40, 25–40 and < 25 mm respectively for earthworms, and 30–40, 20–30 and < 20 mm respectively for Tipulid larvae and caterpillars. The left-hand columns represent percentages of species in the diets, based on 192 and 784 items eaten by parents and chicks respectively. For differences between earthworms and Tipulids, $\chi^2 = 10.59$, $P < 0.01$, and between Tipulids and caterpillars, $\chi^2 = 16.03$, $P < 0.001$. Mean ranks for Tipulids and earthworms combined (large = 3, medium = 2, small = 1) are 1.64 and 2.24 for adults and chicks, respectively (Safriel 1967).

ever, predation on chicks was also significantly greater than on eggs in the Semipalmated Sandpiper (*Calidris pusillus*), whose chicks are fully precocial and hence expected to be visited by their parents less often than are the chicks of Oystercatchers. Furthermore, in the Crested Lark (*Galerida cristata*) which, like most shorebirds, is a ground-nester in open habitats, predation on the fully altricial parent-fed young is, contrary to expectation, lower than on the eggs, which are undoubtedly

Table 9.3 Chick losses and egg losses in various Oystercatcher species[1].

	% mortality		χ^2	% predation		χ^2
	Eggs	Chicks		Eggs	Chicks	
(a) *Haematopus ostralegus*						
Mellum[2]	64.8	86.4	58.03***	15–25	?	?
Ythan Valley[3]	49.7	52.1	0.25 ns	22.7	42.7	25.89***
Skokholm[4]	52.4	65.9	6.57*	38.1	54.8	15.89***
Skokholm[5]	34.3	52.2	14.06***	?	?	
NW England[6]	39.1	36.7	16.60***	?	?	
Greece[7]	55.1	44.4	4.42*	44.4	38.8	0.02 ns
Holland[8]	41.2	77.9	185.01***	?	?	
(b) Other *Haematopus*						
H. palliatus[9]	?	75.4		?	28.6	
H. moquini[10]	73.3	80.0		72.1	48.0	
H. bachmani[11]	54.5	66.7		?	?	

* $P<0.05$; *** $P<0.001$.

1 Only sites with more than one year of data are included. Sites arranged in a decreasing order of sample size.
2 Data of 1952 to 1958 combined, for 1018 eggs and 358 chicks (Schnakenwinkel 1970).
3 Data of 1966, 1967 and 1968 combined, for 429 eggs and 208 chicks of inland-breeding birds (Heppleston 1972).
4 Data of 1965 and 1966 combined, for 343 eggs (U.N. Safriel, unpublished information) and 135 chicks (Safriel 1985).
5 Data of 1963 and 1964 combined, for 283 eggs (U.N. Safriel, unpublished information) and 186 chicks (Harris 1967).
6 Data of 1967 to 1972 combined, for 197 eggs and 120 chicks (Greenhalgh 1973).
7 Data of 1980 and 1981 combined, for 178 eggs and 80 chicks (V. Goutner, unpublished information).
8 Data of 1984–1990, 'Residents' and 'Leapfrogs' combined, for 926 eggs and 545 chicks on Sciermonnikoog (Ens *et al.* 1992)
9 USA (Nol 1989).
10 South Africa (Hockey 1983a).
11 Canada, 167 clutches of 3 yrs combined (Hartwick 1974).

Table 9.4 Per cent predation on eggs and chicks, related to the degree of precociality. Species and groups are arranged in order of an increasing degree of precociality.

	% Mortality			% Predation		
	Eggs	Chicks	χ^2	Eggs	Chicks	χ^2
Crested Lark[1]	42.2	34.5	45.9**	63.2	36.9	11.1*
Great Skua[2]	30.1	9.0	180.1**	10.6	3.3	51.8**
Oystercatcher[3]	50.6	67.6	129.6**	26.0	45.8	65.1**
Semipalmated Sandpiper[4]	28.1	56.4	48.9**	28.1	56.4	48.9**

* $P < 0.01$; ** $P < 0.001$.
[1] *Galerida cristata* (Shkedy and Safriel 1992).
[2] *Stercorarius skua* (Cramp and Simmons 1983).
[3] Data from Table 9.3.
[4] *Calidris pusilla* (Safriel 1975; U.N. Safriel, unpublished information).

'visited' less frequently than the young. In the colony-nesting semi-precocial Great Skua (*Stercorarius skua*), eggs were again predated significantly more often than chicks (Table 9.4). The Crested Lark and the Great Skua are similar in that their chicks are sedentary, while both the Oystercatcher and the Semipalmated Sandpiper have mobile chicks. Thus, the available data from these between-species comparisons of the relative rates of predation on eggs to young do not support our hypothesis that feeding trips are risky. But even if they are eventually shown to be so, the behaviour of the mobile chick seems to be more critical than that of its parents in determining predation rates.

Within-species comparisons

In this section, we compare the behaviour of Oystercatchers that use different breeding tactics in the same area. 'Residents' both nest and forage within the same territory and so can keep the chicks close to where they themselves feed. In coastal sites, such as Schiermonnikoog in The Netherlands, the chicks are fed on marine food (Ens *et al.* 1992). In inland sites, such as the Ythan Valley, Scotland, they are necessarily fed on terrestrial food (Heppleston 1971b), even though they may be close to the shore, as are 'terrestrial feeders' on Skokholm, Wales (Safriel 1985). Chicks of residents follow the foraging parent quite early in their development, and well before fledging. 'Leapfrog' Oystercatchers nest in one territory but forage in another. Adults thus keep the chicks away from where they themselves forage; examples are provided by the 'coastal breeders' in Heppleston's (1971b) study in Scotland and most 'limpet feeders' on Skokholm (Safriel 1985). They transport marine food to the chicks which only start following the parents to the foraging grounds rather late in their development, usually after fledging.

Food item size

Both resident and leapfrog parents should maximize item size in order to minimize the cost of transport. But compared to residents, leapfrogs make more frequent and longer transportation trips. Both central-place foraging theory (Orians and Pearson 1979) and our predation-risk hypothesis predict that, as a result, leapfrogging birds should transport larger and more nutritious items than residents. One index of item size is the ratio of the parent to chick prey size. Since items eaten by parents have no associated transportation cost, they would be predicted to be smaller than those transported to young. In the following, we will compare the ratio of parent to chick item size in leapfrog and residents to evaluate whether leapfrog chicks do indeed receive more nutritious items than resident chicks.

(a) Leapfrogs and limpets Skokholm leapfrogs transported to chicks limpets that were 1.75 times larger in volume than those they ate themselves (Fig. 9.5). The larger limpets were also more likely to be female (rather than male or neuter), to be *Patella vulgata* (rather than *P. aspera*), and to have ripening gonads (Safriel 1967). The energy content per g ash free dry mass of a female *P. vulgata* with ripe gonads, and thus of a large limpet (25.5 ± 0.5 kJ, $n = 14$), is greater than that of a male, neuter, *P. aspera* with unripe gonads, and thus of a small limpet (24.0 ± 0.6 kJ,

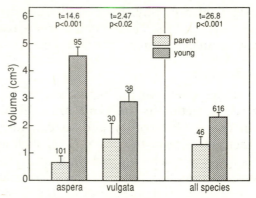

Fig. 9.5 The left-hand columns show sizes and species of limpets taken by Skokholm leapfrog Oystercatchers. The data show the mean sizes, measured as volume (cm³ ± SE), of limpets from three meals from each of seven parents and of all limpets brought to three broods over five days. The right-hand columns refer to one pair raising one chick, and comprise two meals of the parents and all the limpets brought to the chick until it fledged, with *P. aspera*, *P. vulgata* and *P. intermedia* combined. The numbers show the sample sizes with the significance of the difference between the prey of parents and young shown at the top. The linear regression equation for a limpet dry mass is logarithm_{10} dry mass (mg) = 0.131 + 0.654 × logarithm_{10} volume (mm³); the SE of the regression coefficient is 0.079, and R^2 is 0.77 (Safriel 1967).

$n = 8$, $t_{20} = 5.97$, $P < 0.001$). Therefore, as predicted, the limpet-fed chick obtained more (*circa* 1.8 times) energy per item than its limpet-feeding parent. Black Oystercatcher (*Haematopus bachmani*) young are also fed larger limpets than those eaten by their parents (Hartwick 1976).

We propose that the benefit of the large limpets to chicks is not in their high nutritious value *per se* but in minimizing the number of costly and risky transportation trips that have to be made. Otherwise, the highly nutritious large limpets should also be eaten by the parents, unless they carry a cost which makes them unprofitable for parents, a cost that is balanced only by the gain to their young.

On Skokholm, limpets of the sizes brought to young Oystercatchers comprised only 20% of the population (Safriel 1967), hence a large time-cost was indeed invested in searching for them. But it is also costly to dislodge limpets. Whereas small limpets can be dislodged at all times, a large limpet can be dislodged only when it relaxs its grip or moves. Apart from when it is raining, this happens on Skokholm only when limpets are touched by the receding tide. Compared with small limpets, large limpets can only be taken during a small fraction of the time for which they are exposed. This period of availability is greatest during neap tides when the tide remains for longer at shore-levels rich in large limpets; accordingly, birds brought to their young twice as many limpets on neap as on spring tides (Safriel 1967). That the birds have difficulties in dislodging non-moving large limpets is evident from the greater proportion of fractured shells found in anvils of limpets eaten by the parents themselves (41% of 600 limpets) compared with those found around young birds (7% of 269 limpets, $\chi^2 = 145.8$, $P < 0.001$). This difference occurred because a relaxed or moving limpet is attacked at the gap between the surface of the rock and the edge of the shell, and is thus dislodged without damage to the shell. Small limpets, on the other hand, are opened by hammering on the shell.

Usually, a foraging parent ate small limpets itself until it encountered a large one that could be dislodged, which was then transported to the young. When not foraging for the young, the parent appeared to pay more attention to the more common small limpets. This is reflected in the feeding rates, which were 1.5 times greater when parents foraged for themselves than when they hunted for the young (1.48 and 0.98 limpets per minute, respectively; $t_{136} = 2.91$, $P < 0.02$). Thus, feeding the young on the much more common, available and easier to obtain small limpets could have provided the young with more food than when feeding them with large limpets. Since this did not happen, we propose that, as assumed by our hypothesis, feeding the chicks with large limpets is an adaptation for reducing transportation costs. Though flight is costly, it is possible that the major cost over the short distances involved in frequent transportation trips is predation risk to the young rather than energy loss to the parent.

(b) Residents and terrestrial food As already discussed, Skokholm residents mainly feed on Tipulid larvae, earthworms and moth caterpillars. Residents too transport the food to their young, though over shorter distances than leapfrogs, and therefore items brought to chicks should be larger than those eaten by parents. As predicted, on Skokholm the chicks ate larger Tipulids and earthworms than did the adults (Fig. 9.3). The parents hardly ate moth caterpillars themselves: these rare prey, which had the highest energy content, were nearly all fed to the young, as would be predicted by our hypothesis.

An additional prediction of our hypothesis is that, if habitats differ in the size of the food items they provide, parents foraging for their young should select those habitats in which large items prevail. Whereas large limpets are scattered among the much commoner small limpets all over the intertidal rocks, and are visually selected by the birds, the terrestrial food is mostly concealed underground, and habitats may differ in the size distribution of their soil fauna. Thus, size selection in terrestrial food could be effected by habitat selection, allowing another test of the hypothesis.

On Skokholm, parents feed in bogs before the young hatch and again after they have fledged but, during the fledging period itself, they feed in dryer habitats, which occupy the greater part of their breeding territories. This change in feeding habitat affects the comparison between the sizes of prey eaten by parents themselves and those they take to the young. The number of prey species in the dry habitats is lower than in the bogs (Fig. 9.6a). The major prey species *Tipula paludosa* is more than seven times more numerous in bogs than in dry habitats (mean numbers of individuals m^{-2} ± SD are 139.6 + 81.6 [$n = 10$] and 19.1 ± 14.1 [$n = 13$], respectively; $t = 3.73$, $P < 0.02$). *Lumbricus rubellus*, the commonest earthworm in dry habitats, is scarcer there than the commonest earthworm in bogs, *Allolobophora chlorotica* (Safriel 1967). However, though less numerous, the prey in the dry habitat are larger. Thus, the mean length of adult *L. rubellus* is 87 mm whereas that of *A. chlorotica* is only 50 mm (Safriel 1967). Compared with those in bogs, *T. paludosa* in dry habitats are 3.7 larger in May and 2.7 times larger in July (May–bogs: 5.7 ± 5.7 mg, $n = 48$; dry habitats: 21.0 ± 13.4 mg, $n = 8$; $t = 5.39$, $P < 0.001$; June - bogs: 20.9 ± 8.3 mg, $n = 98$; dry habitats: 57.7 ± 25.5 mg, $n = 14$; $t = 10.70$, $P < 0.001$). In the same months, Tipulids and earthworms are 1.9 and 2.4 times larger, respectively (Fig. 9.6b), and are richer in energy (Fig. 9.6c), in dry habitats than in bogs. Thus, although by foraging in dry habitats the search time per prey item increases, the reward per captured item also increases.

As parents foraged in the drier habitats only during the chick-rearing period, our hypothesis correctly predicts that, when feeding young, Oystercatchers should forage where the size of the food items is large, even if their density is low. But even when foraging in such habitats, our

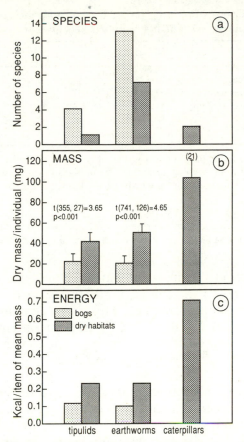

Fig. 9.6 Terrestrial food in bogs, where parents forage for themselves, and in the dry habitats, where most foraging is done for the young, on Skokholm during months of chick rearing (May–early August). (a) number of species, all samples combined; (b) mg dry mass per individual prey. Vertical lines show the SEs while the numbers in brackets give the sample size. The significance of the difference between habitats is also shown. (c) Energy content of an individual prey of a mean size (Safriel 1967).

hypothesis predicts that the foraging tactics should differ according to whether food is to be transported to the young or consumed on the spot by the parent. As predicted, hunting efficiency on Skokholm was 0.5 times lower when a parent foraged for the young than when it hunted for itself (Table 9.5). Though apparently concealed, the birds distinguished small from large terrestrial items, and ignored the small ones when foraging for the young.

To conclude, the analysis of the ratio of chick to parent food item size reveals that, as predicted, both resident and leapfrog parents provide the chicks with large items, but at the cost of increased search time. One way to compensate for this would be for the foraging parent to transfer the

Table 9.5 Efficiency of hunting on terrestrial food, measured as the mean number of items per minute ± SD. The figure in brackets gives the number of items.

	Skokholm, Wales[1]	Tipperne, Denmark[2]
Adult forages for itself	2.54 ± 1.25 (55)	
Hiding chick receives item	1.21 ± 0.71 (27)[3]	1.23 ± 0.17 (4)
Following chick receives item	3.44 ± 1.45 (14)[4]	3.70 ± 0.78 (9)
t	6.61	6.15
P	< 0.001	< 0.001

[1] Each observation lasted until the foraging bout terminated. When the last item was given to the young, all items taken in this bout, including the one given to young, were used for calculating hunting efficiency when foraging for young. In each observation, 2–28 items were captured (Safriel 1967).
[2] Food is intertidal worms (Lind 1965).
[3] Foraging for itself is significantly more efficient than foraging for hiding chick; $t = 5.11$, $P < 0.001$.
[4] Foraging for itself is significantly less efficient than foraging for following chick; $t = 2.31$, $P < 0.05$.

transported item to its mate on watch; the latter could then prepare and present the food to the chicks, while the former would be immediately released to resume foraging. This has actually been observed in Skokholm residents (Safriel 1967), in Schiermonnikoog leapfrogs and residents (where most often females gave the food to the male), as well as in Black Oystercatcher leapfrogs on Cleland Island, Canada (Hartwick 1976).

Our hypothesis also predicts that the ratio of the nutritional value of transported items to that of items eaten by parents should be larger in leapfrogs than in residents. By combining data from Fig. 9.3 and Fig. 9.6c, we calculated that the average item hunted in dry habitats and transported to a resident chick contained 1.3 times more energy than one eaten by the parent (1.247 vs. 0.925 kJ). As this value is less than the ratio of 1.8 found for leapfrogs, the prediction is supported. However, the comparison can only be made during the early part of the chick-rearing period. Leapfrog parents cannot take their chicks to the shore until they fledge, because of the rough terrain and the high density of predators there (Safriel 1985). In contrast, residents take their chicks to bogs within the territories even before they fledge. When this happens, food items are no longer given in the bill, but are extracted from the soil and placed on top for the young—which follow their parents—to pick up. This results in a 2.8 fold increase in the hunting efficiency of the foraging parent, relative to foraging in dry habitats and transporting items to a hiding chick, and in a 1.4 increase as compared to a parent foraging for itself (Table 9.5). We suspect that this increase in hunting efficiency is due to adding small items to the diet. Hence terrestrial items taken by a following young must be much smaller than those that are transported.

Are items brought by leapfrogs indeed more nutritious than those brought by residents during the early part of the fledging period? Transported food items on Skokholm, either by leapfrogs throughout the fledgling period or by residents in the first part of that period only, differ in their energy content, depending on whether they are limpets (mainly taken by leapfrogs) or terrestrial species (taken mainly by residents). The most nutritious terrestrial food items are moth caterpillars (2.989 kJ per item) and the weighted mean content of terrestrial items brought to young is 1.210 kJ per item. Using data from Fig. 9.5 and Safriel (1967), the mean dry mass of a limpet brought to young is 215 mg and has an energy content of 5.350 kJ. Thus, on Skokholm, an item brought to young by a leapfrog parent is nearly 4.5 times more nutritious than an item brought by a resident parent. As predicted by our predation-risk hypothesis, leapfrogs transport highly nutritious food items. Furthermore, though they make longer trips, they presumably make fewer transportation trips than residents, in view of the more nutritious prey items that they transport.

Growth rate and diet

If the larger food items brought by Skokholm's leapfrogs exactly compensate for their longer transportation trips, the young of both types should have similar growth rates. In fact, chicks of Skokholm residents (fed mainly on terrestrial food) weighed 4–5% more than the average chick, while chicks of leapfrogs (fed on limpets) weighed 7–10% less. When aged 28 days, 74% of the variance in chick weight was attributed to the parent's feeding tactics, irrespective of brood size (Safriel 1985). Thus, although a limpet is on average 4.4 times richer in energy than a terrestrial item given to young, limpet-fed young grow more slowly; in other words, the leapfrog parents brought a large, but still an insufficient, number of limpets. One possible beneficial consequence of this, which is not included in our model, is that by doing so, they may reduce their effort and thus increase their chances of future breeding (Drent and Daan 1980). Apparently, the evidence from Skokholm is against this, as on average, breeding effort was positively, not negatively, correlated with the parents' survivorship (Safriel *et al.* 1984). But then again, this may simply mean that high quality birds have both high fecundity and high survivorship (Safriel *et al.* 1984; Ens *et al.* 1992). At any rate, whatever may be the adaptive significance of the leapfrogs' feeding strategy, on Skokholm, their chicks did grow more slowly than those of resident pairs.

This appears to be a general finding in Oystercatchers. In Schiermonnikoog, leapfrog parents spent on average 1766 seconds per tide in transportation trips whereas resident parents spent only 551 seconds (Ens *et al.* 1992). Yet, compared with the growth rate of chicks fed *ad libitum* in the laboratory, growth was slowest in the chicks of leapfrog parents (Fig. 9.7). Chicks of residents feeding on marine food grew faster while

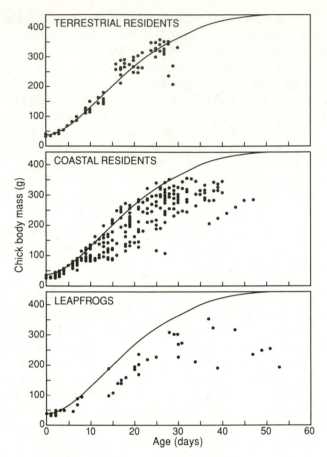

Fig. 9.7 Growth of chicks surviving to fledging on Schiermonnikoog in 1984–1985 compared with a Gompertz curve fitted to the growth of chicks raised in captivity and provided food *ad libitum* in terrestrial residents, coastal residents and leapfrogs (A. Brenninkmeijer, unpublished information).

those of terrestrial residents, which—as on Skokholm—fed their young on terrestrial food, grew at the maximum possible rate, as determined in the laboratory. Altogether, the mean growth rate of 196 leapfrog chicks was 10.2 grams per day, significantly lower than the 11.2 grams per day recorded in 112 resident chicks (t = 3.570, P < 0.001; calculated from Fig. 9.8). Similar results were obtained in the Ythan Valley, Scotland, by Heppleston (1971b). Inland chicks fed on earthworms and Tipulids received, on average, 1.9 times more energy per day than coastal chicks fed mainly on mussels. We therefore conclude that the chicks of leapfrogs grow more slowly than the chicks of residents, in both coastal and inland habitats. The larger prey size transported by the leapfrog parents does not compensate entirely for the fewer transportation trips that are made.

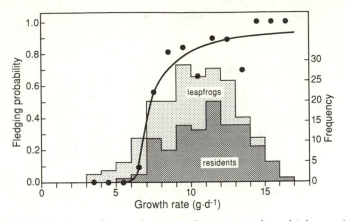

Fig. 9.8 The distribution of growth rates of Oystercatcher chicks on Schiermon-nikoog (cumulative histogram) and the probability of fledging as related to growth rate (points and curve) (Kersten and Brenninkmeijer 1995).

Length of fledging period and growth rate

The most likely consequence of reduced growth rate in the leapfrog chicks is that the chicks take longer to fledge. Fledging periods were vari-able (Fig. 9.9), and as expected, age at fledging was negatively correlated with growth rate (age = −2.0 x growth rate ± 53.6, n = 49, r^2 = 0.68; A. Brenninkmeijer, unpublished information; Fig. 9.10). On Schiermon-nikoog, mean daily gain was 11.7 grams for chicks fledging at under 30 days of age and 6.1 grams for those fledging when older than 40 days (Kersten and Brenninkmeijer 1995). Similarly, on Skokholm, the mean daily weight increase was 12.9 grams in chicks fledging at less than 32 days and 10.7 grams in those fledging at more than 35 days (Fig. 9.10). On Schiermonnikoog, the longest fledging periods were 41 days and 53 days (Kersten and Brenninkmeijer 1995) and in Skokholm 29–38 days and 30–39 days, respectively, for residents and leapfrogs (U.N. Safriel, unpublished information). Thus, whatever the local average rate of chick growth, the average leapfrog chick grew more slowly and fledged later than the average chick of resident birds.

Starvation and predation

We would expect these differences in growth rate and fledging period to result in survivorship in leapfrog chicks being lower than in residents (Fig. 9.8), either because of higher predation, caused by the longer fledging period and greater number of parental trips, or because of starvation. This is indeed so (Fig. 9.11). On Skokholm, a terrestrial feeder produced 3.3 times more fledged young than a limpet feeder; in Aberdeenshire, an inland breeder produced 1.9 times more fledged young than a shore-breeder; and on Schiermonnikoog, a resident produced 3.2 more fledged

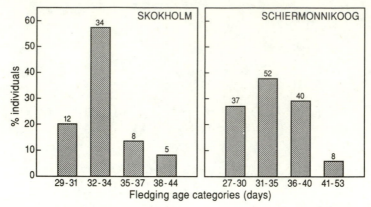

Fig. 9.9 Frequency distribution of fledging ages on Skokholm and Schiermonnikoog; the numbers show the numbers of individuals in each category. On Skokholm, only one diseased chick fledged at 44 days of age (Safriel and Harris 1985). Data for Schiermonnikoog from Kersten and Brenninkmeijer (1995).

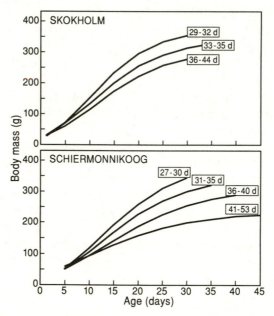

Fig. 9.10 Growth curves of Oystercatcher chicks of different fledging ages, fitted by Gompertz equation, using Ricklefs' (1967) graphical method. (*Top*) 49 chicks on Skokholm in 1965–1966. *k*-values for *A* = 466 (Drent and Klaassen 1989) are 0.089, 0.075 and 0.076 for the three fledging-age classes, respectively. Linear regression coefficients ± SE and number of chicks (n) for the three fledging-age classes are 12.88 ± 0.38 (20), 11.16 ± 0.46 (19), and 10.66 ± 0.47 (7), respectively. (*Bottom*) 139 chicks on Schiermonnikoog in 1985–1991 (Kersten and Brenninkmeijer 1995). *k*-values for *A* = 466 are 0.071, 0.047, 0.045, and 0.027, for the four fledging-age classes, respectively.

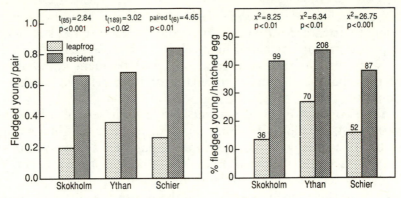

Fig. 9.11 Differences between fledging rates in leapfrog and resident Oystercatchers, expressed as the number of young fledged per pair (left) and the per cent of young fledged per hatched egg. Skokholm data from Safriel (1985) (leapfrogs are limpet feeders; residents are terrestrial feeders). Ythan data from Heppleston (1971b), Table 6 (leapfrogs and residents are shore and inland breeders, respectively). Schiermonnikoog data are from Ens *et al.* (1992) for 1984–1989 and Kersten and Brenninkmeijer (1995), based on 548 hatched eggs. The significance of the difference between leapfrogs and residents is shown along the top. The number on the top of columns gives the number of nests. In Skokholm, 11.1% of hatched resident and leapfrog eggs died of causes other than predation; $\chi^2 = 1.16$, no significant difference between residents and leapfrogs.

young than a leapfrog (Fig. 9.11a). These differences were due entirely to differences in the survival of the young rather than to differences in egg survival (Fig. 9.11b).

Differences in survival between the chicks of leapfrog and resident parents were due to several factors. On Skokholm, the difference over much of the fledging period was associated with predation. The limpet-fed young of leapfrog parents on Skokholm were predated 1.6 more often than terrestrial-fed chicks, while the proportions of those dying from other causes were similar (Fig. 9.11). Mortality other than through predation occurred immediately after hatching. Mortality later in the fledging period was mostly due to a viral disease (Safriel 1982; Safriel and Harris 1985). On Schiermonnikoog, leapfrog chicks grew better when supplied with extra food, and most mortality of old chicks was due directly to starvation (Ens *et al.* 1992). But even where chicks were predated, malnutrition undoubtedly played a part. In field experiments on Skokholm (Safriel 1981) and Ravenglass (Norton-Griffiths 1969), chicks starved to death when predators were excluded. Without protection from predators, these chicks would undoubtedly have been predated before starving (Safriel 1985). In these experiments, as in the wild, a social hierarchy among siblings develops in which the dominants grow faster than subordinates. Subordinates are more restless and careless, especially when under-weight, as also occurs in the American Black

Oystercatcher (*H. palliatus*) (Groves 1984), and are consequently more likely to attract the attention of predators (Safriel 1981, 1985). It would appear that leapfrog subordinate chicks are more prone to predation, not only because of the relatively long and frequent transportation trips of their parents, but also by virtue of their relative carelessness when hungry.

This conclusion is supported by a number of observations. On Skokholm, the chicks of residents spent most of the time within eyesight of both parents, whereas the leapfrog parent was always out of sight of the chicks while it foraged on the shore (Safriel 1985). In the Ythan Valley (Heppleston 1971b), inland and coastal parents spent 95.5% and 83.5% of their time respectively in the nesting territory, with their chicks remaining unattended for 0.5% and 3.8% of the time respectively ($\chi^2 = 388.3$, $P < 0.001$, and $\chi^2 = 5.51$, $P < 0.05$, for both measures). Shore-breeding parents had 17.5 times more aggressive interactions with predators than inland parents, apparently because of the higher density of predators (Heppleston 1971b). But predators may also have made more attacks against Oystercatchers in shore territories if chicks were more careless there and the frequent transportation trips drew their attention to the whereabouts of the chicks. On Skokholm, male parents were more effective than females when on watch (Safriel 1967), but leapfrog males spent less time on watch than resident males; terrestrial- and limpet-feeding males brought, respectively, an average of 16% and 39% of the food items eaten by the young. Thus, the limpet-eating leapfrog males spent relatively more time on provisioning their young than on protecting them.

These observations provide further support for the idea that leapfrog chicks survive less well than those of residents because they are more vulnerable to predation. However, it is not only the frequent visitation by parents that attracts predators' attention, but also the reduced quality of the protection along with the carelessness of the chicks themselves when they are hungry and require more frequent feeding. As we noted earlier, the number of feeding trips can be reduced, and the nutritional condition of the young improved, if the individual food items brought to young are highly nutritious. However, even though leapfrog chicks receive highly nutritious items, they nonetheless survive less well than residents chicks, being in poorer condition and more vulnerable to predation. There must, therefore, be some other feature of the residents' breeding tactics that makes their chicks more successful than those of leapfrogs, and this is now explored.

From being parent-fed to self-feeding

Besides a possible increase in noise level that may attract predators, there is nothing in the behaviour of unfledged altricial chicks that affects their

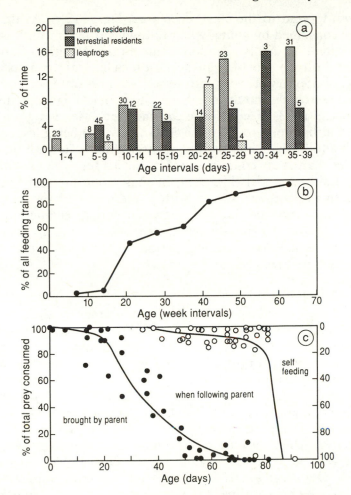

Fig. 9.12 The development in the following behaviour of Oystercatcher chicks as they grow older. (a) Per cent of the total time budget that the chick spent in following the feeding parent in marine residents, terrestrial residents and leapfrogs (M. Klaassen, unpublished information). The number of records is shown above the bars; no columns = no observations. (b) Per cent of the total number of parents' 'feeding trains' at Ravenglass, Cumberland carried out when the chick was following the parent (Norton-Griffiths 1969). (c) Per cent of total prey consumed by the chick when food is transported to them by the parent, when the chick follows the parent and when the chick feeds itself (M. Klaassen, unpublished information).

predation risk as they grow. But unfledged Oystercatcher chicks, though fed by parents, become more mobile and improve their alertness as they increase in size. Gradually, increased vigilance and strength enable chicks to cease hiding and to follow the foraging parent closely (Fig. 9.12a). Accordingly, the proportion of feeding sequences in which the young follows a parent increases with age, while the proportion in which an item is transported to the chick decreases (Fig. 9.12b). As the proportion of prey consumed by young when following a parent increases (Fig. 9.12c), the parent gradually abandons central-place foraging. Captured items are offered on the spot so transportation time, with its associated predation risk, disappears. Theory states that, with no transportation or predation cost, searching should not be limited to just large items (Lessells and Stephens 1983). Feeding rate can then increase, and growth rate accelerate. Furthermore, to increase efficiency when being less selective, parents with a following chick should also select habitats with high food density, rather than areas containing rarer, but larger, prey. According to Safriel and Kaiser's model (Fig. 9.1), in these conditions of high food density and small food items, and with a reduction of the parent/young prey item size ratio, the threshold between parent-feeding and self-feeding is readily crossed. The sooner central-place foraging is abandoned, the shorter should the fledging period be and the faster the parent- to self-feeding threshold can be crossed.

Feeding with parents

Are these expectations actually borne out by field studies and do residents and leapfrogs differ in any of these respects? Field studies confirm the increased efficiency associated with feeding an accompanying chick compared with transporting food to it. When hunting for a following young, Skokholm parents offered the young 88.1% of the items they captured, compared with only 44.1% when they transported food to hiding young (t_{293} = 4.90, P < 0.001; Safriel 1967). On Skokholm and in Tipperne, Denmark, parents were 2.8–3.0 times more efficient, i.e., less selective, when they were hunting for a following young than for one that was hiding some distance away (Table 9.5). Similarly, following Black Oystercatcher chicks were fed 4.4 more limpets and 1.3 times more flesh per unit time than were hiding chicks (Hartwick 1976).

As would be expected from the model, unfledged chicks of coastal residents living next to a mudflat with high food density on Schiermonnikoog increased the time they spent following adults more rapidly than terrestrial residents living in meadows far from the shore and with a lower food density. Leapfrogs that defended a mudflat distant from the breeding ground spent even less time in following (Fig. 9.12a). After fledging, between the age of 70 and 80 days, nearly all food was obtained by resident and leapfrog chicks alike while following (Fig. 9.12c).

Differences between the growth rates of chicks being reared in these different circumstances became obvious at the age of 10–15 days (Fig. 9.10), and were perhaps related to the differences in the rates of acquiring the following habit and abandoning central-place foraging. In Schiermonnikoog, the survival of the chicks of resident and leapfrog parents hardly differed before they reached the age of 10 days. Later, when only resident chicks were able to increase the time spent following

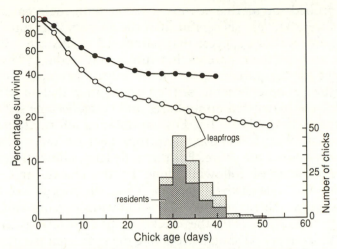

Fig. 9.13 Chick survivorship for residents and leapfrogs in relation to age, Schiermonnikoog 1985–1991 (Kersten and Brenninkmeijer 1995). The cumulative histogram shows the fledging age for residents and leapfrogs.

parents, their mean survival stabilized at circa 40% at around their 20th day. In contrast, leapfrog chicks simply could not follow the parents when they flew to the mudflats, and their survival dropped to less than 20% by their 50th day (Fig. 9.13). Clearly, delayed self-feeding is associated with reduced chances of survival, and represents a severe penalty that is suffered by leapfrog Oystercatchers.

Becoming independent

When and how does self-feeding replace parent-feeding? Self-feeding requires a certain proficiency in manoeuvring the bill and detecting the food. Oystercatcher chicks have the smallest optic lobes among nine shorebird species examined, probably because they do not hunt for themselves (Nol 1986). Self-feeding would thus be expected to await the development of the optical lobes, although the rate of the development may depend on overall growth rate.

Unfledged chicks often 'exercise', or practice, self-feeding by collecting items from the surface and by probing. In Schiermonnikoog, chicks 'foraged' for grit and occasional insects (B.J. Ens, unpublished information) for approximately 1% of their time during their first few days and for 10%–15% of the time from the age of 15 days and until fledging (M. Klaassen, unpublished information). In England, chicks 10 days old collected beetles and spiders, when 'foraging' or when following, and picked larvae from inside cows' dung pats after the parents had broken the crust (K.B. Briggs, personal communication). In Denmark, parents bored holes in the sand and chicks older than 35 day picked up Nereid worms from the hole (Lind 1965). At Ravenglass, chicks aged 15 to 30 days probed for marine worms together with their parents, though the chicks were far less successful than the parents (Norton-Griffiths 1968). These observations suggest that parent-fed chicks are 'trained' by themselves and by their parents for future self-feeding. The kind of food brought by the parents also has an effect. Parent-fed chicks that find items by probing became self-fed soon after fledging, whereas chicks which were parent fed on molluscs and crabs became self-fed much later (Norton-Griffiths 1969; B.J. Ens unpublished information).

Most mollusc feeders are leapfrogs because their shellfish prey live some way down the shore. For them, fledging allows a dramatic increase in the amount of following done by the chicks, and an end to the leapfrog life style of their parents. Thus, the shorter the fledging period of leapfrogs, the earlier they start following parents and begin to learn feeding techniques. For the chicks of both leapfrogs and residents, a short fledging period should accelerate the shift from parent- to self-feeding. Mass at fledging increases with growth rate, and since fledging age decreases with growth rate, fledging mass is higher when the fledging period is shorter (Fig. 9.10). Although chicks that fledged with a large mass were more likely to return to the natal areas than chicks that fledged with a small mass, this trend was not significant (Kersten and Brenninkmeijer 1995). This may have been due to small sample sizes, and furthermore, mass at fledging may also determine the rate of acquiring self-feeding skills and the onset of full independence.

The evolutionary stability of parent-feeding in Oystercatchers

The fundamental reason why parent-feeding in Oystercatchers is advantageous over the semi-precocial and the self-feeding precocial states found in other birds lies in the adaptations for exploiting very large food items. But this rather rare state could prove to be evolutionary unstable. Environmental changes, such as the enrichment of suitable habitats due to the intensification of agriculture and their expansion at the expense of other habitats unsuitable for Oystercatchers, may select for more common and possibly more stable states along the altricital–precocial gradient.

Where predation pressure has increased recently, leapfrog Oystercatcher chicks become so restricted in their movements that in some cases they resemble semi-precocial gulls (Safriel 1967). However, the Oystercatcher may not be able rapidly to adopt the coloniality or multiple-prey loading that enable gulls to rear successfully their semi-precocial young. On the other hand, it would seem to be but a short step for residents breeding inland to adopt successfully the more common state of self-feeding that occurs in most Charadrii. If so, the recent trend for breeding Oystercatchers to spread inland may be strengthened still further, and may even reach the point at which this species becomes primarily an inland, rather than coastal, breeder.

Summary

The Oystercatcher chick is precocial but nonethless depends on parental feeding, at least until fledging. This is a very rare state among birds. It is intermediate between the more common states of precociality with restricted mobility and full parental feeding, as found in gulls, and full mobility with partial parental feeding, as occurs in several groups, including Charadriiformes. This chapter investigates how the Oystercatcher may have come to occupy such a unique position in the adaptive landscape.

Although Oystercatcher chicks hide most of the time, the food-transporting parent may attract the attention of predators to the hiding places. The evidence for this is still equivocal, but self-feeding would be risky too; though cryptic, movement by the foraging chick does make it conspicuous. A mathematical model that explores the trade-off between these two types of predation risk predicts that when parents feed their young with items that are very much larger than the young could obtain for themselves, parent-feeding is less risky than self-feeding over a very wide range of prey sizes and densities. We propose that it is the difference in bill size between parent and chick, together with feeding specializations that enable parents to capture very large items, which make parent-feeding advantageous in Oystercatchers.

Once the trade-off in predation risk between parent-feeding and self-feeding has determined the overall strategy, the tactics of rearing the young are determined by the trade-off in predation risk between the length of the fledging period and the number of transportation trips made by parents. Predation risk may be reduced by either minimizing the length and number of transportation trips, or by shortening the fledging period via efficient food provisioning. The Oystercatcher's foraging tactics should be further explored as an optimal solution to the problem of maximizing the chicks' growth rate, while minimizing their predation risk as they grow.

As predicted by the model: (a) Oystercatcher chicks are fed much larger items and grow faster than self-fed Charadrii chicks, and their fast growth shortens their fledging period, thus increasing their survivorship; and (b) chicks of Oystercatcher pairs that nest some way from where they feed ('leapfrogs') are fed with larger, more nutritious items than chicks of pairs that nest and feed in contiguous areas ('residents'). Yet, leapfrog young still grow more slowly and have a longer fledging period than do chicks of residents. Inspite of the presumed smaller number of trips, and in accordance with the expected effect of a long fledging period, leapfrog chicks suffer higher mortality than residents. This is due to predation, or to starvation where predators are uncommon, or to a combination of the two; starving chicks run a higher risk of being predated.

As chicks grow, food transportation to a hiding chick is replaced by the fast provisioning of young that accompany the foraging parent. Later on, parent-feeding is gradually replaced by self-feeding. Both these shifts, that result in a shorter transportation time, occur earlier in residents than in leapfrogs. The fact that resident chicks survive better than leapfrog chicks suggests that transportation is costly, possibly both in terms of overall energy provision and in terms of predation risk, hence a reduction in transportation time is highly adaptive. It is proposed therefore, that the unique strategy of young-rearing in Oystercatchers may not be evolutionarily stable under the current environmental changes that affect Oystercatcher habitats.

10 Haematopus ostralegus *in perspective: comparisons with other Oystercatchers*

Philip A. R. Hockey

Introduction

The primary aim of this book is to provide a scientific interpretation of our knowledge of the Eurasian Oystercatcher, one of the best-studied shorebirds in the world. There are, however, 11 other species of oyster-catchers worldwide. None has been as well studied as the Eurasian Oystercatcher, but some have been studied in different ways and to address different questions. The aim of this chapter is to provide an overview of how facets of the biology of *Haematopus ostralegus* compare or contrast with other oystercatcher species, and to consider what the reasons for these similarities or differences might be.

Taxonomy and distribution

Oystercatchers occur along many of the world's coasts, and inland in some areas, but do not breed in central Africa, southern Asia or at very high latitudes. Despite their panmictic distribution, and the existence of two colour phases, one pied and one black, there has been little morpho-logical divergence within the genus, and the systematics of *Haematopus* are not fully resolved. In general, discrete populations of black morphs have been treated as full species, Old World pied morphs have been treated as subspecies of *H. ostralegus*, and New World pied forms, with the exception of *H. leucopodus* of South America, as subspecies of *H. palliatus*. One species, *H. unicolor* of New Zealand, is polymorphic, containing black, pied and intermediate colour morphs.

Attempts have been made to unravel the phylogeny of the Charadri-iformes based on the colour patterns of the downy young (Jehl 1968), character compatibility analysis (Strauch 1978) and protein electro-phoresis (Christian *et al.* 1992). Cracraft (1981) attempted a phylogenetic classification of recent birds, and Sibley and Ahlquist (1990) have sub-sequently made major revisions to bird classification based on DNA–DNA hybridization. All of the above studies conclude that the Haematopodidae and the Recurvirostridae (avocets and stilts) are closely related. Cracraft included them within the superfamily Haematopodoidea and Sibley and Ahlquist placed them within the tribe Haematopodini. While it is not the purpose here to attempt a taxonomic revision of the Haematopodidae, we need some baseline classification in order to compare behaviour and life-history characteristics between taxa. What follows is a brief evaluation of recent taxonomic treatments of oystercatchers and a summary of broad-scale distribution patterns.

Nearctic and Neotropical Oystercatchers

With the exceptions of *H. leucopodus* and *H. ater*, Peters (1934) classified all New World Oystercatchers as subspecies of *H. ostralegus*. Subsequently, Heppleston (1973) added *H. leucopodus* as a subspecies of *H. ostralegus*. Baker (1973) ascribed specific status to *H. bachmani*, and classified all pied forms except *H. leucopodus* as subspecies of *H. palliatus*. Justification for separating New and Old World pied forms at the specific level was provided by Wetmore (1965) on the basis of eye colour and dorsal coloration. Baker (1973) recognized five subspecies of *H. palliatus*, including *H. p. frazari*, which is a hybrid swarm between *H. bachmani* and *H. palliatus* and thus has no valid name. It appears, however, that none of the mainland or West Indian *H. palliatus* subspecies recognized by Baker is sufficiently separable on mensural grounds to satisfy the seventy-five per cent rule of Amadon (1949) and, for the purpose of this chapter, all are referred to the nominate subspecies. The race on the Galapagos Islands *H. p. galapagensis* is sedentary and very isolated from congeners; its subspecific status probably should be retained.

Palearctic Oystercatchers

Peters (1934) and Heppleston (1973) classified all Palearctic oystercatchers as subspecies of *H. ostralegus*, and Hockey (1982) subsequently presented evidence for the specific status of the probably extinct *H. meadewaldoi* of the Canary Islands. There are three populations of pied oystercatchers in the Palearctic which have disjunct breeding and non-breeding distributions. The modern consensus (e.g. Cramp and Simmons 1983) is that these populations are subspecifically distinct—*H. o. ostralegus* in the west, *H. o. longipes* in the central regions and *H. o. osculans* in the east.

Afrotropical Oystercatchers

Haematopus moquini of southern Africa is the only oystercatcher which breeds in the Afrotropics. Since *H. meadewaldoi* has been accorded specific status, *H. moquini* should be treated binomially, not trinomially as by Baker (1973).

Australasian Oystercatchers

The taxonomy of Australasian Oystercatchers has been reviewed by Baker (1974a, 1975b, 1977) and McKean (1978). Regarding the Australian forms, *H. longirostris* is phenetically similar to New World pied oystercatchers, differing from Palearctic *H. ostralegus* subspecies in wing pattern and lack of an eclipse plumage. On these grounds it probably warrants specific status (Baker 1977; McKean 1978). The black oystercatchers of Australia present more of a taxonomic problem. Generally treated as subspecies of *H. fuliginosus* (nominate *H. f. fuliginosus* and *H. f. ophthalmicus*), they are phenetically similar to one another (Baker 1977). However, McKean (1978) points out that *H. f. ophthalmicus* differs from *H. f. fuliginosus* in bill shape and in having a prominent fleshy red orbital ring. Additionally, the ranges of the two forms are discrete, with *H. f. ophthalmicus* having a tropical, and *H. f. fuliginosus* a subtropical distribution, with no evidence of hybridization between them. In this review they are considered specifically distinct.

Concerning the New Zealand forms, there is no good reason to question the conclusions of Baker (1974a) that *H. unicolor* and *H. chathamensis* be accorded specific status and that *finschi* be classified as a subspecies of *H. ostralegus*. In addition to being morphologically similar to Palearctic *H. ostralegus* forms, *H. o. finschi* shares with them the inland breeding habit.

On the basis of the above, the following classification of oystercatchers is used in this review.

Nearctic and Neotropical Regions
> *H. bachmani* [1]
> *H. ater* [1]
> *H. leucopodus* [3]
> *H. p. palliatus* [3]
> *H. p. galapagensis* [3]

Palearctic Region
> *H. meadewaldoi* [1]
> *H. o. ostralegus* [3]
> *H. o. longipes* [3]
> *H. o. osculans* [3]

Afrotropical Region
> *H. moquini* [1]

Australasian Region
> *H. fuliginosus* [1]
> *H. ophthalmicus* [1]
> *H. unicolor* [2]
> *H. longirostris* [3]
> *H. chathamensis* [3]
> *H. o. finschi* [3]

1 = black form, 2 = variable, 3 = pied.

At the species level, Sibley and Monroe (1990) differ from this classification in according full species status to *finschi*, but not to *chathamensis* (= *unicolor*) or *ophthalmicus* (= *fuliginosus*).

Perspectives on distribution: Eurasian or Gondwanaland origins?

The breeding and main nonbreeding ranges of the 16 taxa listed above are illustrated in Fig. 10.1 and Fig. 10.2. Records of vagrancy are not included. With the exception of Palearctic *H. ostralegus* subspecies and, to a lesser extent *H. bachmani*, *H. leucopodus* and *H. o. finschi*, the world's oystercatchers are essentially non-migratory. The four northern hemisphere migrants all have their breeding ranges centred north of 40°N, and the most northerly breeders, *H. o. ostralegus* and *H. o. longipes*, migrate the farthest (Fig. 10.3). *H. o. longipes* and *H. o. osculans* are the only oystercatchers whose breeding and nonbreeding ranges are totally disjunct. In the Old World, there are no breeding oystercatchers in central Africa and southern Asia, although they breed to the north and south of these areas. There is no such latitudinal polarization in the New World. The reason for this distributional gap is unclear, especially as there is no comparable gap in the distributions of the closely related Recurvirostridae.

Larson (1957) proposed that oystercatchers originated as dark-plumaged birds in Eurasia. Emigrants from this stock moved south during the Pliocene and established themselves as isolated new species in Australia, New Zealand, southern Africa and South America. Subsequently, most of the original northern stock mutated to pied forms and a second

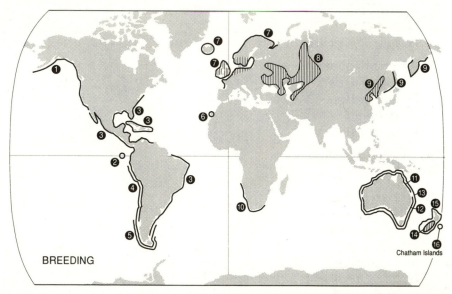

Fig. 10.1 Breeding distributions of the world's oystercatchers. Numbers refer to the following taxa. 1 = *H. bachmani*, 2 = *H. p. galapagensis*, 3 = *H. p. palliatus*, 4 = *H. ater*, 5 = *H. leucopodus*, 6 = *H. meadewaldoi*, 7 = *H. o. ostralegus*, 8 = *H. o. longipes*, 9 = *H. o. osculans*, 10 = *H. moquini*, 11 = *H. ophthalmicus*, 12 = *H. fuliginosus*, 13 = *H. longirostris*, 14 = *H. o. finschi*, 15 = *H. unicolor*, 16 = *H. chathamensis*.

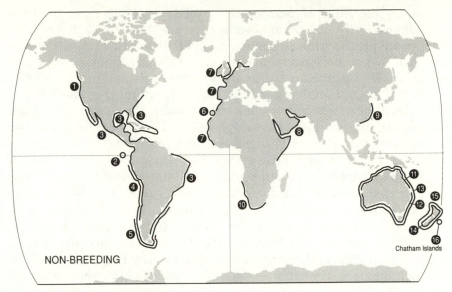

Fig. 10.2 Nonbreeding distributions of the world's oystercatchers. Numbers refer to the same taxa as Fig. 10.1.

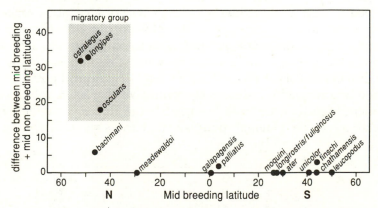

Fig. 10.3 Migration distance (° latitude) of oystercatcher species and subspecies as a function of mid-breeding latitudes. Migration distance was calculated as the difference between the mid-latitudes of the breeding and nonbreeding ranges.

southerly movement, this time of pied forms, occurred during the Pleistocene. Pied birds settled as secondary immigrants, sometimes in areas where black forms had already settled, but were able to coexist owing to evolutionary specialization that had taken place in the older black forms.

The mechanistic assumptions in this hypothesis cannot be verified, but the argument seems flawed on several grounds, not least that of niche contraction by black taxa in the absence of congeneric competitors. It seems likely that *H. leucopodus*, a pied South American species which is behaviourally unique in several respects (Miller and Baker 1980), is a prototypical oystercatcher which split off early from the main lineage (A.J. Baker personal communication). Apart from behavioural differences, *H. leucopodus* has its own taxon of feather louse, lacks brownish dorsal coloration, has a yellow, not red, orbital ring and has underwing coverts of a different colour to the belly. The latter two features are unique among the Haematopodidae. The oldest oystercatcher fossil known, from Lower Pliocene deposits in North Carolina, shows some close affinities with present-day *H. leucopodus* (Olson and Steadman 1979). Apart from *H. leucopodus*, all New World Oystercatchers, black and pied, have brown dorsal coloration and all Old World species have black dorsal coloration. This cannot be explained by Larson's Pliocene–Pleistocene migration model, and his initial assumption of a northern origin of the Haematopodidae is called into question.

The diversity of oystercatcher species increases from north to south. Including the probably extinct *H. meadewaldoi*, only four species of oystercatcher breed in the northern hemisphere, in contrast to the 10 which breed in the southern (Fig. 10.4). Two of the three extant northern hemisphere species are migratory, whereas none of the southern hemisphere species moves more than a few degrees of latitude between the breeding and nonbreeding seasons. These observations suggest that, contrary to Larson's hypothesis, the Haematopodidae have a Gondwanaland origin. Consistent differences between New and Old World forms in dorsal plumage and eye colour suggest that these characters diverged following the separation of South America from the rest of Gondwanaland and, if pied plumage is the primitive character, that melanism arose independently in the two lineages. If the Recurvirostridae are as closely related to the Haematopodidae as Cracraft's (1981) and Sibley and

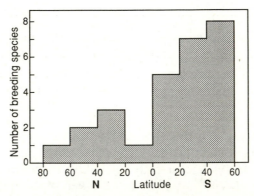

Fig. 10.4 The number of oystercatcher species breeding in 20° latitude bands. Any species breeding in more than one latitude band is recorded as present in each of the bands where it occurs.

Ahlquist's (1990) studies conclude, it would be predicted that a comparable pattern of distribution and diversity should exist within this family. Although the Recurvirostridae are not as diverse as the Haematopodidae, six of the seven species breed in the southern hemisphere, whereas only three species breed in the northern. Of the 12 oystercatcher species, 50% breed in Australasia, as do four (57%) of the Recurvirostridae. Baker's (1977) contention that *H. o. finschi* of New Zealand is a Palearctic element in the Australasian fauna should perhaps be inverted, and the three subspecies of *ostralegus* which breed in the Palearctic could be considered as southern elements in the Palearctic fauna.

The coasts of southern South America are the only part of the world where three species of oystercatcher (*H. leucopodus, H. ater* and *H. palliatus*) occur sympatrically. Two species co-occur in central South America, Australia and New Zealand. In these latter areas, one species is a black or polymorphic rocky-shore specialist and the other is pied and forages preferentially on soft substrata. In southern South America, these rôles are filled by *H. ater* and *H. palliatus* respectively, while *H. leucopodus* generally forages in inland and marshy habitats. In areas where only one species occurs, there is no general rule as to whether it will be a rocky- or sandy-shore specialist. *H. palliatus* and the three Palearctic *H. ostralegus* subspecies generally, but not invariably, forage on soft substrata, whereas *H. bachmani, H. p. galapagensis, H. moquini, H. meadewaldoi* and *H. chathamensis* prefer rocky shores (Table 10.1).

Habitat partitioning by sympatric species in New Zealand is slightly different to other areas. It appears that *H. unicolor* is undergoing a shift in habitat usage. At present, it feeds on rocky shores to a greater extent than the sympatric *H. o. finschi*, but it has been suggested that this is due to competitive exclusion of *H. unicolor* from soft substrata by the rapidly increasing numbers of *H. o. finschi* (Baker 1974a). The very long bill of *H. unicolor*, approximately equal in length to that of *H. o. finschi*, places this species with the soft substratum foragers (Table 10.2).

The problem of sympatric speciation

Although geographic patterns of species diversity suggest that oystercatchers have a southern origin, and there is some niche differentiation (sandy *versus* rocky habitats) between sympatric species, there is a difficulty in defining the mechanism for sympatric speciation, particularly as some species occasionally hybridize (e.g. *H. bachmani* with *H. palliatus*, and *H. fuliginosus* with *H. longirostris*). The genus is morphologically conservative but is catholic with respect to habitat choice. Because oystercatchers are predominantly coastal, and coastlines are essentially linear, unidimensional habitats, the probability of new isolates becoming established beyond an existing species' border is perhaps lower than for species which occupy two-dimensional terrestrial habitats. Oystercatchers do exhibit fidelity to natal site. It is conceivable that sympatric speciation

Table 10.1. Movements, habitat choice, age of first breeding , mate fidelity and sex rôles during breeding of the world's oystercatchers.

Species	Migratory?	Nonbreeding flocks?	Nonbreeding habitat	Breeding habitat	Breeding age	Mate fidelity	Sex roles
H. bachmani	No[1]	2–50[2]	Mudflats, rocky shores[3]	Rocky shores, islands[3,1]	3[4]	Long term[5,6]	Equal investment[5,7]
H. p. palliatus	Only in NE of range[8,9]	Yes[9]	Mudflats, sandy beaches[10]	Saltmarshes, sandy beaches[11]		Long term[12]	Complementary[13]
H. p. galapagensis	No[4]		Coastal[14]	Coastal[14]			
H. leucopodus	Partly[15,16]	Yes[15,17] 100s–1000s[5]	Beaches, adjacent upland, freshwater ponds[10,17,18]	Beaches, inland grass flats, freshwater lagoons[16,17,18]			
H. ater	Mostly no[17], some to Isla Grande[16]	Small groups 4–5[18] largest 8[17]	Shingle and rocky coasts[16,17]	Pebble and shell beaches[19]		At least through year[18]	Both defend young[20]
H. o. ostralegus	Yes[21]	Feed 1–10 Roost 25–100, up to 1000s[20]	Mudflats, saltmarshes, sandy and rocky beaches[10]	Saltmarshes, sand and shingle beaches, riparian, agrarian 22	4–5[23] 3.5[20]	Long term[20]	Both tend young[20]
H. o. longipes	Yes[24]	Small groups <40[25,26]	Mudflats, sandy and rocky shores[26]	Inland riparian, fresh and saline[24] lakes		'live in pairs'[7]	Both feed young[27]
H. o. osculans	Yes[24]		Coastal[28]	Sea coasts, lakeshores [24]			
H. meadewaldoi	No[29]		Rocky and sandy shores [29]	Rocky and sandy shores [29]			
H. moquini	No[30]	Small groups[30]	Rocky and sandy shores[30]	Rocky and sandy shores [9]	3–4[46]	Long term[36]	Both incubate and defend nest/young[32]

H. longirostris	No[21,33]	100s[21]	Mudflats, beaches, estuaries[21,34]	Saltmarshes, sandy beaches, dunes, pastures[34]	4[35] 4–6[33]	Long term[33,34]
H. fuliginosus	No[21]	Small groups[36]	Mudflats, sandy beaches[34]	Sand/shingle beaches, rocky shores and islands[34]		Yes[40] — Both feed young[37]
H. ophthalmicus	No[34]					
H. unicolor	No[21]	Occasionally, up to 150[38]	Rocky and sandy shores[37]	Sandy beaches, dunes[34]	3[39]	Through year[39] — Both incubate, involved in parental care[40,41,42,43]
H. chathamensis	No[39]	Stay on territory or flock[39]	Mostly rocky shores, also sandy beaches[39]	As non-breeding		
H. o. finschi	Yes[39]	Large flocks[38]	Coasts and estuaries, not rocky areas[21,38]	Inland, grass fields, riparian, agrarian[21,38]	2–3[44]	Questionable[39]

Data sources. 1: Webster 1941a, 2: Kenyon 1949, 3: Hartwick and Blaylock 1979, 4: Dircksen 1932, 5: Purdy and Miller 1988, 6: Harris 1967, 7: Purdy 1985, 8: Bent 1927, 9: Baker and Cadman 1980, 10: Burger 1984, 11: Lauro and Burger 1989, 12: Nol 1984, 13: Nol 1985, 14: Lévêque 1964, 15: Jehl 1978, 16: Humphrey *et al.* 1970, 17: Woods 1975, 18: Johnsgard 1981, 19: Zapata 1967, 20: Cramp and Simmons 1983, 21: Hayman *et al.* 1986, 22: Briggs 1984b, 23: Harris 1970, 24: Dement'ev and Gladkov 1969, 25: Bundy *et al.* 1989, 26: Meinertzhagen 1954, 27: Gavin *et al.* 1962, 28: Etchécopar and Hüe 1978, 29: Collar and Stuart 1985, 30: Hockey 1983b, 31: Summers and Cooper 1977, 32: Hall 1959, 33: Newman 1989, 34: Lane and Davies 1987, 35: Blakers *et al.* 1984, 36: Reader's Digest 1976, 37: Wakefield and Robertson 1988, 38: Baker 1973, 39: Falla *et al.* 1979, 40: Baker 1974a, 41: Braithwaite 1950, 42: Jones 1979a, 43: Watt 1955, 44: Warheit *et al.* 1984, 45: Sibson 1966, 46: P. A. R. Hockey, unpublished information.

Table 10.2. Summary biometrics of the world's oystercatchers.

Species	Sex	Mass Range (g)	Mass Average (g)	Wing length Range (mm)	Wing length Average (mm)	Culmen length Range (mm)	Culmen length Average (mm)
H. bachmani	M	555–648	607[1]	241–267	251.7[2]	64–75	69.1[2]
	F	618–750	689[1]	251–268	257.1[2]	70–83	75.6[2]
H. p. palliatus	M		567±113[3]	241–270	254.4[4]	76.5–89	83[4]
	F		638±42[3]	244–280	259.5[4]	84–104	92.3[4]
H. p. galapagensis	M	No data		241–259	250.9[2]	78.7–86	82.1[2]
	F			241–259	253.5[2]	77.5–87.6	82[2]
H. leucopodus	M	585–610	597.5[5]	237–259	251.9[4]	68–76	73.2[4]
	F	585–700	632[5]	244–260	251.5[4]	72–85	79[4]
H. ater	M	700–708	702[5]	253–269	261.5[4]	68–74.5	71[4]
	F	585–700	642.5[5]	251–280	267.7[4]	72–84	79.5[4]
H. o. ostralegus	M	425–805	594[6,7]	241–270	255.0[8]	67–86	75.8[8]
	F	445–820	608[6,7]	240–276	255.7[8]	70–90	81.2[8]
H. o. longipes	M		623[9]	246–261	256.4[8]	69–91.8	78.2[8]
	F	384–498	436[9]	247–273	263.4[8]	66.5–101	88.7[8]
H. o. osculans	M			unsexed: 230–280[8]			98[10]
	F		518[10]			91–99.2	96.0[10]
H. meadewaldoi	M	no data		259–262	261[10]	72.5–77	74.8[10]
	F			250–257	254[10]	79–81	80.0[10]
H. moquini	M	582–757	666±35[12]	265–286	276.5[12]	57.7–69.5	63.0[12]
	F	646–800	722±34[12]	265–289	279.5[12]	67.4–76.9	71.7[12]

	Sex						
H. longirostris	M	410–775	635	258–291	275	61.0–83.5	71.3[13]
	F	500–782	662	261–295	281	67.6–90.0[13]	80.6[13]
H. fuliginosus	M	600–825	740	271–310	290[13]	66.4–90.5	73.0[14]
	F	640–950[5]	779	271–300	285[14]	69.5–85.8	80.8[14]
H. ophthalmicus	M	610–620		264–290	274[14]	65.0–78.0	67.9[14]
	F	800		273–289	281[14]	78.4–80.1	79.2[14]
H. unicolor (black)	M		678[15]		270[15]		78.8[15]
	F		724[15]		275[15]		88.5[15]
SH. unicolor (pied)	M		717[15]		267[15]		83.6[15]
	F		750[15]		271[15]		94.1[15]
H. unicolor (intermediate)	M		710[15]		271[15]		84.4[15]
	F		779[15]		277[15]		95.3[15]
H. chathamensis	M		540[16]		251±6.2[16]		67.8±2.7[16]
	F		640[16]		266±8.5[16]		76.8±5.6[16]
H. o. finschi	M	518–615[16]	538[15]	247–259	256[16]		81.0[16]
	F	538–622[16]	559[15]	245–263	261[16]		90.7[16]

Data sources. 1: Dunning 1993, 2: Ridgway 1919, 3: Nol *et al.* 1984, 4: Blake 1977, 5: Johnsgard 1981, 6: Harris 1967, 7: Dare 1977, 8: Glutz von Blotzheim *et al.* 1975, 9: Cramp and Simmons 1983, 10: Dement'ev and Gladkov 1969, 11: Hockey 1982, 12: Hockey 1981b, 13: Marchant and Higgins 1993, 14: McKean 1978, 15: Baker 1975a, 16: Baker 1975b.

could occur given a rocky/sandy habitat mosaic in which the patches of each habitat type were sufficiently large and the populations inhabiting such patches were sedentary, as is the case with most southern hemisphere species.

To the human eye, the black and white plumage of pied forms is less conspicuous in a sandy beach habitat than is black plumage, and the reverse is true on rocky shores. It is therefore possible that a selective advantage against predation promoted the evolution of melanism in populations living in rocky areas, and that this advantage could be suffi-cient to explain the independent evolution of melanism in the Old and New World groups. This hypothesis would explain why the only pied rocky-shore specialists are *H. p. galapagensis* and *H. chathamensis*, which occur on isolated island groups lacking indigenous predators. An analysis of habitat choice by the polymorphic *H. unicolor* in New Zealand may shed some light on the validity of this hypothesis. If the crypsis theory has substance, black and dark *H. unicolor* (*sensu* Baker 1973) should be disproportionately common on rocky shores, whereas pied and pale birds should favour sandy beaches. One potentially complicating factor here is that predation pressure on these birds is believed to be very low, and the selective advantage in melanism may be concomitantly low and so difficult to demonstrate.

Morphological characteristics

The morphological similarity of oystercatchers worldwide is one reason why their classification has been so contentious. As a general rule, black and variable species are heavier than pied species (Table 10.2), but the ratio of wing length to body mass is no greater for migratory species than for sedentary ones. There is consistent sexual dimorphism in all species in both mass and bill length, with females being heavier and having longer bills than males. Table 10.2 indicates that weight dimorphism in *H. ater* may be very small, but these figures are based on a sample of only five birds. Although the bill length of an individual oystercatcher can vary during its lifetime (Hulscher, Chapter 1, this volume), there are consistent differences in bill morphology between species which normally forage on soft substrata and those which forage on rocky shores (mean bill length: soft substrata, males = 82 ± 9 mm, females = 88 ± 6 mm; rocky shores, males = 71 ± 6 mm, females = 78 ± 4 mm: Mann-Whitney test, $P < 0.01$ and < 0.02 in males and females respectively).

Within species there is an overlap in bill length between the sexes. However, within pairs of *H. moquini*, the female's bill is almost invari-ably longer and more pointed than the male's. There are dietary differ-ences between the sexes, and it has been proposed that these within-pair differences allow pairs to defend smaller territories than would be the case if the diets of both pair members were the same (Hockey and Underhill 1984).

Habitat choice

Most oystercatcher species are coastal breeders, nesting on sandy and rocky shores and in saltmarshes. Extensive inland breeding is restricted to *H. ostralegus* and *H. leucopodus* and, to a lesser extent, to *H. longirostris*. Subspecies of *H. ostralegus* breed inland in both the Palearctic and New Zealand, the behaviour being most pronounced in *H. o. longipes* and *H. o. finschi*. All species which do breed away from the coast fall within the soft substratum group of foragers, consistent with a breeding season diet of earth-dwelling invertebrates. During the nonbreeding season, most of these birds move to the coast (Figs 10.1 and 10.2).

None of the rocky shore specialists, which includes all of the black forms, breeds away from the coast, although these species may show some habitat shift between the breeding and nonbreeding seasons. For example, *H. bachmani* and *H. fuliginosus* increase their use of soft substrata during the nonbreeding season. At least in the case of *H. bachmani*, this is due in part to avoidance of winter storms on exposed rocky shores (Hartwick and Blaylock 1979); this may, indeed, also explain why the principal foraging sites of *H. ostralegus* in winter are estuarine mudflats. *H. moquini* is found on both rocky and sandy shores throughout the year, but during the breeding season there is a movement away from mainland rocky shores to sandy beaches and rocky offshore islands. This movement is, to some extent, predation-mediated; offshore islands are predator-free, and sandy beaches allow birds to select nest sites with good all-round visibility for predator detection.

During the nonbreeding season, *H. leucopodus*, *H. ostralegus*—especially the subspecies *ostralegus* and *finschi*—and *H. longirostris* all gather in flocks which may exceed 100, and in some cases 1000 birds. In contrast, black species which forage on rocky shores rarely form feeding flocks in excess of 10 birds, with the exception of *H. bachmani*, which occurs in flocks of up to 50 in soft-substratum foraging habitats during the nonbreeding season (Table 10.1). Several species aggregate at high-tide roosts. For example, although the majority of *H. moquini* are territorial throughout the year, they gather at high tide in roosts of up to 120 birds. As this species rarely experiences temperatures below the lower threshold of its thermoneutral zone, the primary function of this behaviour in *H. moquini* seems to be avoidance of predation. There are three lines of evidence for this. Roosts are sited either where they cannot be reached by terrestrial predators, such as offshore rocks, on promontories adjacent to such rocks, or in areas with good all-round visibility (Hockey 1985). During the breeding season, breeding birds cannot join such roosts, and it is at this time of year that adult mortality due to predators is highest (Cooper *et al.* 1985). Most predators of *H. moquini* are nocturnal mammals, and night-time roosts are larger than day-time ones. Communal roosting by *H. ostralegus* in winter, by contrast, may confer a dual advantage, serving both as an anti-predator behaviour and as a means of conserving energy (heat) through close huddling. The thermoregulatory rôle of *H. ostralegus* roosts is supported by their being much more closely packed than roosts of *H. moquini*.

Although oystercatcher species can be crudely classified as hard or soft substratum specialists, all species exhibit some plasticity in habitat choice. In other words, the degree of specialization at the species level is not pronounced, presumably because it is not morphologically constrained. Nonetheless, several studies of *H. ostralegus* have shown that there is considerable specialization in feeding behaviour at the individual level (Sutherland *et al.*, Chapter 3, this volume) and oystercatcher populations are best viewed as being composed of specialist rather than generalist individuals.

Site fidelity

Individual *H. ostralegus* are highly site-faithful to wintering (Ens and Cayford, Chapter 4, this volume) and breeding (Hulscher *et al.*, Chapter 7, this volume) areas. Indeed, long-term site fidelity is almost certainly widespread in oystercatchers, occurring as it does in the most migratory (*H. ostralegus*) and the most sedentary (*H. moquini*) species. Coupled with the long-term pair bonds that are typical of most, and possibly all, oystercatchers (Table 10.1), such high fidelity provides the potential for high offspring philopatry. In fact, this does appear to be strong in *H. ostralegus* (Ens *et al.*, Chapter 8, this volume) and other oystercatchers. On the offshore islands of south-western Africa, juvenile *H. moquini* disperse away from the islands at the end of the breeding season to the mainland coast, occasionally as far as 650 km from their birthplace (Hockey 1983a,b). Many return, however, to breed on the islands 3–5 years later. During an outbreak of avian cholera in 1991–1992, eight birds were recovered that had been ringed as chicks between four and 10 years previously; all were recovered on their natal islands, where they were presumed to be breeding. In recent years, convincing evidence has been presented that migration has a strong genetic component (Berthold 1990). Through assortative mating, such inheritance would lead to the establishment of 'demes' of birds with fixed movement patterns and there is evidence, although no genetic proof *per se*, that such inheritance occurs in oystercatchers.

Diet and foraging

Diet breadth

The few detailed studies of the diets of oystercatchers other than *H. ostralegus* indicate that they take a similarly wide variety of prey species, ranging in morphology from amphipods and crabs to bivalves, poly-

Table 10.3   A summary of principal Oystercatcher prey types

| Species | SUBSTRATUM | | Source |
	Rocky	Soft	
H. bachmani	Mussels, limpets, chitons, goose barnacles		Webster 1941a Legg 1954 Hartwick 1976 Morrell *et al.* 1979
H. p. palliatus	Snails, limpets, crabs		Levings *et al.* 1986
H. ater	Mussels, limpets		Woods 1975
H. leucopodus	Limpets, mussels	Polychaetes, earthworms	Woods 1975
H. moquini	Mussels, limpets, snails Polychaetes	Bivalves	Hockey and Underhill 1984
H. longirostris		Bivalves, polychaetes	Lane and Davies 1987
H. fuliginosus	Limpets, mussels, chitons	Amphipods	Lane and Davies 1987
H. unicolor	Mussels, limpets, snails	Bivalves	Baker 1974a
H. o. finschi	Chitons, small gastropods	Bivalves	Baker 1974a
H. chathamensis	Chitons, small gastropods		Baker 1974a

chaetes, echinoderms, ascidians and even fish (Baker 1974a; Toft 1982; Hockey and Underhill 1984; Levings *et al.* 1986). In Panama, Levings *et al.* recorded 30 prey species eaten by *H. palliatus*. Baker recorded 27 species in the diets of *H. o. finschi* and *H. unicolor*, Chafer (personal communication) recorded 24 species eaten by *H. fuliginosus*, and Hockey and Underhill (1984) reported 52 prey species eaten by *H. moquini*. Although dietary diversity may be high at the prey species level, when viewed in terms of prey morphology, the dominant dietary components vary little between oystercatcher species. Molluscs dominate the diets of all species which feed on rocky shores, particularly mussels, limpets, snails (coiled gastropods) and chitons. On soft substrata, bivalves and polychaetes dominate (Table 10.3). Such dietary conservatism within the genus is predictable given the limited morphological variation between species.

Can diet explain distribution?

The diet does not, however, explain the puzzling absence of breeding oystercatchers in the Old World tropics. Crabs are abundant in tropical estuaries and lagoons, where they are the major prey of several 'crab specialists', such as the Crab Plover (*Dromas ardeola*) and Greater Sand Plover (*Charadrius leschenaultii*). Although oystercatchers are not normally regarded as crab specialists, the European Shore Crab *(Carcinus maenas)* is an important prey of *H. ostralegus* during summer. Likewise, *H. ostralegus* eats Fiddler Crabs (*Uca* spp.) on the coast of Guinea-Bissau during winter, although their preferred prey here is the large Bloody Cockle (*Anadara senilis*) (Hulscher, Chapter 1, this volume). *H.*

o. ostralegus and *H. o. longipes* do migrate in small numbers as far south as the equator, but occur only as vagrants in the southern tropics. In southern Africa, the northward limits to the range of *H. moquini*, both in the west and east, coincide with a transition from temperate to sub-tropical biogeographical provinces. *H. moquini* is common on rocky shores of the eastern Cape Province, where it eats mostly the Brown Mussel *(Perna perna)*. But although this mussel is also common along the Natal coast, oystercatchers are absent. A study of the biology of *H. moquini* at the limits of its range could provide useful insights into the limiting habitat requirements of oystercatchers.

Temporal changes in diet

Many migratory oystercatchers, belonging principally to races of *H. ostralegus*, are forced to change their diet seasonally when they switch habitat; for example, from inland fields to estuaries. Among resident coastal species, such dietary switches are likely to be less pronounced, but good seasonal data are lacking to test this supposition. In resident oystercatchers, we might expect seasonal prey-switching to occur in response to the reproductive condition of prey animals just as, on Schiermonnikoog, the importance of *Macoma* in the diet of *H. ostralegus* decreases in late summer as the meat content of the clams decreases. Temporal diet switches may also be mediated by a change in prey activity or burrowing depth, as occurs in *H. ostralegus* on Schiermonnikoog (Hulscher, Chapter 1, this volume). There is also evidence that this occurs in *H. moquini* where there is a diurnal diet shift. Mussels predominate in the diet during the day but, by night, the proportion of the limpet *Patella granularis* taken increases. This limpet forages during low tide at night, and active limpets are then more easy to dislodge from the rocks than during the day when they are clamped tightly to the rock face to minimize desiccation (Hockey and Underhill 1984).

Oystercatchers respond to changes in the relative abundance of their prey species, as has been abundantly documented for *H. ostralegus* (Hulscher, Chapter 1, this volume). Predictably, this happens in other oystercatcher species. In the late 1970s, the diet of *H. moquini* at islands off the south-western Cape, South Africa, was dominated by the indigenous mussel *Aulacomya ater* and the limpet *Patella granularis*. In the early 1980s, there was an accelerating invasion by an alien Mediterranean mussel *Mytilus galloprovincialis*. By the austral summer of 1987–1988, at two islands, this species made up between 42% and 92% of the oystercatchers' diet. By the early 1990s, this proportion had stabilized at 65–70%. By the summer of 1992–1993, *Aulacomya*, the dominant dietary component 10 years previously, had disappeared totally from the diet, reflecting its competitive exclusion from the shore by the faster-growing, more fecund and desiccation-tolerant *Mytilus* (Hockey and van Erkom Schurink 1992).

Prey handling

Because of similarities in prey morphology in different parts of the world, different species of oystercatchers are faced with the same problems in gaining access to prey. With limpets and chitons there are strong adhesive forces to be overcome, bivalves must be prised or hammered open, and the flesh has to be extricated from coiled gastropods, or they have to be smashed open, as by *H. palliatus* (Levings *et al.* 1986). Factors such as prey size and shell strength play a rôle in determining the range of attack techniques that are effective, but several studies have reported very similar prey attack and handling techniques used by oystercatchers in many parts of the world. But there are some small differences. For example, *H. ostralegus* and *H. unicolor* regularly smash open mussel shells (Hulscher, Chapter 1, this volume; Baker 1974a), whereas *H. moquini* almost invariably attacks only gaping mussels, gaining access by stabbing between the valve margins and severing the posterior adductor muscle (Hockey 1981a). Feare (1971) found that *H. o. ostralegus* attacks limpets at the anterior shell margins, whereas *H. moquini* attacks limpets at the posterior margins (Hockey 1981a). At present, however, the reasons for these differences have not been studied in depth.

In South Africa, male *H. moquini* take far more whelks and limpets than do females which, as in *H. ostralegus*, eat more polychaetes and small unshelled items. Although both sexes of *H. moquini* take mussels in approximately equal proportions, sexual dimorphism in the diet is probably widespread in oystercatchers worldwide because of the marked differences in bill structure between males and females (Sutherland *et al.*, Chapter 3, this volume*)*.

Prey selection and optimal foraging

Size selection by oystercatchers within a prey species is usually unimodal (Hockey and Underhill 1984; Ward 1993). Within a prey species, the upper limit to the size of animal taken by *H. ostralegus* seems normally to be set by the bird's ability to dislodge or handle the prey item, whereas the lower limit may be constrained by handling ability (dexterity) or, more usually, by profitability (Ward 1993; Zwarts *et al.*, Chapter 2, this volume).

The lower size rejection threshold of *H. ostralegus* varies with average intake rate (Zwarts *et al.*, Chapter 2, this volume). *H. moquini* also experiences short-term variability in the availability of different prey sizes when feeding on the Wedge Clam (*Donax serra*). Small clams predominate high on the shore, and thus are available almost throughout the tidal cycle, whereas large clams are only found low on the shore. The oystercatchers do include some small clams in the diet, but it is probable—although not proven—that these are eaten only on the flowing tide. The birds spend more time foraging on the flow than the ebb tide, indicating that they do not search for small clams when these are first available, but wait for the large, profitable clams to become available. Whether small clams are eaten may thus depend on the satiation level of birds during the flowing tide (Ward 1993).

Prey characteristics other than size affect prey choice in oystercatchers, of course (Hulscher, Chapter 1, this volume). Just as hammering *H. ostralegus* select thin-shelled mussels (*Mytilus edulis*), *H. moquini* attacks limpets with a particular shell shape (Hockey and Branch 1983). They preferentially attack limpets at the posterior shell margin and make the correct attack decision most regularly on markedly pear-shaped limpet species. The most abundant limpet on the shore, and the one eaten most often by these oystercatchers, is *Patella granularis*. This species is much closer, on average, to being elliptical than pear-shaped. In this case, oystercatchers selectively remove the most pear-shaped individuals, leading to a change in 'average' shell shape in populations of *P. granularis* which are heavily exploited by oystercatchers.

Avoidance of parasitized prey

The infective stages of cestodes and trematodes are probably widespread in oystercatcher marine prey (Goater 1988) and it has been shown that, when taking clams, *H. ostralegus* is able to distinguish and avoid prey that are heavily infected by parasites (Hulscher 1982). In New Zealand, *H. o. finschi* preys extensively on the estuarine cockle *Chione stutchburyi*, which is the intermediate host of the trematode *Curtuteria australis*. The incidence of metacercariae in cockles is 64%, but the incidence in oystercatchers, at 50% (Allison 1979), is low in comparison. This suggests that *H. o. finschi* may also be partially successful in rejecting parasitized bivalves. Similarly, in the Swartkops estuary, South Africa, the mudprawn *Upogebia africana* is the most important prey of all large invertebrate-eating birds except *H. moquini*, which eats almost exclusively the razor shell *Solen capensis*. They only occasionally eat *Upogebia* even though the prawns are readily available to birds when many prawns crawl out of their burrows and on to the surface of the mud at low tide. It is suspected that this otherwise highly maladaptive behaviour—almost all surfacing prawns are eaten by predators—is parasite-mediated. By avoiding *Upogebia*, oystercatchers therefore reduce their risk of infection (Martin 1991).

Oystercatchers are not always able to recognize parasitized prey, however. Dogwhelks *Nucella lapillus* are intermediate hosts of the castrating trematode parasite *Parorchis acanthus*, which produces gigantism and a reversal to juvenile behaviour in *Nucella* (Feare 1967). Dogwhelks form winter aggregations where they are free from predation by *H. ostralegus*, which only attack whelks distributed singly on the open shore. Infected whelks enter winter aggregations later than healthy ones and are therefore susceptible to predation by oystercatchers for longer. Oystercatchers appear unable to detect which whelks are parasitized and the behavioural changes induced in the intermediate host are thus of survival value to the parasite (Feare 1971).

Episodic events and food shortages

Oystercatchers worldwide are sometimes affected by sudden and dramatic changes in their feeding conditions. The continental population of *H. ostralegus* is subjected to severe cold spells in winter (Hulscher *et al.*, Chapter 7, this volume). Similar episodic events can occur even within the more mesic climes of Africa and central South America. In Chile and Peru, an important prey of *H. palliatus* on sandy beaches is the clam *Mesodesma donacium*. Near Lima, Peru, *Mesodesma* was eliminated by the 1982–1983 El Niño, and had not recolonized the area three years later (Arntz *et al.* 1987). Similarly, in early 1988, a massive flood in the Orange River, South Africa, caused a major mortality of rocky-shore invertebrates. Within 10 km of the river, all potential oystercatcher prey were killed, and invertebrate mortalities occurred as far as 140 km away (Branch *et al.* 1990). A similar, near total mortality of oystercatcher prey occurred in St Helena Bay, South Africa, in February 1994 as a result of an anoxic 'black tide'.

The major difference between the European and the African and South American episodic events is the number of birds affected. In winter, *H. ostralegus* congregate in estuaries. Comparable seasonal changes in distribution do not occur in more mesic regions, where oystercatchers remain more evenly spaced along the shore throughout the year. There is no direct evidence that localized prey extinction in Africa results in any mortality of oystercatchers; birds simply move out of the affected areas. The only substantial prey-related mass mortalities of *H. moquini* have occurred as a result of Paralytic Shellfish Poisoning (Hockey and Cooper 1980). But even these are not, on an absolute scale, comparable to the mass mortalities that sometimes occur in wintering oystercatchers in continental Europe where the numbers of birds involved may exceed the world population of *H. moquini*; more than 5000 *H. ostralegus* died in the Delta area of The Netherlands during one cold spell alone (Meininger *et al.* 1991). On a relative scale, however, the impacts on *H. moquini* may be greater. A single outbreak of Paralytic Shellfish Poisoning in May 1978 may have killed as much as 6% of the world population, and an outbreak of avian cholera in 1992 could have killed 4%. In long-lived species such as oystercatchers, episodic events such as mass deaths through starvation, poisoning or disease may have significant, but as yet unquantified, impacts on the species' long-term population dynamics.

Displays and vocalizations

The displays of *H. ostralegus* have been far better studied than have those of any other oystercatcher, although there are fairly detailed accounts of the behaviour and vocalizations of *H. ater*, *H. palliatus*, *H. leucopodus* and *H. moquini* (Miller and Baker 1980; Baker and Hockey 1984). *H. leucopodus* is behaviourally as well as morphologically enigmatic, but the similarities in the behavioural and vocal repertoires of other oystercatchers are further evidence for evolutionary conservatism within the genus.

The piping display

This is a conspicuous component of the behavioural repertoire of all species and functions both in territory establishment and defence. In *H. ostralegus*, but to an unknown extent in other species, it also plays an important rôle in establishing and asserting dominance hierarchies outside the breeding season (Ens and Goss-Custard 1986).

The precise form of the display varies between oystercatchers. Old World taxa hold their tails horizontally in this display, whereas in the New World, *H. ater* and *H. palliatus* raise the tail obliquely, at least for some of the time, and *H. leucopodus* has a particularly striking display, with the tail raised vertically and the undertail coverts fanned and fluffed. Piping vocalizations of *H. leucopodus* differ from those of other taxa studied (*H. ater*, *H. palliatus*, *H. bachmani*, *H. o. ostralegus*, *H. fuliginosus*, *H. chathamensis*, *H. o. finschi*, *H. unicolor* and *H. moquini*) in having an extremely narrow-band width. Miller and Baker (1980) suggest that this is an adaptation for long distance sound propagation as the spectrum of a narrow band call changes little with distance relative to a wide band call with the same centre frequency; *H. leucopodus* breeds on windy, open grasslands. Similar long range vocal displays occur in other wader species which nest at low densities in open habitats, such as *Numenius* spp. (Skeel 1978).

The piping displays of *H. ater* and *H. palliatus* are much more similar to one another than either is to the sympatric *H. leucopodus*. *H. palliatus* and *H. ater* regularly hybridize where they co-occur (Jehl *et al.* 1973). There is no evidence that either species hybridizes with the behaviourally distinct *H. leucopodus*, apart from one putative *H. ater* × *H. leucopodus* hybrid (Jehl 1978).

The piping vocalizations of *H. ater*, *H. palliatus* and *H. leucopodus* are very similar to their respective alarm calls, suggesting that piping is a highly ritualized form of these calls (Miller and Baker 1980). By contrast, piping calls of *H. moquini* include not only alarm calls, but also distraction-lure calls (Baker and Hockey 1984). The piping calls of *H. moquini* are wide band, but are structurally much more similar to *H. ostralegus* than to New World Oystercatchers, as too are the flight calls of *H. moquini* (Baker and Hockey 1984). This provides further indirect evidence in favour of an independent evolution of black and pied taxa in the Old and New Worlds following the break-up of Gondwanaland.

Anti-predator behaviour

Distraction displays of oystercatchers have not been as well studied as piping displays, and the significance of apparent differences between species should be inferred cautiously. False brooding, or adopting a brooding posture away from the nest to distract predators, is widespread in waders and probably occurs in all oystercatcher species. In addition, rodent runs, injury feigning and distraction-lure and threat displays are well described for *H. o. ostralegus* and *H. moquini*. The distraction-lure display of *H. moquini* is more elaborate than that of *H. ostralegus*, involving rapid vocalizations and exaggerated body posturing.

It is tempting to suggest that the most elaborate distraction displays will be found in those species which have an evolutionary history of high predation pressure on eggs and young. However, there are a number of observations that are counter to this hypothesis. Crouching distraction displays are well developed among *H. o. ostralegus* on the Faeroe Islands and in *H. chathamensis,* yet neither is exposed to mammalian predators (Miller and Baker 1980), and ground-based distraction displays are unlikely to be particularly effective against visual avian predators of eggs and small chicks, such as gulls and corvids. *H. moquini* has been observed to drive the diurnal Small Grey Mongoose *(Galerella pulverulenta)* away from the nest, rather than resort to distraction displays. *H. ostralegus* makes similar attacks, but also uses distraction-lure and threat displays against mammalian predators (Cramp and Simmons 1983). In the case of *H. moquini*, most predators are nocturnal hunters and most mortality of adults occurs at night. It is not known whether these birds use visual displays at night but, assuming that nocturnal predators are 'stealthy' rather than 'scary', one could predict that that such displays would be ineffective. Thus, there is little evidence to suggest that the displays reflect the intensity of predation. Rather, with the exception of *H. leucopodus*, displays and vocalizations are conservative within the genus. It is likely, therefore, that they serve common functions with common evolutionary origins, and do not represent convergent characteristics. Several displays have homologues in the Recurvirostridae (Cramp and Simmons 1983), further supporting the close relatedness of the two families.

Life history characteristics

Longevity

H. ostralegus can live in excess of 40 years (Ens *et al.*, Chapter 8, this volume). There have been no comparable long-term studies of other species, although some are now in progress. *H. palliatus* reaches at least 10 years old (Rydzewski 1978), and *H. moquini* 18 (P.A.R. Hockey, unpublished information). On the assumption of a positive association between survivorship and body mass, it would be expected that survivorship of most species will at least equal that of *H. ostralegus*.

Greenberg (1980) demonstrated that Nearctic passerine species which breed at high latitude or altitudes have a higher survivorship than those which breed in more mesic, less seasonal habitats. He was able to explain these differences on the basis of reproductive opportunity related to the length of the breeding season. Species which breed at low latitudes have a relatively long breeding season and thus the potential for multi-broodedness. Species breeding at high latitude have to maximize non-breeding survivorship by longer distance migration to more mesic winter quarters to counteract the constraints of a short breeding season. Such reasoning cannot be extended directly to predict differences in longevity between oystercatcher species, because all oystercatchers are single brooded. However, it could be argued on the same basis that survivorship among species which lay relatively small clutches of 1–2 eggs (e.g. *H. p. galapagensis*, *H. leucopodus*, *H. ater* and *H. moquini*) may be greater than those which make a larger investment in reproduction, such as *H. ostralegus*, *H. palliatus* and *H. longirostris*. At present, however, the data to test this expectation are not available.

Clutch size

In many groups of birds, including plovers, sandpipers and gulls, clutch size tends to be larger at high than at low latitudes. This pattern is not repeated in the oystercatchers. The mid-latitude positions of the breeding ranges of the northern hemispheric *H. ostralegus* and the southern hemispheric *H. leucopodus* are almost identical, but the two species represent extremes of clutch investment within the group (Table 10.4). Again, *H. bachmani* has a breeding range spanning 33° of latitude (Fig. 10.1). However, there is no latitudinal variation in either egg size or clutch size (L'Hyver 1985), although there is a two week difference in the timing of clutch initiation between the north and south of the breeding range. Among oystercatchers, there is a tendency for those species which breed at relatively high latitudes to have a shorter laying period than low-latitude breeders. However, intraspecifically, these differences are not pronounced, even in species with wide latitudinal ranges, including *H. bachmani* (L'Hyver 1985) and *H. o. ostralegus* (between 50°N and 64°N—Harris 1967; Heppleston 1972; Väisänen 1977).

Although no oystercatchers are multi-brooded, all species probably lay replacement clutches; as many as three in the case of *H. palliatus* (Nol 1984). There is evidence for replacement clutches containing fewer eggs than initial clutches in *H. ostralegus* (Harris 1967; Heppleston 1972), *H. palliatus* (Nol *et al.* 1984) and *H. ater* (Nol 1984), but not in *H. moquini* (Hockey 1983a). Replacement clutches laid by *H. bachmani* are smaller than initial clutches in Alaska (Webster 1941b), but not in British Columbia (L'Hyver 1985). Although clutch size of *H. bachmani* does not vary between initial and replacement clutches in British Columbia, eggs in replacement clutches of *H. bachmani* tend to be smaller than those in initial clutches (L'Hyver 1985). Smaller eggs in replacement clutches may be a strategy to reduce the incubation period (Runde and Barrett 1981).

Timing of egg laying

The timing of egg laying affects reproductive success in single-brooded species of birds. Several studies have shown that the probability of an *H. o. ostralegus* egg leading to a fledged chick decreases strongly during the course of a breeding season, i.e. late layers are less successful. However, young *H. o. ostralegus* tend to lay later than older birds, and experience, rather than the timing of breeding *per se*, may thus be the primary correlate of breeding success. However, the timing of egg laying by individual *H. ostralegus*, *H. bachmani* and *H. palliatus* is correlated between years (Nol 1984; L'Hyver 1985), suggesting that age differences may not be the only variable necessary to explain the seasonal decrease in breeding success. Nol (1984) suggests that social facilitation plays a rôle in determining the timing of egg laying in *H. palliatus*, as breeding synchrony is most pronounced when breeding density is high.

Risk, investment and limitation

The first-laid egg of *H. ostralegus* is only partially incubated and is at high risk from predators; on Schiermonnikoog, 40% of first-laid eggs are lost before the second egg is laid (Ens *et al.*, Chapter 8, this volume). The first egg of *H. palliatus* is not incubated at all, and is certainly at high risk. The second-laid egg of *H. palliatus* is the largest in the clutch. It has been suggested that this pattern evolved because the first egg has a high chance of being lost to predators and, if the second and third eggs hatch asynchronously, the third egg may be abandoned due to disturbance by predators (Nol 1984). The female *H. palliatus* thus places maximum investment in the egg most likely to hatch. In *H. moquini*, where hatching of modal two-egg clutches is closely synchronous, and incubation commences after laying of the first egg, there are no size differences between first- and second-laid eggs (Hockey 1983a). Both eggs of *H. moquini* are at equal risk from predators (Hockey 1983a), and investment per egg within a clutch is equal. Although insufficient data exist to test the hypothesis, it can be predicted that within-clutch variance in egg investment will be greater among species which lay three or four eggs than those which lay one or two. The reason for this is the longer time taken to clutch completion by the former which therefore places differential risks of loss (or late hatching) on individual eggs, depending on their position in the laying sequence.

Table 10.4. Breeding parameters of the world's oystercatchers.

Species	Clutch size:			Eggs:		Modal clutch %of female Mass	Incubation period	Fledging period	Fledging success (young per year/%)	Causes of failure
	Range	Mean	Mode	Size	Mass					
	(mm)	(mm)	(mm)	(mm)	(g)		(d)	(d)		
H. bachmani	2-4 [1] 1-3 [2]	2.69 [1] 2.1 [3]	3 [1] 2 [3]	57.0 × 38.9 [3]	46.8 [3]	21.9	24-33 (27) [3] 14.6	40 [2,4]	0.19-0.31 [2] 0.5-1.1 [5]	Predation (gulls,crows), storms, man, seals [2,5,45]
H. p. palliatus	1-4 [1,6]	2.40 [2] 2.56 [6] 2.8 [7]	3 [7]	56.3 × 39.7 [6,8]	49.3 [8]	23.2			0.24-0.39	Predation, high tides flooding, storms [6,7,9]
H. p. galapagensis	1-2 [10]		2 [11]							Predation (mockingbird) [11]
H. leucopodus	1-2 [12]		2 [13,14]	59.0 × 40.0 [15] 57.5 × 40.2 [12]	51.3 [15] 50.3	16.1				
H. ater	1-2 [12]	1.89 [9]	2 [13,16]	62 × 41 [15] 60.3 × 40.8 [9]	54.7 [9]	17.2			0.05 [9]	Predation (gulls, skuas) [27]
H. o. ostralegus	2-5 [17]	2.7 [18,19]	3 [17,20]	57 × 40 [17] 56.3 × 39.9 [20]	46 [20] 48.5	23.3	23.5-34.5 (27) [20]	33 [21]	0.57 [18] 0.61 [19]	Man, disturbance, agriculture [18]
H. o. longipes	2-4 [20,22]		3 [20]							
H. o. osculans	2-4 [23]		3 [23]	58.4 × 31.9 [20] 57.2 × 38.1 [24]	44.9	26.0		25-35 (27) [22]		
H. meadewaldoi										
H. moquini	1-2 [25,26] rarely 3	1.81 [27] 1.74 [25]	2 [25]	60.7 × 41.0 [25]	55.8 [25]	15.5	27-39	35-40 [25] (32) [25]	20% [26]	Predation (gulls, mammals), storms, hatching, man disturbance [25,26,28]

H. longirostris	1-4 [29]	2.08 [30]	2 [30]	59 × 41 [28]	53.7	16.2	28-32 [29]	42-49 [29]	0.27-0.89	Predation (ravens), man, domestic animals, high tides [37]
H. fuliginosus	1-3	1.8 [30]	2 [31]	63.1 × 42.8 [30] 65 × 44 [15]	57.8 [30] 68.1	14.8				Predation (gulls) [31]
H. ophthalmicus										
H. unicolor	1-5 [30]	2.3 [32]	3 [33]	62 × 43 [15] 58.8 × 40.9 [34]	58.5 [15] 49.4 [34]	19.7	25-32 [28]	42-49	0.64 [30]	
H. chathamensis	2-3 [30]			56.9 × 40.5 [34]	46.0 [34]	14.4	25.1±1.2	48	0.47 [30]	
H. o. finschi	2-3 [33]	2.3 [9]	2	56.0 × 38.6 [34]	44.2 [34]	15.8	24-28 [33]	35 [33]	0.70 [30]	Agriculture [30]

1. Data sources. 1: Webster 1941b, 2: Hartwick 1974, 3: L'Hyver 1985, 4: Purdy 1985, 5: Paine *et al.* 1990, 6: Baker and Cadman 1980, 7: Nol 1989, 8: Nol *et al.* 1984, 9: Nol 1984, 10: Harris 1967, 11: Lévêque 1964, 12: Johnson and Goodall 1965, 13: Woods 1975, 14: Johnsgård 1981, 15: Schönwetter 1962, 16: Zapata 1967, 17: Bent 1927, 18: Heppleston 1972, 19: Briggs 1984b, 20: Dement'ev and Gladkov 1969, 21: Cramp and Simmons 1983, 22: Gavrin *et al.* 1962, 23: Caldwell and Caldwell 1931, 24: Etchécopar and Hüe 1978, 25: Hockey 1983a, 26: Jeffrey 1987, 27: Summers and Cooper 1977, 28: Cooper *et al.* 1985, 29: Lane and Davies 1987, 30: Marchant and Higgins 1993, 31: Wakefield and Robertson 1988, 32: Baker 1975b, 33: Falla *et al.* 1979, 34: Baker 1974a.

2. Egg weights cited without reference are calculated using the formula $W(g) = Kw.(LB^2/10^3)$ (Hoyt 1979), where L = egg length in mm, B = egg breadth in mm and Kw is a constant. Kw was calculated from 25 freshly laid *H. moquini* eggs and = 0.541 (± 0.008 SD).

Local differences in the clutch size of *H. o. ostralegus* have been explained on the basis of variation in foraging success and hence the ability of adults to feed chicks (Safriel 1967; Heppleston 1972; Briggs 1984b). Similarly, the clutch size of *H. palliatus* is smaller in Argentina than in Virginia and Nol (1984) suggests that food limitation may explain this difference, the rate of food acquisition being higher in Virginia than in Argentina. Whilst lowered food availability can lead to clutch size being reduced below the species' mode, there is no evidence among oystercatchers that an increase in food availability can have the opposite effect, although it can increase the number of young that fledge from a clutch.

There may also be density-dependent variation of clutch size within species. For example, *H. bachmani* breeding at relatively high density on Cleland Island, British Columbia, lay fewer three-egg clutches than do *H. bachmani* in British Columbia as a whole (L'Hyver 1985). Similarly, the average clutch size of *H. moquini* breeding at high density (5 pairs per ha) on offshore islands in south-western Africa (1.74) is slightly smaller than when all clutches, including island clutches, are averaged (1.81) (Summers and Cooper 1977; Hockey 1983a). The average clutch size of *H. moquini* at offshore islands is the smallest recorded in any oystercatcher (Table 10.4), and, at 5 pairs per ha—including both breeding and feeding territories—*H. moquini* also reaches the highest breeding density of any oystercatcher species studied to date. Mean clutch size of *H. ater* (1.89) and *H. bachmani* (2.04) breeding at high density (respectively 0.5 and 2.12 pairs per ha) are also small relative to the average (Groves 1984; Nol 1984). In seven study populations, average clutch sizes of *H. palliatus* and *H. o. ostralegus* ranged from 2.31 to 2.80 eggs (Nol 1984). In six of these seven study populations, breeding density was < 0.28 pairs per ha. At one site, however, *H. o. ostralegus* bred at an average density of 1.47 pairs per ha and laid an average of 2.70 eggs per clutch. But, overall, there are some gounds for believing that reproductive output is reduced at high breeding densities.

Because of high predation rates, accurate determination of clutch size, defined as the number of eggs laid as distinct from the number surviving, can be difficult, and clutch size is probably regularly underestimated (Ens *et al.* 1992). The suggestion that clutch size is inversely related to the density of breeding birds thus requires careful substantiation, although it does conform to the predictions of the Fretwell–Lucas density-dependent fitness model (Fretwell and Lucas 1969). One mechanism for explaining reduced clutch size at high breeding density is that predators, such as gulls, are attracted to areas with high densities of breeding birds. Egg loss may thus increase proportionally and, along with it, the risk of underestimating true clutch size. Thus, whether the birds actually lay fewer eggs at high breeding densities, rather than fewer being found by research workers, remains to be established.

Breeding behaviour and parental investment

Mating and rearing system

All species of oystercatchers are predominantly monogamous, although bigamy and extra-pair copulations occur (Ens *et al*, Chapter 8, this volume). They also form long-term pair bonds; in *H. o. ostralegus*, the same pair may occupy the same breeding territory for up to 20 years. Long-term pair bonds are retained both by those species which spend the whole year on one territory and by those which are migratory and flock in the nonbreeding season. Worldwide, there is evidence for long-term pair bonding in *H. bachmani, H. palliatus, H. o. ostralegus, H. moquini* and *H. longirostris* (Table 10.1), but the existence of long-term pair bonds in *H. o. finschi* has been questioned (Baker 1974a).

Being long-lived, mate-faithful, site-faithful and having contiguous territories are attributes which may predispose oystercatcher populations to becoming saturated during the breeding season. Removal experiments show clearly that this is often the case in *H. ostralegus* and that a surplus of birds exists which are capable of breeding but lack a mate, a territory, or both (Ens *et al.*, Chapter 8, this volume). Observations on islands off South Africa, where *H. moquini* breeds at high densities, also indicate that territory and mate vacancies are rapidly claimed. If competition for territories is intense, delayed breeding is to be expected (Ens *et al.*, Chapter 8, this volume); in the case of *H. ostralegus*, one male bred for the first time at 14 years old! In both *H. ostralegus* and *H. moquini*, some females breed for the first time at three years old, and some males at four years old. On the assumption of an equal sex ratio at hatching, it has been suggested for *H. ostralegus* that earlier breeding by females is a consequence of a higher mortality of subadult females than males, perhaps because females are inferior competitors for preferred prey (Ens *et al.*, Chapter 8, this volume). An alternative hypothesis is that there is a disproportionately high mortality of females prior to independence because they are larger than males and have a greater energy demand, which parents may be unable to satisfy.

Although oystercatchers are mostly monogamous and mostly mate-faithful, there are exceptions. Polygyny and divorce also occur. The life-history implications of such exceptions and the selective pressures that may drive their occurrence are discussed in detail for *H. ostralegus* in Ens *et al.* (Chapter 8, this volume). As yet, no other species of oystercatcher has been studied in sufficient depth to make meaningful interspecific comparisons. The norms of the breeding behaviour of most oystercatcher species are poorly described, and interpretation of the exceptions is a goal for the future.

When the probability of retaining the same mate in future breeding attempts is high, or when the probability of finding a new mate of equal or greater quality is low, behavioural complementarity between members of the pair would be expected to evolve (Trivers 1972). However, a more basic condition for the evolution of apparent cooperation is that one parent cannot raise the young alone. Indeed, oystercatchers frequently breed in areas where the risk to eggs and young from predators is high. Additionally, because they are 'precocial 5' birds (Safriel *et al.*, Chapter 9, this volume) which experience strong competition for territories and have an extended period of parental provisioning, one adult cannot guard and feed young simultaneously except when young follow adults onto the feeding grounds. One would therefore expect both behavioural complementarity and apparent cooperation between mates.

In the case of *H. ostralegus*, both sexes contribute almost equally to territory defence, incubation and the brooding and feeding of chicks (Ens *et al.*, Chapter 8, this volume). The behaviour of male and female *H. palliatus* in the pre-laying period is also very similar, although females spend more time foraging than males, as is the case in *H. ater* (Nol 1984). However, during incubation, males are involved in more territorial interaction than females and consequently make a smaller contribution to incubation. In the population studied by Nol (1984, 1989) in Virginia, feeding and nesting territories were separated by 100–1000 m, demanding that adults fly to the foraging area to obtain food for chicks. During this chick-feeding stage, males provision young more frequently than females; this has been attributed to the relatively low wing loading of the males. Although there is rôle differentiation between the sexes at various stages of the breeding cycle, the net energy investment by the two sexes is very similar, with female energy expenditure being approximately 6% greater than that of the male. This equal cumulative energy expenditure suggests that there is no differential mortality of the sexes during the nonbreeding season. If there was a difference, the sex with the higher probability of not surviving until the next breeding season should invest a greater proportion of its residual reproductive effort to ensuring fledgling survival. This prediction of equal survival of the sexes is supported by the return rates to the breeding grounds; over a five-year period male return rate was 84.4% and female return rate was 84.6% (Nol 1984).

Rôle differentiation by *H. bachmani* occurs at several stages of the breeding cycle. Most oystercatchers make several nests before laying. The male *H. bachmani* plays the dominant rôle in nest building, but the final choice of nest site is probably made by the female. Male-dominated nest building has also been reported for *H. ater* (Nol 1984), and in several other species of monogamous shorebirds, including avocets (Gibson 1978). The construction of numerous nests in the pre-laying stage may serve to strengthen the pair bond as nest building usually occurs only when both members of the pair are present. As in *H. palliatus* and *H. ater*, female *H. bachmani* spend more time (6% more) foraging than males in the pre-laying period, whereas males spend more time in aggressive interaction. High levels of male aggression at this time occur in *H. o. ostralegus*, *H. unicolor* (Jones 1979a) and *H. ater* (Nol 1984), but not in the population of *H. palliatus* studied in Virginia.

This high level of aggression appears to fulfil a dual function, one purely territorial, and the other anti-cuckoldry. Because of the extended breeding season, both sexes are susceptible to cuckoldry, although the risk to males is highest in the pre-laying period and the attack rate on conspecifics is correspondingly high. Although difficult to detect in purely observational studies, cuckoldry does occur rarely. Both sexes of *H. ostralegus* engage in extra pair copulations (*EPCs*) early in the breeding season (Ens *et al.*, Chapter 8, this volume). Once the female starts egg laying, the frequency with which she attempts *EPC* decreases, and she makes no attempts after clutch completion. Males rarely copulate with any females—including their own mates—after clutch completion, but on the rare occasions when they do so, it is often an *EPC*. Possible cuckoldry has been also observed in *H. palliatus* (Nol 1984) and *H. bachmani* (Purdy 1985). The lack of high aggression levels by male *H. palliatus* in Virginia may be related to a low risk of cuckoldry. In this area, the experimental removal of birds did not result in rapid replacement, suggesting that there were few or no 'floaters' in the population. In a similar experiment with *H. o. ostralegus*, where cuckoldry occurs sufficiently often for differences in its frequency during the breeding season to be detected, mates were rapidly replaced (Harris 1970).

It has been proposed that the ultimate objective of *EPCs* by *H. ostralegus* is an attempt to either 'sample' a potential new mate or to change mates (Ens *at al.*, Chapter 8, this volume). However, lost mates are usually replaced within 24 hours in *H. moquini* breeding on offshore islands where densities are high. Cuckoldry would therefore be expected here but, in fact, it has not been observed. Although *H. moquini* has not been studied in the detail of *H. ostralegus*, divorce happens at a much lower frequency than the 8% reported for *H. ostralegus* on Skokholm Island. The reason for the low incidence of divorce and cuckoldry in *H. moquini* may be that the majority of pairs are sedentary and defend fixed territories throughout the year. The opportunity to occupy vacant positions may thus arise at any time; such opportunities for *H. ostralegus* arise only during the breeding season, so this behaviour is compressed into a short period and so is, perhaps, more easily detected. Whether neighbour-pair familiarity among sedentary species also plays a rôle in reducing the frequency or intensity of interaction between adjacent territory holders, as it apparently does in wintering Grey Plovers *(Pluvialis squatarola)* (Turpie 1994), is not known.

Incubation and chick rearing

Eggs are usually laid at 1–2 day intervals. There is clearly a disadvantage to nidifugous young in asynchronous hatching (Ashkenazie and Safriel

1979a), and, at least in some species of oystercatchers (*H. o. ostralegus*, *H. bachmani*, *H. palliatus*), either the first egg is not incubated at all, or it is only incubated intermittently. Continuous incubation only begins after the second egg has been laid, thus exposing the first egg to a potentially high predation risk (Harris 1970). However, some protective steps can be taken. *H. moquini* is susceptible to terrestrial mammal predators on the mainland and to Kelp Gulls (*Larus dominicanus)* and Sacred Ibises (*Threskiornis aethiopicus)* on islands. The first-laid eggs are covered and are no more or less susceptible to predators than are second-laid eggs, unlike the first-laid egg of *H. ostralegus* (Hockey 1983a). After clutch completion, incubation is continuous. This pattern is typical of monogamous charadriiforms, even in warm climates. Probably, incubation serves as much to protect eggs from predators as to keep them warm because Purdy (1985) found that unattended eggs of *H. bachmani* were rapidly predated by gulls or crows.

Among monogamous charadriiforms which breed inland, the male generally incubates during the day and the female at night. However, food availability to coastally-breeding oystercatchers is usually determined by the tidal cycle and no such diet patterns of incubation have been recorded. Participation by male *H. bachmani* in incubation increases as incubation progresses—a pattern repeated in several monogamous shorebirds (Ashkenazie and Safriel 1979a; Miller 1985), but not all (Jehl 1973; Howe 1982). There are several theoretical reasons why direct male investment in incubation should increase during the incubation period. In terms of past investments, the value of the eggs increases constantly (Trivers 1972), the future investment expectations of the male increase (Dawkins and Carlisle 1976) and the opportunity for cuckolding other females decreases as the breeding season progresses (Maynard Smith 1977). It may also serve to counterbalance the greater investment made by the males in nest building and aggression early in the season. Incubating male *H. bachmani* sometimes leave the nest to attack an intruder—when this occurs, the female takes over the incubation. As incubation proceeds, the duration of incubation bouts by both sexes increases. A similar pattern has been reported for Semipalmated Sandpipers (*Calidris pusilla)* (Ashkenazie and Safriel 1979b) and some gulls (Drent 1970; Maxson and Bernstein 1984). Although these patterns hold true in general for *H. bachmani*, there is inter-pair variation in rôle differentiation. This implies that the contribution made by one mate is adjusted to the contribution of the other.

Young *H. bachmani* in British Columbia are brooded almost continuously for the first few days after hatching. Thereafter, the incidence of brooding is apparently related to low temperatures, but brooding does not continue after the chick is 21 days old. The intensity of brooding by *H. palliatus* in Virginia is much less, which may reflect higher ambient temperatures and a lower predation risk. In both *H. bachmani* and *H.*

palliatus, the majority of brooding is done by the female, freeing the male to spend more time in territorial defense and in provisioning chicks. Once the female is freed from brooding duties, investment by both sexes in chick feeding is equal. Helbing (1977) reported that female *H. bachmani* brought more food items to chicks than did males, but did not consider the possibility of dietary differences between the sexes. Purdy (1985) suggests that there is a tendency for males to bring larger food items than females. There are sexual differences in the diet of several oystercatcher species, and it is likely that these are manifest in different prey spectra delivered to chicks by males and females.

Although both parents provision chicks in all species studied, there is evidence that adults may be stressed during this period. Chick mortality, often apparently due to starvation, has been recorded in *H. ostralegus* (Safriel *et al.*, Chapter 9, this volume), *H. palliatus* (Nol 1985), *H. bachmani* (Groves 1984; Purdy 1985) and *H. moquini* (P.A.R. Hockey, unpublished information). At least in *H. bachmani*, some of this starvation is due to the establishment of a sibling hierarchy and death of the subordinate chick (Groves 1984). Indirect evidence for food limitation is provided by the response of *H. moquini* to invasion of the shore by an alien mussel *Mytilus galloprovincialis*. During the 1980s, the midshore regions of the south-western Cape, South Africa, previously dominated by limpets, gradually became blanketed with thick beds of *Mytilus*. Invertebrate standing stocks on the shore increased dramatically, and *Mytilus* came to dominate the diet of oystercatchers. During this period, there was a steady increase in the reproductive output of the oystercatchers (Hockey and van Erkom Schurink 1992). Slow growth and delayed fledging carry an additional burden in as much as the chick is exposed to predation risk for longer.

The constraints that may have given rise to parent-feeding in oystercatchers are discussed in detail by Safriel *et al.* (Chapter 9, this volume). They conclude that the behaviour evolved in environments with large food items that chicks could not catch or handle themselves. They also predict that the increasing prevalence of inland breeding in *H. o. ostralegus*, but presumably also in *H. o. finschi*, will lead to self-feeding of chicks becoming widespread because of the relative abundance of small prey, such as insect larvae, in terrestrial habitats.

However, an additional hypothesis for the unique adaptive position occupied by oystercatchers can be put forward, based on the suggestion that oystercatchers originated in Gondwanaland. Most of the world's oystercatcher species breed in the southern hemisphere, predominantly on the open coast, and several breed on exposed rocky coasts; this was probably the common, ancestral condition. Worldwide, only a small proportion of waders breed on the open coast, although several breed among coastal dunes and use largely or partially terrestrial foraging habitats. Most coastally-breeding oystercatchers forage exclusively in the intertidal zone. The potential for the evolution of self-feeding in chicks is therefore constrained not only by the scarcity of small and accessible prey, but also by tidally-imposed limitations on foraging time and the risk of being swept off the shore by waves, which is very considerable along rocky shores. Neither of the latter constraints is imposed on inland-breeding oystercatchers, nor indeed on most breeding waders. Hence an explanation for the unique rearing strategy adopted by oyster-catchers may lie simply in the nature of the terrain in which most species have evolved.

Breeding success and population trends

Breeding success in oystercatcher populations worldwide is low (Table 10.4), and to a large extent reflects mortality during the egg stage rather than the mortality of dependent chicks. For example, eggs hatched in only 14% of 229 nests of *H. palliatus* in Virginia, but of the nests which did produce hatchlings, 69% subsequently fledged at least one chick. In Argentina, only 25% of *H. palliatus* nests hatched and Nol (1989) esti-mates that only 5–10% of pairs fledge young in any one season. Hatch-ing success of *H. ater* reported by Nol (1984) from Argentina is even lower than that of *H. palliatus*; eggs hatched in only one of 20 nests, and only one chick fledged. Hatching success of *H. moquini* in the presence of predators is low (12%) in relation to predator-free island populations of *H. bachmani* (25–46%) and *H. o. ostralegus* (44–82%) (Harris 1967; Hartwick 1974; Hockey 1983a). Predators and storms are the main causes of egg loss in most species (Table 10.4) and, because the intensity of both factors varies locally, there may be substantial differences in fledging success in the same species in different areas. For example, the fledging success of *H. bachmani* varies from 0.19–1.1 young per pair. Fledging success also varies between years: at one site, a population of *H. palliatus* produced from 0–0.5 young per pair over four years (Nol 1989) (Table 10.4).

Most chick mortality occurs in the first week after hatching; for example, 87.5% of deaths occur then in *H. moquini* (Hockey 1983a) while, in *H. o. ostralegus*, it is 61–62% (Heppleston 1972). Overall chick mortality rates vary from 31–80% in *H. o. ostralegus* (Goss-Custard *et al.*, Chapter 13, this volume) and 63–72% in *H. bachmani* (Hartwick 1974). After chicks reach seven days old, mortality of young *H. moquini* is less than 30% (Hockey 1983a). In most studies, predation has been identified as a major cause of chick deaths.

The importance of predation for egg and chick survival was clearly illustrated in a 'natural' experiment at Marcus Island in the south-western Cape, South Africa (Cooper *et al.* 1985). This 11 ha granitic island sup-ports a breeding population of *ca* 55 pairs of *H. moquini*. In 1976, the island was joined to the mainland by a 2 km long causeway as part of harbour developments. This causeway provided access for terrestrial mammals, including predatory species, to the island. Between 1976 and 1982, eight species of mammalian predators were recorded there, of which four species occurred regularly. All of these four species are preda- tors of birds, their eggs and young. In addition to killing 28 adult *H. moquini*, equivalent to 25% of the resident population, these predators had a profound effect on the reproductive output of the oystercatcher population between June 1979 and April 1982. In three consecutive breeding seasons, a total of only five chicks fledged from the island, a production of only 0.03 young per pair per year, quite insufficient to maintain the population in the long term. Even assuming no mortality between fledging and breeding, a pair would have to remain mated and breed for 67 seasons to replace themselves! In 1982, an effective preda- tor-proof wall was constructed across the causeway, and the remaining predators on the island were trapped. In the following breeding season, at least 15 chicks (0.3 young per pair) fledged, and in subsequent years breeding success has been at least as high. Predators had thus reduced breeding success by an order of magnitude. Incidentally, the breeding success on the adjacent mainland is normally as low as it was on Marcus Island when predators were present. This implies that chick production on predator-free offshore islands (0.3–0.6 young per pair per year— Urban *et al.* 1986) may thus be crucial for maintaining the mainland breeding population.

As with breeding success, population trends in oystercatchers vary from place to place. For example, numbers of *H. bachmani* in lower California have decreased steadily since the beginning of the century, whereas in other places numbers have remained unchanged (Kenyon 1949). On San Nicolas Island, numbers have increased exponentially in recent years (D. R. Lindberg, unpublished information). During the pre- sent century, numbers of *H. o. ostralegus* and *H. o. finschi* have increased markedly, in concert with an increased use of inland habitats (Sibson 1966; Cramp and Simmons 1983). In contrast, numbers of *H. o. longipes*, *H. o. osculans*, *H. meadewaldoi* and *H. fuliginosus* have decreased (Dement'ev and Gladkov 1969; Hockey 1987; Lane and Davies 1987). *H. meadewaldoi* is now presumed extinct (Hockey 1987). Some species and subspecies have small total populations, e.g. *H. p. gala- pagensis* (<100 pairs—Harris 1974), *H. o. osculans* (Dement'ev and Gladkov 1969), *H. moquini* (*ca* 5000 birds—Hockey 1983b), *H. ophthalmicus* (<1000 birds—Lane and Davies 1987), *H. unicolor* (2000–3000 birds—Hayman *et al.* 1986) and *H. chathamensis* (*ca* 50 pairs—Baker 1973). The population of the latter species is increasing slowly (King 1981).

For some species, it is possible to deduce the causes of population trends. *H. chathamensis* was reduced to low numbers by introduced predators, and the same potential threat faces *H. p. galapagensis*. The demise of *H. meadewaldoi* probably stemmed from a combination of habitat-induced natural rarity and exploitation of shellfish resources by man (Hockey 1987); intense shellfish collection elsewhere, for example in Chile and South Africa (Siegfried 1994), may locally depress oyster-catcher populations. Coastal development and the recreational use of off-road vehicles have also undoubtedly had some local impacts, particularly in developed countries. This is thought to have led to a population decrease in *H. p. palliatus* and may be responsible for a 40% drop in the numbers of *H. fuliginosus* around Hobart since the 1960s. Off-road vehicles cause disturbance to *H. longirostris* in Australia, but the ultimate effects of this are unknown (Lane and Davies 1987).

In other species, causes of population trends are uncertain. There are, for example, no obvious reasons why the numbers of *H. o. longipes* should have decreased in central Siberia or why *H. o. osculans* also should have decreased over the past century. Reasons for the dramatic population increases in *H. o. ostralegus* and *H. o. finschi* since the mid-20th century, following decreases of both in the 19th century, are likewise unclear. The increases parallel increasing use of inland breeding habitats, but the causes of the habitat shifts in the first place have not been established.

Summary

There is much morphological and behavioural conservatism within the Haematopodidae. The only species which may be considered as morphologically and behaviourally aberrant, and probably represents the primitive condition, is *H. leucopodus* of southern South America. There are several species for which basic biometric and biological data are lacking, but it seems unlikely that these will produce any major surprises. And although there are many detailed and excellent accounts of various aspects of oystercatcher ecology and behaviour, evolutionary relationships within the group are not well understood. A thorough systematic study of oystercatchers is required before we can reasonably explain the evolution of sympatry in black and pied taxa in some parts of the southern hemisphere.

Relative to the Charadrii as a whole, a high proportion of oystercatchers are sedentary. Within the genus, *H. ostralegus* is the most migratory taxon, and breeds at higher northern, and thus more ice-prone, latitudes than any other species, except *H. bachmani*. Northerly breeding and migration in *H. ostralegus* is paralleled by a relatively large clutch size. The distribution of the genus as a whole is predominantly southerly, and *H. ostralegus* probably is one of the more recently evolved taxa. *H. ostralegus* of western Europe and New Zealand are currently the only populations undergoing major increases and occupying new breeding habitats. This may reflect greater behavioural vagility of a recently evolved taxon. However, thorough systematic studies are needed to determine which species represent relatively primitive and derived conditions. For example, *H. ostralegus* shares the inland breeding habit with *H. leucopodus*, thought to be a prototypical oystercatcher.

Although oystercatchers are among the best-studied waders, some aspects of their ecology have received surprisingly little attention. For example, detailed studies of rôle differentiation between sexes have never included measurements which test for differences in the field metabolic rates of males and females. Although the numerical impact of oyster-catchers on their prey has been assessed for some species, very few studies have addressed what this means for the population dynamics of prey organisms and the possible consequences for littoral community structure. Indeed, in general, studies have concentrated on the rôle of environmental factors in shaping the behaviour of the birds rather than considering the rôles the birds themselves play as agents of selection. Sedentary species of oystercatchers are particularly tractable in this regard, but the impacts of migratory or nomadic species which form flocks during the nonbreeding season could also be dramatic. Dietary segregation between sympatric species has never been well quantified, and the tropical/subtropical allopatry between *H. ophthalmicus* and *H. fuliginosus* begs an explanation.

Part II

Population ecology

11 Oystercatchers and man in the coastal zone

Rob H.D. Lambeck*, John D. Goss-Custard and Patrick Triplet

Introduction

The coastal zone of western Europe, an important breeding area and the principal wintering range of the Oystercatcher, has a dense human population. Consequently, the potential impact of man on the breeding, staging and wintering areas is large. There is a whole array of possible interactions. These may be direct, as from competition for food with fisheries, from shooting and from the abrupt loss of feeding areas due to land reclamation, or more indirect and gradual through changes to the ecosystem, such as those induced by eutrophication and pollution. Man can disturb but also protect areas. In this chapter, an overview will be given of present knowledge on the possible effects of the main human activities on Oystercatchers, with emphasis on the nonbreeding season.

Eutrophication and pollution

The impact of man on the quality of estuarine and coastal waters is large. Well-known environmental problems are related to the input, by air and water, of nutrients, heavy metals, oil-products and organochlorines, such as pesticides and polychlorinated biphenyls (PCBs). However, the range of modern chemical products is much larger, and the ecological effects of many of them are still poorly known. It is, moreover, not feasible to include them all in monitoring schemes. Although the significance of atmospheric deposition should not be underestimated, the obvious main sources in the estuarine environment, directly as well as by river input, are the run-off from agricultural land, discharges of untreated sewage, and the effluents of oil-refineries, chemical factories and purification plants. Shipping can also have effects, sometimes dramatically in the case of spillage accidents but mostly less conspiciously, as with waste of oil compounds and the release of antifouling products, such as tributyltin.

Organics and nutrients

Discharges of enormous amounts of organic waste occur in areas with,

* This is publication number 2000 of The Netherlands Institute of Ecology

for example, pulpmills, potato-flour and straw-board factories, or near cities with unpurified sewage. Increased loads of organic matter can induce several environmental changes which have consequences for the range and abundance of Oystercatcher prey. High amounts of organic suspended matter may decrease: (a) the primary production in the water by reducing light penetration (Eilers and Peeters 1988); and (b) the filtration capacity of filter-feeding molluscs by clogging their gills (Widdows *et al.* 1979; McLusky and Elliott 1981). Accretion of organic matter changes the chemistry of the sediment, and, under extreme circumstances only, its physical properties. Mineralization may seriously affect oxygen concentrations in the sediment and even in the water, depending on the amount of organics discharged and on the local hydrodynamic conditions (van Es *et al.* 1980; Weston 1990). An increased amount of organic matter in combination with a deterioration of oxygen conditions in the sediment reduces the number of benthic species present, induces a shift from benthic filter-feeders, such as Edible Cockles (*Cerastoderma edule*) and Edible Mussels (*Mytilus edulis*), to deposit-feeders, and from relatively large species to smaller opportunist species with short generation times. When conditions deteriorate still further, only a few tolerant species survive. Finally, permanent anoxia and, consequently, nearly azoic conditions can prevail (Pearson and Rosenberg 1978; Gray 1992).

Such drastic changes in the benthos would be highly unfavourable to Oystercatchers, which usually specialize on the larger bivalves (Hulscher, Chapter 1, this volume). However, semi-enclosed bays are more susceptible to large inputs of organics than well-mixed estuaries, with anoxia first occurring in the deeper parts before spreading to the shallower areas, especially in summer (Rosenberg 1985). In open tidal areas, anoxia sufficient to destroy life only develops around discharge pipes (Essink 1978). In estuaries, areas most affected are those with the highest sedimentation rates, which are normally situated in the more brackish zones. Fortunately, Oystercatchers mainly occupy the marine, and so the least affected parts of an estuary (Wolff 1969).

Discharges may even have a positive effect on Oystercatchers. Mineralization of organic inputs increases nutrient levels and, along with fertilizers in agricultural run-off, effluents from, for example, sewage works and atmospheric deposition, contribute to the so-called eutrophication of estuaries and coastal waters. The availability of nutrients mostly limits primary production (Granéli 1987; Radach and Moll 1990), although such complex relationships are not necessarily manifest in the field and, moreover, may vary between areas, not least because of the prominent role played by the local hydrodynamic conditions (Granéli *et al.* 1990; Radach and Moll 1990; Gray 1992). Nevertheless, the result of eutrophication may be an increase in both the length and intensity of phytoplankton or phytobenthos blooms (Cadée 1986). This, in turn, may improve the growth of some macrozoobenthos and, consequently, improve the feeding conditions for waders (van Impe 1985), including Oystercatchers. Indeed, the increase of zoobenthic biomass in the Waddenzee—the Dutch sector of the Wadden Sea—from 1970 to 1990 has been ascribed to eutrophication (Beukema and Cadée 1986; Beukema 1991).

However, there can also be detrimental effects from discharges, depending on the magnitude of nutrient increase. An increase in productivity can be accompanied by a shift in algal species composition, which may not benefit the food species of Oystercatchers. Thus, Beukema and Cadée (1991) recently showed that the growth rate of the Baltic Tellin (*Macoma balthica*) depends on the abundance of diatoms, whereas an increased abundance of a dominant flagellate had no effect. Beukema (1991) also found an increasing proportion of small-sized species, such as polychaetes, accompanying the increase in zoobenthic biomass. Such a gradual shift in benthic community structure, and even the disappearance of some prey species, may not matter to Oystercatchers or other waders as long as the biomass of their particular prey species continues to increase, or the increase in the alternative prey exceeds any losses in their normal food supplies (van Impe 1985). More important is the increased risk that blooms of toxin-producing algae will occur, resulting in die-offs of zoobenthos and fish (Gray 1992). In general, large blooms of phytoplankton species increase the risk of oxygen depletion in the sediment during the mineralization of the dead algae sunk to the bottom. Mortality due to algal bloom induced anoxia can also occur in intertidal areas. It frequently occurred, for example, in the Somme Estuary in France in the 1980s, resulting in the cockle populations in some sectors being greatly reduced. Desprez *et al.* (1992) claimed that this had no effect on the number of Oystercatchers because their feeding behaviour became in general more opportunistic and part of the population switched to other feeding habitats and other prey. However, these authors only related seasonal peak numbers of Oystercatchers to cockle abundance. Triplet and Etienne (1991) found reduced numbers in midwinter in bad cockle years, suggesting that some of the Oystercatchers had departed in such years. Since the passage of migrants and coldweather immigrants from the north have a great influence on Oystercatcher numbers in the Somme, a more thorough analysis of existing long-term census data is warranted using, perhaps, bird-days per winter as the measure of Oystercatcher abundance.

Another change induced by eutrophication is an increased abundance of macroalgae, chiefly of the sea lettuce *Ulva* sp. and *Enteromorpha* sp. (Nicholls *et al.* 1981; Soulsby *et al.* 1985; Raffaeli *et al.* 1989), dense mats of which can locally smother the sediment or mussel beds. The composition and abundance of the benthic fauna may change through inhibition of the settlement of larvae, anoxia, increased shelter for crabs and other predators (Hull 1987; Ólaffson 1988) and better conditions for epifaunal grazers, such as the mudsnail *Hydrobia ulvae* (Nicholls *et al.* 1981). Most wader species avoid dense algal mats on mudflats, probably because they create a physical barrier to feeding or because increased levels of hydrogen sulphide deter the birds (Nicholls *et al.* 1981). However, there is no firm evidence that algal mats actually affect wader numbers: in estuaries with an increase of algae, wader populations have also increased (Nicholls *et al.* 1981; Raffaelli *et al.* 1989). As long as the algae occur in just local patches, their influence is apparently low (Raffaelli *et al.* 1989). Indeed, any effect may be confined to the summer, since algal mats break up during autumn, especially in the more exposed marine parts of estuaries (Soulsby *et al.* 1985) that are favoured by Oystercatchers. In a study of the relation between Oystercatcher densities and mussel bed characteristics in the Exe estuary, Goss-Custard *et al.* (1992) found no effect of *Fucus* cover on bird numbers, although only one survey of the algae was available.

Contaminants

Several studies have shown that most of the heavy metals and organochlorines taken up each year by waders originates from west-European estuaries (Parslow 1973; Goede 1985; Goede and de Voogt 1985; Lambeck *et al.* 1991; Becker *et al.* 1992). Nonetheless, there are only a few cases where wader mortality can be directly related to such pollution. Alkyl lead in an industrial effluent killed about 1500 waders, chiefly Dunlin (*Calidris alpina*), in the Mersey estuary in 1979 but, as Oystercatchers hardly occur there, they were not affected (Bull *et al.* 1983). Alkyl lead is likely to be as toxic as inorganic lead (Osborn *et al.* 1983). Sub-lethal levels may result in altered organ weights and decreased energy reserves and, consequently, a lower chance of surviving harsh conditions (Goss-Custard *et al.*, Chapter 6, this volume). Goede and de Voogt (1985) suggested that some of the waders in the western Waddenzee may already be suffering from lead intoxication, because blood concentrations of an indicator metabolite were similar to those in mammals that had been affected. However, in contrast to mammals, waders can accumulate lead in feathers and eliminate it during moult, whilst salt gland excretion is possibly a second additional elimination mechanism (Evans *et al.* 1987). Perhaps because of this, lead levels for the kidney of these Waddenzee birds (Goede and de Voogt 1985) were far below those in waders that have actually died from poisoning (cf. Bull *et al.* 1983), suggesting that any lethal degree of intoxication was rather unlikely. Similarly, the levels of lead in Oystercatchers from the Burry Inlet, Wales, were also far below toxicity levels known from other bird species (Hutton 1981). A special case of lead intoxication arises from the ingestion of the pellets discharged by guns. In contrast to wildfowl, this also seems not to be of significance in Oystercatchers because of a sample of 105 birds found dead in The Netherlands, only one Oystercatcher had apparently died from lead intoxication (Smit *et al.* 1988).

In birds, the highest levels of cadmium are found in the kidney, with intermediate ones in the liver and low levels in other organs and feathers (Goede and de Voogt 1985; Evans *et al.* 1987; Stock *et al.* 1989). Cadmium is excreted via the urine, the nasal salt glands and possibly the preen glands (Evans *et al.* 1987). Loss rates for Oystercatchers are unknown, but the average half-life for renal cadmium in adult Dunlin was found to be about one year (Blomqvist *et al.* 1987). The concentration of cadmium in Oystercatchers shot in autumn in the Burry Inlet increased with age, rising from 4 µg g^{-1} DW in the kidney of juveniles to 21 µg in subadults and 43 µg in adults (Hutton 1981). Levels in females were significantly higher than in males, probably due to an increased uptake during egg production resulting from calcium–cadmium synergism. Oystercatchers that starved to death during severe frost in the German Wadden Sea contained lower concentrations than were recorded in the Burry Inlet, but subadult levels were similar or slightly higher than adult ones, while males had higher levels than females (Stock *et al.* 1989). Whether starvation affects the total level of cadmium and its distribution within the bodies of Oystercatchers is unknown. Although cadmium levels found in Oystercatchers are generally higher than in other wader species (Evans and Moon 1981; Goede and de Voogt 1985; Blomqvist *et al.* 1987), the highest levels found in the Burry Inlet sample were still below the concentrations known to cause histological kidney lesions in birds (Hutton 1981). As with lead, there is thus little evidence to suggest that cadmium levels are currently high enough to threaten Oystercatchers.

Mercury, mostly in the form of methylmercury, is another potential threat to Oystercatchers, since it is especially accumulated by molluscs, their main prey (Phillips 1977). Contamination with this heavy metal may lead to a decreased reproductive success in birds (Goede 1985; Becker 1989; Lewis *et al.* 1993). Pollution of an area with mercury will be reflected in higher concentrations in the eggs of locally breeding birds, as demonstrated for the Elbe Estuary by Becker *et al.* (1992). Although relatively high, levels in the eggs of Elbe Oystercatchers remain considerably below those in eggs of fish-eating birds such as the Common Tern (*Sterna hirundo*) that may approach critical values (Becker 1989; Becker *et al.* 1992). The levels of mercury recorded in organs and feathers of Oystercatchers from the Burry Inlet (Hutton 1981), and in other wader species from the Waddenzee (Goede 1985), were regarded as being low. However, these values are possibly unrepresentative because the birds were collected in early autumn, before the metal had had much time to accumulate in the birds. Thus, Parslow (1973) found in Knot (*Calidris canutus*), Dunlin and Redshank (*Tringa totanus*) in the Wash a roughly ten-fold increase in the liver concentrations of mercury over the winter, an accumulation that was apparently lost during the following breeding season or soon afterwards. Although females excrete mercury via the eggs, moult is probably the main cause of these seasonal fluctuations in mercury concentration. Feathers contain a high proportion of the total body load and this is shed when feathers are lost. Part of the mercury remaining in the body will then be redistributed to the newly formed feathers (Parslow 1973; Lewis *et al.* 1993). Indeed, mercury levels in new feathers from Knot and Bar-tailed Godwit (*Limosa lapponica*) from the Waddenzee were relatively high (Goede 1985). Birds living in a marine environment may, indeed, be better adapted to elevated concentrations of mercury (Parslow 1973), and possibly other heavy metals, than strictly terrestrial birds. The more powerful osmoregulation of coastal birds may contribute to a higher loss rate, as does an easier uptake of metal antagonists, such as zinc for cadmium and selenium for mercury (Hutton 1981; Goede 1985; Evans *et al.* 1987; Stock *et al.* 1989).

In contrast to most heavy metals, the increasing use in agriculture of pesticides, particularly chlorinated hydrocarbons such as DDT and dieldrin, has become a serious threat to coastal birds since the 1950s. This was clearly demonstrated by a well-known incident that took place in The Netherlands around 1965. Discharges from a processing plant near Rotterdam, heavily contaminated with insecticides, reached the Wadden Sea via the residual northward current along the Dutch coast. Accumulation of these toxic compounds resulted in a population crash of several breeding species, especially terns and Eider Ducks (*Somateria mollissima*) (Koeman 1971; Swennen 1972), even as far north as Germany (Becker and Erdelen 1987). The effects on Oystercatchers are less well-documented, although presumably a mass mortality did not occur. But perhaps because of poor breeding success, the breeding population in the German Wadden Sea showed a decline in that period, abruptly interrupting the steady increase in numbers that had occurred from 1950 to 1980 (Becker and Erdelen 1987). Eiders, which take similar foods to Oystercatchers, suffered a high mortality because females fast during breeding and were intoxicated by pesticides released from their fat reserves (Swennen 1972). A ban on this specific discharge and other, more general legislative measures, allowed the populations to recover. Nowadays, residues of pesticides in both eggs (Becker 1989) and organs of adult Oystercatchers (Everaarts *et al.* 1991) from the Wadden Sea are low.

Another category of synthetic chlorinated hydrocarbons, PCBs, is mainly transported to the coast by rivers that drain industrialized areas. PCB congeners are very lipophilic and are thus easily taken up by all aquatic organisms, including Oystercatcher prey (Langston 1978; Duinker *et al.* 1983; Hummel *et al.* 1990). Knowledge of the PCB structure (Fig. 11.1) is important, because large differences occur in the metabolizability and toxicity of individual congeners (Boon *et al.* 1989). Unfortunately, their analysis is complicated. Some 132 congeners, out of a theoretical maximum of 209, seem to occur in the natural environ-

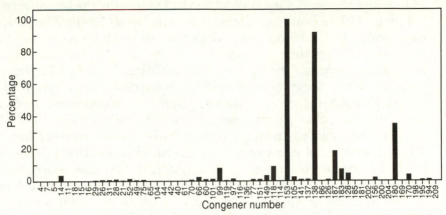

Fig. 11.1 Pattern of PCB congeners in the pentane extractable lipids of Oyster-catcher eggs from the inner Westerschelde estuary (SW Netherlands). In this analysis, 59 congeners could be identified using standard mixtures as reference. Numbering of congeners is according to international (IUPAC) nomenclature. Concentration of each detected congener is expressed in per cent of the dominant congener number 153, and the pattern shown is the average of chromatograms of 10 different eggs. Sequence of congener number on the x axis is according to sequence of elution from chromatography column (R.H.D. Lambeck, unpublished information).

ment, although the number assayed in pollution studies is lower and, moreover, variable. Because of this and differences in methods, it is difficult from the literature to compare the levels found in a specific organism.

PCBs are known to kill birds, but environmental levels seem generally to be too low to provide the high dosage required, apart from a few local cases (e.g. Koeman *et al.* 1972). A more likely effect is of a reduced reproductive success arising through embryonic deformation and egg-shell thinning on the one hand, and a decreased parental attentiveness on the other, such as has been found in a colony of an inland breeding tern species in the USA (Kubiak *et al.* 1989). For Oystercatchers, the risk of taking PCBs varies regionally. The summed concentration of 59 PCB congeners in eggs collected along the inner Westerschelde estuary in the South-west Netherlands was, on average, 43 µg g^{-1} pentane extractable lipids (PEL). This is 3.5 times higher than on the island of Texel in the west of the Waddenzee (R.H.D. Lambeck, unpublished information), where PCB-contamination is already regarded as a matter of concern because of its deleterious effect on seal reproduction (Reijnders 1986). Although Becker *et al.* (1992) found an increase in PCBs in eggs from the German Wadden Sea during the 1980s, there is no evidence so far that reproductive success of Oystercatchers has anywhere been affected.

Based on the sum of 22 major congeners, Oystercatchers from the western Waddenzee had an average total liver concentration of 4.5 µg g^{-1}

PEL in juveniles and 8 μg in older birds. Figures for the brain were about 0.4 μg g^{-1} PEL in both age groups (Everaarts *et al.* 1991). The bulk of the total body burden is, however, deposited in the adipose tissue. Oyster-catchers build up their fat reserves in late autumn and the total adult body mass increases by 50–60 g (Goss-Custard *et al.*, Chapter 6, this volume). When these reserves are metabolized, a redistribution of PCBs occurs through the body. Concentrations in vital organs, such as the brain and liver, thus strongly depend on the condition of a bird. Oyster-catchers that starved to death during harsh winter weather lost nearly 40% of the normal midwinter mass (Lambeck *et al.* 1991; Goss-Custard *et al.*, Chapter 6, this volume). As a result, the average concentration of PCBs in a sample of severe weather victims had increased by a factor 35 in the liver, and in the brain by a factor of 56 in juveniles and 120 in adults (Lambeck *et al.* 1991). Despite this large increase, concentrations still remained far below the toxic levels reported for other bird species. It thus seems unlikely that PCBs contributed to the death of ten thousand Oystercatchers in the Dutch Delta region during three successive winters with severe spells (Lambeck *et al.* 1991).

Another contaminant that has attracted much attention during the last decade is tributyltin (TBT), the active agent in anti-fouling paints used on ships. Related organotin compounds are also used as fungicides in agriculture. Nowadays, TBT affects the reproduction of many marine gastropods worldwide, and has adverse effects on the survival of both larval and adult bivalves, cockles and mussels being amongst the more vulnerable species (Beaumont *et al.* 1989). TBT may thus cause a decrease in the food stocks of Oystercatchers in areas situated near busy shipping or marinas, although there is no direct evidence for this. The use of this contaminant is, however, declining owing to an entire or par-tial ban in several countries.

Oil spills are one of the most widespread environmental threats. Waders are regularly smeared by oil that has drifted ashore, and some die. But compared to the mass mortalities that occur in seabirds (Camphuysen 1989), the importance is small. Nevertheless, the large oil spills remain a 'sword of Damocles' hanging over many of the important feeding areas for Oystercatchers and other waders.

Shellfishery and bait-digging

Despite their name, Oystercatchers rarely feed on oysters these days. These bivalves are only taken regularly on oyster farms in the Baie de Morlaix in Brittanny and around the Isle de Ré in South-west France. Here, *Crassostrea gigas* is the main prey for 10–25% of the local Oyster-catcher population which accounts for 8–11% of the oyster mortality, an amount regarded as low (Lunais 1975). Cockles and mussels figure much more importantly in the relationship between man and Oystercatcher.

Bait-digging for Lugworms (*Arenicola marina*) and Ragworms (*Nereis* sp.), is another human activity which affects an Oystercatcher food source, although these worms are usually not their main prey and indirect effects, such as disturbance from the bait-diggers themselves, may be of more importance.

Cockle fishery

Cockle abundance can vary considerably between years, because of large annual variations in spatfall and the predation of juveniles, along with a vulnerability to severe winters and storms (Beukema *et al.* 1993). Cockles are fished in several countries but this does not necessarily increase the annual mortality significantly because total landings are often much smaller than natural mortality, particularly in a severe winter. However, in many places fishing is done mostly in autumn, thus decreasing food stocks for the Oystercatcher before the winter arrives. Moreover, the size distribution of cockles eaten by man and Oystercatchers largely overlap, so a fishery can potentially affect the feeding conditions for the birds.

Traditional fishing by hand-raking is now uncommon in Europe, except in the Somme estuary. In The Netherlands, cockle fishing this way considerably increased during the 1980s, but the number of licences issued has recently been reduced. A more sophisticated method used to be employed in the Wash. Over high tide, a vessel would rotate around its anchor, so that the turbulence from the propellor would blow the cockles out of the sand. By drawing in the anchor ropes at intervals, a characteristic pattern of 'blowing rings' was created in the sediment and the exposed cockles were collected by hand over low water (Franklin and Pickett 1978). Nowadays, most fishing is done in Europe by hydraulic dredging from a boat at high tide. A metal basket, with bars 1–2 cm apart and fitted with sledges and a blade, is pulled over the bottom. The blade removes the upper 5 cm of the sediment which is then flushed by water jet into the basket, allowing sediment and smaller animals to pass through while the rest is pulled on board for processing. A modern Dutch vessel has a dredge on both sides and a 1 m blade and, aided by a draught of only 0.6 m, can dredge about 4 ha of intertidal flat a day.

Since blowing can remove the sand to a depth as much as 12 cm, the indirect consequences for Oystercatchers of both blowing and dredging may be even greater than the direct removal of large cockles. It may induce short- or long-term changes in the cockle population and in the benthic community in general, either directly by killing vulnerable species and juveniles or indirectly by physically disturbing the sediment. Because of injury, increased susceptibility to scavengers and displaced sediment, many of the undersized animals that are too small to be of commercial interest and thus allowed to pass the sieving baskets may be unable to resettle. In fact, it was shown in the Thames estuary that 25% of the undersized, usually juvenile cockles, were killed by dredging

(Franklin and Pickett 1978), while estimates of the mortality in the Dutch Wadden Sea varied from 10–50% (Anonymous 1987). After the experiment, however, the natural mortality in undredged plots was apparently higher than in dredged areas because the differences in density equalized in about six months. Movements of cockles into the dredged areas were probably of minor importance in this equalization process. Nor was the potential of the area to receive spatfall and to support cockles affected, even when the fishing was carried out shortly before spatfall. The infilling of dredge trails mostly occurred within eight months (Franklin and Pickett 1978). Since the burying depths of other benthic species taken by Oystercatchers, such as *Nereis* and *Macoma*, depends on their age, type of sediment and time of the year, the mortality of these alternative prey was variable, ranging from nearly zero to several tens of per cent, the highest losses occurring where fishery tracks overlapped. Measurable effects of the fishery gradually disappeared over 1–4 years (Anonymous 1987). However, as sessile animals living in tubes were mostly unable to resettle after dredging, it seems that frequent dredging can have a long-term effect on benthic community structure.

The extent of a fishery is therefore crucial (cf. Anonymous 1987). Total Dutch landings increased spectacularly from the mid-1970s, having now reached 7–10 million kg of cockle meat, while the fleet of large modern vessels fishes about 4000 ha each year. In years of high cockle numbers, only a few per cent of the whole stock is removed. But in the poor years of 1984–1986, for example, which followed severe winters, the percentages for the Waddenzee increased to 17, 36 and 31% per annum (Anonymous 1987). Because the mortality of cockles from severe winters is much less in the Dutch Delta, the fishing fleet moves there in such years. Given the great capacity of the vessels and the much smaller size of the area compared to the Waddenzee, the impact of fishing in the Delta can be large; in 1987, for example, the fishery achieved record national landings, despite the virtual absence of cockles from the Waddenzee. The overall density and biomass of adult cockles on a 16 km² intertidal flat in the western Oosterschelde declined by 90% and 70%, respectively, between March and November 1987. The fishery was thought responsible for at least 40% of this loss, implying that vessels had removed about 80% of the late summer stock of cockles (Lambeck *et al.* 1988).

More recent and precise data from the Oosterschelde demonstrated a natural mortality of adult cockles of about 10% between August and December 1989. In contrast, about 68% had disappeared in fished areas, with as much as 85% disappearing from the best cockle banks. Overall, 58% disappeared from the tidal flats that were studied, implying a mortality of 48% due to the fishery. Since the areas lying high in the intertidal zone are practically inaccessible to the vessels, the loss of cockles in 1989 due to fishing in the Oosterschelde as a whole lies between 29 and 38% (van Stralen *et al.* 1991).

That year, it was uneconomical for the fishery to collect cockles below densities of 30 m^{-2} (van Stralen *et al.* 1991). The lowest densities at which Oystercatchers can forage profitably presumably lie in the same range, however (Horwood and Goss-Custard 1977; Sutherland 1982c). In years with a low stock and high market prices, it is profitable for the fishery to take undersized cockles, including animals as small as 16–18 mm long (Lambeck *et al.* 1988). Since the minimum profitable size for Oystercatchers is around 15 mm (Zwarts *et al.*, Chapter 2, this volume), the birds cannot really escape the effects of fishing by taking small cockles. However, the present Dutch regulations impose a minimum size to fished cockles which is relatively favourable to birds (M. van Stralen, personal communication). Because the best cockle areas are nearly completely fished out and the growth and flesh content of cockles from the higher parts of the intertidal zone is relatively low, the cockle fishery is easily able to reduce by over one half the food supply of the Oystercatcher in this major wintering area.

In The Netherlands, cockles comprise 50% (Oosterschelde) to 75% (Waddenzee) of the food intake of Oystercatchers, with similar values being reported for some British and French estuaries (Triplet and Etienne 1991; Goss-Custard *et al.* 1996a). The birds may consume up to several tens of per cent of the stock (Goss-Custard *et al.* 1996a). Studies in the Burry Inlet of overwinter changes in cockle densities in netted areas, that excluded both the birds and fishing, and in fenced areas, that excluded only fishing, suggested that Oystercatchers were responsible for a considerable part of the sometimes large mortality of second-winter cockles that occurred there. Since these losses took place just as the cockles entered the size range that could be legally fished, they were viewed as serious to the fishery, so Oystercatchers were regarded as a pest (Davidson 1967). This resulted in an extensive culling programme (Prater 1981). After a re-examination of the data, however, Horwood and Goss-Custard (1977) concluded that Oystercatchers could not have been the sole cause of the large losses of second-winter cockles. In years of medium or high cockle abundance (>400.m^{-2}), when as much as 80–90% of the stock disappeared over the winter, the birds themselves could only have removed 22–34%. The evidence suggested that illegal fishing was responsible for much of the remainder of the loss. The exclosure experiments which concluded that the birds were responsible had been flawed. Excluding the fishery by fencing an area creates a very rich patch of cockles compared with the surroundings, thus attracting Oystercatchers until they have grazed the stock down to the densities occurring around the fenced area. Accordingly, the original experiment exaggerated for the Burry Inlet as a whole the impact that Oystercatchers had on the second-winter cockles. Nonetheless, an equilibrium model of the fishery estimated that, even without illegal fishing, Oystercatchers reduced the long-term yield to the fishery by up to 40%. Indeed, in years of low stocks of second-winter cockles, their short-term effect could be much greater (Horwood and Goss-Custard 1977; Goss-Custard *et al.* 1996a).

In this case there is thus a genuine conflict of interest between birds and the fishery. Much depends, however, on the minimum size of cockle that is allowed to be fished and on local conditions. For example, in the Somme estuary, the minimum legal size is 30 mm, and this can be enforced since all fished cockles have to be screened for health purposes. The local Oystercatchers mostly eat smaller cockles. Given a considerable natural winter mortality of cockles, accompanied in some years by a very high summer mortality, Oystercatcher predation here is not regarded as making the situation worse and is thought to have no serious impact on fishery.

By contrast, a conflict of interests does exist in The Netherlands where the usual minimum fishable cockle size is 20–22 mm. In the Oosterschelde, the effect of the fishery on Oystercatchers first became clear in 1987. The intensive fishery that year reduced the adult cockle stock on the two intertidal flats of the western sector to an average density of 35 m^{-2}. This made it difficult for Oystercatchers to meet their energy demands, and average adult body mass in midwinter was 30 g lower than in previous years. In contrast, the weights of birds feeding in the nearby eastern sector, where cockles remained abundant, remained at normal levels (Lambeck 1991). But despite the lower weights of birds in the western part of the Oosterschelde, the mortality of Oystercatchers was not unusually high as the winter was extremely mild and wet, so the birds were able to supplement their intake by increased inland feeding at both low and high tide. Nonetheless, the difficulties the birds were experiencing were clearly demonstrated during three fine frosty days in December, when unusually large numbers of birds emigrated to France; normally, such a response only occurs after much longer spells of really severe weather (Lambeck 1991; Hulscher *et al.*, Chapter 7, this volume).

Due to the combined effects of very poor spatfalls in both 1989 and 1990 and of the fishery, the cockle stocks in the tidal flats in the western Oosterschelde reached a level in the winter 1990–1991 that was even lower than it had been in 1987–1988. The local winter population of Oystercatchers promptly decreased from its normal level of 18000 birds to 6000, by far the lowest figure recorded since the census began in 1965 (Lambeck *et al.* 1989). Many Oystercatchers apparently moved into the central part of the Oosterschelde, where cockles were available in the 2000 ha of mussel-culture lots that were protected from the cockle fishery. A large increase, from 400 to 1800 birds, was also recorded in a nearby small Belgian estuary, and more birds than usual were observed along the North Sea beaches. Despite this redistribution of birds, over 1200 starved Oystercatchers were found dead in the Dutch Delta during a cold spell lasting two weeks, this represents some 2% of the midwinter population. This percentage mortality is slightly higher than that

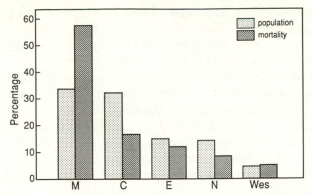

Fig. 11.2 A comparison between the relative size, expressed as a percentage of the total number of Oystercatchers ringed, of the populations originally ringed in four sectors of the Oosterschelde and in the Westerschelde (Dutch Delta) and their relative contribution, as a percentage of the number found dead, to the number of ringed victims of a cold spell in 1990–1991. M = western sector or mouth, C = centre, E = eastern sector and N = northern branch of the Oosterschelde while Wes = Westerschelde (R.H.D. Lambeck, unpublished information).

recorded in 1985, when the weather conditions were even more severe (Lambeck 1991; Meininger *et al.* 1991). Compared with elsewhere in the Oosterschelde, Oystercatchers in the western sector suffered a three times greater rate of mortality (Fig. 11.2). As argued by Goss-Custard *et al.* (Chapter 6, this volume), it is in periods of severe weather that Oystercatchers are most likely to pay the penalty for not having been able to feed at an adequate rate earlier in the winter. In this case, it seems that the poor feeding conditions experienced by the birds before the severe weather arrived, due partly to poor spatfall, had been exacerbated by the commercial cockle fishery.

Dutch policy makers became aware of the conflicts of interest involved and new regulations were made. Several areas, comprising some 25% of the Waddenzee, were permanently closed to fishing. A fundamental assumption made in the legislation concerned the amount of cockle food required by all the bird populations, with Oystercatchers and Eider Ducks being the principal consumers; this amounted to 12.6 million kg fresh meat in the Waddenzee and 3.4 million kg in the Oosterschelde. A similar calculation was made for the stock of natural mussels in the two areas. Each year, a survey must now provide data on the distribution, numbers and biomass of cockles, with only densities of above 50 m^{-2} being regarded as harvestable by birds. In years of a poor cockle stock, at least 60% of the food demand of the birds is required to be protected from fishing and so remain available to the birds. Depending on the situation, the management options can range from the tempory closure of specific areas of the fishery up to a complete ban (Revier 1993). Further fine-tuning can be arranged through a quota system for ships and instructions for reducing the losses of smaller cockles.

Ownership of fishery firms is becoming increasingly international and regulations in one country can easily increase pressure in another, as has been shown by recent developments in the Wash. The allowed maximum size of ships there is much lower than in The Netherlands, but intensive dredging in combination with poor spatfall apparently largely decreased the local cockle stocks during 1992. The effects of this on the Wash Oyster-catchers seem to have been even worse than in the Oosterschelde. Besides a decline in numbers, reduced body mass and more inland feeding, moult was apparently delayed and mortality in autumn was higher than usual (Clark 1993).

Mussel fishery

Mussel beds are very important feeding areas for Oystercatchers. Densities can exceed 100 birds ha^{-1} (Zwarts and Drent 1981) and several studies have shown that Oystercatchers can have a considerable impact on the mussels. Combining data from beds of all qualities, Goss-Custard *et al.* (1980) calculated that up to 30% of the autumn stock of the large mussels taken by Oystercatchers in the Exe estuary were eaten by them over the winter. An estimated 40% of the mussels greater than 40 mm long (*circa* 3 years old) near Schiermonnikoog in the Waddenzee was consumed during the year, the decrease of mussels in isolated clumps being twice as high as that on the continuous banks (Zwarts and Drent 1981). Baird *et al.* (1985) calculated that 12.6% of the annual produc-tion of the three main bivalves in the Wadden Sea tidal flats, including the mussel, was consumed by Oystercatchers whereas, in the Ythan estuary, 15% of the mussel biomass and production was consumed annually (Baird and Milne 1981). Craeymeersch *et al.* (1986) estimated that 40% of the total annual production of a mussel bed in the Oosterschelde was eaten by Oystercatchers; this increased to 60% when only the most im-portant mussel size class (30–50 mm) to the birds was considered. Since only 60% of these mussels were actually accessible to Oystercatchers, the depletion rate over the year was even higher.

Such high rates of impact by Oystercatchers can easily bring them into conflict with mussel fishermen. In most areas, mussel culture relies on the availability of natural banks of small juveniles. Part of this so-called seed is transplanted to usually subtidal lots where growth conditions are optimal, and marketable sizes can be reached within two growing sea-sons. In The Netherlands, mussel fishing is nowadays virtually confined to the Waddenzee, where roughly 20% of the seed comes from the inter-tidal zone while the rest is taken from subtidal banks. Nowadays, large amounts are also delivered to the Dutch from the German and Danish Wadden Seas. Mussel culture used to be confined to the Delta area, which is still the home base for the fishermen and of the trade, but the western Waddenzee has become as important a growing area over recent decades. Although the growth and development of seed banks into mussel beds that are attractive to Oystercatchers is impeded by the fishery, the majority of such banks that are situated in exposed areas would have disappeared anyway within a few months, due to storms and, sometimes, ice. Furthermore, Dutch fishermen have created new banks in the intertidal zone, mostly in the Oosterschelde, where surplus mussels are stored. Although mussels can sometimes be put on one site this way for several years, the turnover at such storage places is usually high, thus providing feeding conditions for Oystercatchers that may be even more favourable than on natural beds because a higher proportion of the mussels is accessible and fewer are overgrown by barnacles and other epifauna (Meire and Ervynck 1986). Indeed, the increased cultivation of mussels during this century may have raised the winter survival of Oystercatchers and contributed to the long-term growth in the population (Cramp and Simmons 1983).

Recent developments may be less favourable, however. In response to a reduced supply of mussels and, in part, an increased predation pressure by Oystercatchers and gulls, mussel farmers are tending to reduce the extent to which they use intertidal plots. After three years of bad spatfall in 1988, 1989 and 1990, most mussels in the entire Wadden Sea were fished away and virtually no naturally established banks were left (Beukema 1993). Nonetheless, Oystercatcher numbers on the Balgzand in the westernmost Waddenzee remained unchanged. Based on zoobenthos data, Beukema (1993) claims that Oystercatchers survived to the end of winter by switching to alternative bivalve prey and Lugworms, whose stocks were thereby reduced to unprecedently low levels. Mass mortality of Oystercatchers was not reported from elsewhere in the Wadden Sea either, but the Eider Duck, which feeds mainly on subtidal mussels, suffered serious winter losses in 1991 and many birds left the area (Swennen 1991; Revier 1993).

The long-term consequences for Oystercatchers of such large-scale changes in mussel culture are poorly understood. The few studies available have dealt only with the local effects of mussel fishing on Oystercatchers. For example, an experimental study showed that, in response to the partial fishing of a mussel bed in June, an increase in the densities of Oystercatchers and Herring Gulls (*Larus argentatus*) occurred, owing to the presence of large numbers of damaged and detached mussels which the birds found easy to take. After a few weeks, bird numbers declined to their usual level, and Oystercatcher intake rates in the fished and unfished areas became similar (Meesters and Münninghoff 1986). This study did not cover the subsequent winter months, however. Similarly, Zwarts and Drent (1981) found only short-term effects when one mussel bed was depleted: the birds simply redistributed themselves over the remaining feeding habitats. Nevertheless, a large-scale disappearance of mussel beds for a long period would be expected to have a much larger adverse impact than was recorded in these two studies of quite local changes. As described by Goss-Custard *et al.*, techniques are now being developed that may allow such effects to be anticipated at both local (Chapter 12, this volume) and global (Chapter 13, this volume) scales.

Bait-digging

Digging for large worms as bait for fishing for sport is widespread around the European coasts. Digging is mostly done by hand but, in some areas, it is also done mechanically by a flat-bottomed boat with specialized equipment. The density of worms, the accessibility of an area and the local market for selling bait determine the amount of exploitation of unprotected tidal flats that takes place; in suitable areas, several tens of bait-diggers can be at work simultaneously. Clearly, the disturbance resulting from this activity potentially threatens the feeding opportunities available to Oystercatchers. Because the sediment is dug up to a depth of 40 cm, bait-digging may also have an effect on the local macrofauna, and thus on predators such as the Oystercatcher.

When calculated for an entire coastal area, the impact of digging on the abundance of a bait species is small; an estimated 1–1.9% of the Lugworm population is removed annually from the Oosterschelde and 0.8% from the western Waddenzee (van den Heiligenberg 1984). Locally, the figure can be much higher. Thus, Blake (1979a,b) estimated the annual removal of Lugworms at 7.8% and of King Ragworms (*Nereis virens*) at 23% in tidal flats near Newcastle, and intensive digging almost depleted the adult Lugworm stock in a bay of the Lindisfarne Reserve in North-east England (Townshend and O'Connor 1993). Because smaller worms are easily missed by the bait-diggers and because adult worms rapidly immigrate into dug areas (van den Heiligenberg 1984; McLusky *et al.* 1983), the local populations of such worms are never endangered. Nevertheless, there is evidence that the benthic community may change in areas heavily frequented by bait-diggers. Temporal changes in microrelief and sediment texture (McLusky *et al.* 1983) may play a part in this as well as differences between species in their vulnerability to digging. For example, mudsnails and Baltic tellins are not affected by digging (McLusky *et al.* 1983), whereas cockles that are left on the surface or buried under more than 10 cm of dug sediment often die, which can result in a considerable reduction in the cockle population in popular digging areas (Jackson and James 1979). This effect may be exacerbated by frequent treading of the surface, which also increases mortality of cockles and some other shallow-living species (van den Heiligenberg 1984). Uncontrolled digging for King Ragworms may also have deleterious effects on natural mussel banks where these worms are relatively common. Because of the larger capacity and greater efficiency of mechanical digging, the unrestricted deployment of such techniques may be harmful to the tidal flat ecosystem. On the other hand, boats will not disturb foraging Oystercatchers, which is inevitable when people dig over low tide. Although opportunistic feeding by Oystercatchers around fresh dug mounds may partly compensate for this, it may be supposed that frequent and large-scale bait-digging will affect the number of Oystercatchers that can feed on a tidal flat, as was demonstrated for some other waders in Lindisfarne (Townshend and O'Connor 1993).

Sports fishing in coastal waters has increased since the war and, along with it, so has bait-digging, much of which is now done on a commercial basis. A conflict of interests arose in the Dutch Delta area where, during 1960–1986, several large dams were built between the islands, opening up much more of the tidal flats to exploitation. However, recent regulations limit bait-digging to specially assigned tidal flats, preferably those that are less important to birds. A similar management policy was pursued in Lindisfarne (Townshend and O'Connor 1993). Another solution to fulfil the still increasing demand for bait comes from the specialized mariculture farms, that hatch King Ragworms. In the long term, such large-scale commercial developments may be crucial to prevent a too large presence of people in feeding areas of Oystercatchers and other waders.

Disturbance

People can disturb feeding or roosting Oystercatchers in many ways. Deliberate actions, such as walking towards a flock, or shooting at birds, are the most obvious examples. Disturbance can also result unintentionally from leisure activities, such as dog-walking, surfing, birdwatching and boating, and from other activities, such as low-flying airplanes and helicopters, bait-digging, seismic research, drilling, military practices and scientific research. The number of disturbances caused directly or indirectly by people in some areas is many times greater than that caused by natural factors, such as predators (Tensen and van Zoest 1983; Kirby *et al.* 1993). Since leisure activities in particular have increased substantially in recent decades, birds are probably more affected nowadays than in the recent past (Prater 1981; Goss-Custard and Verboven 1993).

It is difficult at present to quantify the real impact of disturbances on Oystercatchers, although attempts are now being made by modelling (Goss-Custard *et al.*, Chapter 12, this volume). The response depends on the type of disturbance, possible synergistic effects of other disturbing factors, and on their own previous experience and physiological state (hungriness) and on the behaviour of other birds nearby. For example, the reduced opportunities to habituate may explain why birds roosting in tidal habitats, such as salt marshes, are more easily disturbed than birds on inland sites (Smit and Visser 1993). The frequency of predator attacks and, especially, shooting may also be responsible for the large differences in response found between areas (Meltofte 1982). In the Banc d'Arguin, the Oystercatcher is one of the most wary waders, flying up in response to an approaching walker at distances of 400–500 m (Smit and Visser 1993), and a similar 300–400 m was found in the Somme estuary and in other French estuaries where hunting is allowed (P. Triplet, unpublished information). By contrast, elsewhere in western Europe, its tolerance is in general higher than that of most other species (van der Meer 1985; Kirby et al. 1993; Smit and Visser 1993). For example, Tensen and van Zoest (1983) showed that, when approached, the flight distance of inland roosting Oystercatchers on Terschelling was 60 m whereas it was 95 m in Curlews (*Numenius arquata*). The difference in tolerance between these two species was even larger for birds on tidal flats; average distances of respectively 85 and 211 m were measured in the Dutch Delta area (van der Meer 1985), and 136 m versus 339 m in the Waddenzee (Smit and Visser 1993). Interestingly, response distances considerably declined when a person with whom the birds were familiar, such as a local farmer, caused the disturbance (Smit and Visser 1993). Similarly, a group of people is more disturbing than a single person, as is the presence of a dog (Kirby et al. 1993; Smit and Visser 1993). Indeed, the appearance of a person, the vehicle he is using, and the predictability of his behaviour plays a part; a wind surfer was more threatening to roosting birds, for example, than a person in a kayak (Koepff and Dietrich 1986). Foraging Oystercatchers are, in general, quicker to take wing than are roosting ones, although, again, habituation plays a role; birds close to the seawall of Terschelling had an average flight distance of 79 m as compared to 113 m on the flats further away. The status of birds was probably also important, since territorial birds on mussel beds took flight at on average 77 m, two thirds of that at which birds on the adjacent mudflat flew away (Smit and Visser 1993).

Smit and Visser (1993) further found that airplanes and helicopters are particularly disturbing. Altitude, speed, size and whether they flew straight ahead or made unexpected turns were important in determining the response of the birds. Small planes flying above 300 m alarmed roosting Oystercatchers and other waders in 8% of cases whereas, below that altitude, the frequency rapidly increased and it is usually 100% when the plane passes below 150 m. In comparison, a low-flying military jet within 1200 m had relatively little effect; roosts were disturbed in 5% of the cases, while at low tide, birds did not even stop feeding. In contrast, a low-flying helicopter disturbed Oystercatchers in 27% of cases when within 1500 m and in 73% when within 250 m. Oystercatchers may habituate to planes to some extent when the disturbances they make are predictable and when the birds are not approached too closely. The opposite may also occur as jet fighters are associated with heavy explosions at shooting ranges.

The recovery time for Oystercatchers following disturbances caused by people is longer than those caused by more natural events. In Zeger's (1973) study near Schiermonnikoog, feeding was resumed within 15 minutes when a person walked nearby. This implies an average recovery time of 7.5 minutes because the Oystercatchers returned only gradually to the area. The average time increased to 30 minutes after a severe disturbance caused by several people walking slowly across the area. A few preliminary tests in the Dutch Delta area (van der Meer 1985) and in the Somme estuary (P. Triplet, unpublished information) corroborated these results. A time loss of 10 minutes was found by Smit and Visser (1993) after the passage of a low-flying plane, whereas van der Meer (1985) recorded a value of 30 minutes on two occasions in the Dutch Delta. However, recovery times depend on the hunger of the birds, tidal conditions, and, particularly, on their preceding experience. After being approached for the second time, about half the Oystercatchers left a study area for the remainder of that tidal period (Metcalfe 1989). Similarly, Smit and Visser (1993) found that repeated disturbance by the same plane lengthened the recovery time four-fold.

The area affected by a disturbance is also highly variable, being much smaller around a stationary person than one that moves around. Van der Meer (1985) calculated that the total size of the area that, due to a walker, was either abandoned by Oystercatchers or where their feeding was interrupted, was described by the formula:

$$\text{Area (in m}^2) = \tfrac{1}{2}\pi r_2^2 + 2r_1 h_1 s + 2(r_2 - r_1)h_2 s \qquad (11.1)$$

in which s is the walking speed of the person in metres per second, r_1 is the flight distance for birds in metres (85 m), r_2 is the disturbance distance in metres at which birds stop feeding (121 m), h_1 is the recovery time in seconds for birds that flew away (900 s), and h_2 is the recovery time in seconds for birds that stopped feeding (300 s).

Using the parameter values measured in the Dutch Delta and given in brackets, a single walker disturbs an area of 20 ha in which Oystercatchers stop feeding or fly up; for comparison a light aircraft flying at a height of 150 m would disturb an area of over 15000 ha. For a person such as a bait-digger, remaining on the same spot for a long time, the formula simplifies to

$$\text{Area (in m}^2) = \pi r_3^2 \qquad (11.2)$$

in which r_3 is the distance of approach in metres, this being the shortest distance from that person at which the birds feed normally. Assuming that r_3 is intermediate between r_1 and r_2, a bait-digger prevents Oystercatchers feeding in an area of roughly 3 hectares.

Few data are available on the time actually lost through disturbance, but this clearly depends on its frequency. Two planes removed 12% and 17% respectively of the available feeding time during one low tide period (van der Meer 1985). Averaged over the summer, roosting Oystercatchers

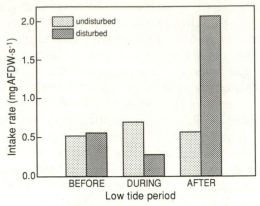

Fig. 11.3 Compensatory feeding by a marked Oystercatcher in its mussel bed territory after an experimental disturbance (Hooijmeijer 1991).

were on the wing on average for 38 seconds per hour due to walking people on Terschelling (Smit and Visser 1993). The effect of such disturbances will be lower in the colder periods of the year when birds are more tolerant and fewer people are present (Smit and Visser 1993). For many Oystercatchers, a certain amount of disturbance is unlikely to be significant because the average bird does not use all its available feeding time anyway, even in winter (Goss-Custard et al. 1977a). Furthermore, there is evidence from cage experiments that Oystercatchers can compensate for lost foraging time by raising their intake rate above normal levels (Swennen et al. 1989), as recent data (Hooijmeijer 1991) on another bird has confirmed (Fig. 11.3). Frequent disturbances, such as shooting, that is so common on tidal flats in many French areas and locally also in Denmark (Meltofte 1982), may have adverse effects on Oystercatchers, especially on the minority that have difficulty in acquiring enough food during severe weather. But so far, the impact disturbance has on this particularly vulnerable section of the population remains to be estimated, although attempts to do so by modelling are in progress.

Clearly, at some point, an increasing frequency of disturbance will make an area unsuitable for Oystercatchers and they will leave. On the feeding grounds, this will increase the density of foraging birds elsewhere, and thus depress intake rates (Ens and Cayford, Chapter 4, this volume). If it only concerns a roosting area, Oystercatchers would normally be expected to find another location in the immediate surroundings (Smit and Visser 1993), but this is not always possible. For example, when high-tide roosts in the Dee estuary became unusable due to leasure activities, Bar-tailed Godwits and Knots started roosting in an adjacent estuary, involving up to an extra 40 km of flight per tidal cycle (Mitchell et al. 1988). Oystercatchers feeding in the Baie d'Authie in northern France cannot roost there because of frequent shooting in combination with there being few roosting alternatives, and so fly to a reserve in the Baie de Somme (F. Sueur, personal communication), adding a 28 km flight per tidal cycle. Such adaptations cost energy and may increase the proportion of birds in relatively poor condition. Whether this reduces their survival or promotes emigration, is as yet unknown.

Shooting

Hunting Oystercatchers has been done for both culling and recreational purposes. Culling was particularly practised in the United Kingdom, when Oystercatchers were regarded as a pest of cockle fisheries; some 16 000 were shot in Morecambe Bay during 1956–1969 and 11 000 in the Burry Inlet in 1973–1974. But because culling failed to stop the decline in local cockle stocks, it has not been tried subsequently (Prater 1981).

Nowadays, legal recreational shooting is only practised in France because in Denmark, the only other northern European country with a relatively large population of hunters, the Oystercatcher has been protected since 1983. Before then, there was an open season from 1 August to 1 January. No direct assessment of the pre-1983 Danish bag of Oystercatchers is available, but an average annual kill of 52 000 birds from 1961–1962 to 1971–1972 has been recorded for an administrative category of ten wader species, including the Oystercatcher (Møller 1978). Because the Oystercatchers presumably did not belong to the five most favoured game species amongst these ten waders (Møller 1978), not more than a few thousand Oystercatchers, the majority juveniles (Lambeck *et al.* 1995), were probably shot each year in Denmark, taking into account the additional losses of wounded birds.

The present shooting season in France varies between Départments, but on average extends from 20 July to 28 February. Hunting occurs along the whole coast, except in reserves. The largest bags are apparently obtained between the Belgian border and southern Normandy, since 80% of the French ring recoveries are reported from there (Triplet *et al.* 1987), even though only a quarter of the French Oystercatcher population winters there. Because many migrants often do not have experience of hunters, it is perhaps likely that their chances of being shot are higher during the first part of their flight along the French coast. But data on the total annual kill are lacking. An estimate of 1000 birds made by Woldhek (1979) is nowadays thought probably to be too low. With an annual mean of nearly 40 recovered foreign-ringed birds from 1978 to 1985, of which presumably 90% were shot (Triplet *et al.* 1987), the annual kill would be 1800 birds on the assumption that 5% of the migrants are ringed and the percentage of shot birds reported is 40%. Since both percentages are relatively high, it is quite likely that even larger numbers are killed. Most victims are juveniles or immatures, probably because they may migrate further south than adults (but see Hulscher *et al.*, Chapter

7, this volume) and are less experienced (Cramp and Simmons 1983; Triplet *et al.* 1987).

In some winters, severe weather to the north causes large numbers of Oystercatchers to move south-westwards into France and shooting mortality increases (Triplet *et al.* 1987; Hulscher *et al.*, Chapter 7, this volume). A recent example illustrates the potential that French hunters have for killing Oystercatchers in such circumstances. On the 14 January 1987, at least 63 000 Oystercatchers were seen migrating southwards along the Dutch North Sea coast (Hulscher 1989; Hulscher *et al.*, Chapter 7, this volume). According to the regular midwinter counts held a few days later, only 80 000 birds stayed behind in the partly frozen Dutch Wadden Sea, some 120 000 or 60% less than normal. Extrapolating to the German and Danish Wadden Sea, Hulscher (1989) suggested that about 235 000 Oystercatchers had taken part in this cold weather movement. However, a more reliable estimate may be 100 000–150 000 birds. Ring recoveries showed that Danish birds were hardly involved (P. Triplet, unpublished information), and the numbers remaining in the Wadden Sea, from which the calculation was made, may have been underestimated as many birds moved to more remote areas with open water (Temme and Gerß 1988; Hulscher 1989). Of the 394 colour-ringed individuals in the 1986 breeding population of Schiermonnikoog in the eastern Dutch Wadden Sea, at least 19% did not return after that winter. Four birds were found dead on Schiermonnikoog while nine were reported shot in France, and three were found on Dutch and Belgian beaches (Hulscher 1989). Assuming that 60% of that population emigrated, some 236 of these colour-ringed birds may have left the Wadden Sea. This allows a crude estimate to be made of the number of victims shot in France, using two alternative calculations, based on the proportion of 20 out of 96 (0.21) disappeared birds that were reported dead. In one, all birds were assumed to have a 0.21 chance of being reported, irrespective of the cause of death. In the second, shot birds were assumed to have a 0.4 chance of being reported while, *ergo*, the chances of birds dying for other reasons being reported must have been 0.15; note that this is certainly too low for areas specially searched for victims, such as Schiermonnikoog and the Dutch Delta area (R.H.D. Lambeck, unpublished information). On the first calculation, 43 of the emigrated birds ($\approx 18\%$) would have been shot, and in the second 23 (10%). Assuming this Schiermonnikoog sample is representative for all 100 000–150 000 involved, a minimum of 10 000–15 000 and a maximum of 18 000–27 000 Oystercatchers were shot in France that winter. The large size of the kill in January 1987 is corroborated for the Calvados coast, Normandy, by Debout (1990) who, on basis of the number of hunters, the number of shots fired by each and their success rate, arrived at a minimum kill of 10 000 Oystercatchers for this region alone. This represents 25–33% of the recorded local influx of 30 000–40 000 birds, ten times the normal population. Most birds died in the first four days after they arrived from the north, after which the shooting season was temporarily closed for the rest of the cold spell. The total number of foreign-ringed Oystercatchers recovered in France during the season 1986–1987 was ten times higher than the annual average of the preceding eight years (P. Triplet, unpublished information), further suggesting that an annual kill in France of 1500–2000 birds in normal years is indeed realistic.

In conclusion, legal hunting is not nowadays normally a substantial mortality factor, except in some winters in which a large number of exhausted and inexperienced birds arrive in France, displaced by extreme weather in their northern wintering areas. In normal years, juveniles and immatures are most at risk from hunters, whereas in cold winters with large-scale movements, adults also suffer large losses. Finally, it should be noted that the extent to which illegal shooting occurs is unknown although, considering the recoveries of birds shot, for example, in Fenno–Scandinavia and Russia, it certainly does occur widely.

Habitat loss

Besides activities that cause the immediate death of birds, the irreversible destruction of feeding areas is potentially the most harmful human influence on Oystercatchers in the intertidal zone. There is a continuing process worldwide of land-claim in estuaries and shallow coastal bays, closely linked to the increasing human population (Davidson *et al.* 1991; Evans 1991). Many British estuaries have lost over 25% of their area during the last three centuries and, in some, virtually all tidal flats have been removed (Evans 1991). Current proposals for land-claim, storage schemes and barrages threaten up to 10% of the remainder (Prater 1981; Davidson *et al.* 1991). In France and southern Europe, the story is similar. In the Seine estuary, for example, over 75% of the tidal flats have disappeared since 1850 (Desprez and Dupont 1985). In The Netherlands, there is a long tradition of accelerating accretion by various simple technologies and reclaiming intertidal areas, usually for agriculture. Nonetheless, the process has not been entirely one way. Many of the most recently built polders were land in historic times; indeed, geological transgression combined with storm disasters created the Wadden Sea only a few thousand years ago. Similarly, the first small polders in the Delta area date from the Middle Ages, yet the estuaries enlarged considerably up until the 17th century (Pieters *et al.* 1991). The pace of land-claim recently has been accelerated by technical developments which allow thousands of hectares to be embanked in one project.

These trends have led to increased demands to predict the effects of further winter habitat loss on the numbers of Oystercatchers and other shorebirds. Other chapters in this volume discuss the attempts being made to make such predictions at the local (Goss-Custard *et al.*, Chapter

12, this volume) and global (Goss-Custard *et al.*, Chapter 13, this volume) scales. This chapter reviews studies which have monitored and investigated the effects of habitat loss on local Oystercatcher numbers.

The first in-depth study of the effect of habitat loss on shorebirds was carried out during 1971–1975 in the estuary of the River Tees in Northeast England where, over the preceding hundred years, over 80% of the original 2400 ha of intertidal flats had been reclaimed for docks and industrial developments. Although the combined effects of further habitat loss and a reduced feeding time were studied in detail on eight species (Evans 1979; Evans *et al.* 1979), Oystercatchers were unfortunately very scarce. A second study of habitat loss and local shorebird numbers was made in the Højer/Rodenäs area in the Wadden Sea, around the Danish–German border. A sea-wall was constructed during 1979–1981 which embanked 800 ha of grazed saltmarsh and 300 ha of the tidal flats in the Danish territory (Margrethe–Kog) along with 525 ha of saltmarsh and 25 ha of tidal flat in the German sector (Rickelsbüller Koog). In the first two years, the number of waders in the German sector remained unchanged, but an average decrease of 76% occurred in the Danish sector. Oystercatchers were not affected, however, because their main feeding areas lay outside the embanked area. Interestingly, population sizes of waders in this part of the Danish–German Wadden Sea as a whole have not changed; there seems only to have been a large-scale redistribution of birds (Laursen *et al.* 1984). An evaluation of possible long-term changes has not yet been made.

A third study, however, was able to focus on the Oystercatcher. It was carried out in the Delta area, the former estuarine complex of the rivers Rhine, Meuse and Schelde in the South-west Netherlands, one of the top areas for migratory and wintering waders in Europe (Wolff 1967; Saeijs and Baptist 1980). After a storm-flood disaster in 1953, it was decided to close four of the six branches of the Delta. The two remaining ones are busy shipping lanes, and here safety was increased by raising the seawalls. The smallest branch, The Veersche Gat–Zandkreek, a bypass of the Oosterschelde, the largest estuary in the Delta (Fig. 11.4), was closed in 1961 and became a brackish lake; however, any consequences for waders were not followed. The much larger Haringvliet–Hollandsch Diep was cut off from the North Sea in 1970 and became a fresh water lake while retaining its function as a main outlet of water from the Rhine and Meuse. However, this brackish and rather silty estuary had previously not been important for waders, except Avocets (*Recurvirostra avosetta*) and, to a minor extent, Dunlins and Curlews (Wolff 1967). More crucial was the loss of 55 km² of tidal flats due to the closure of the Grevelingen estuary in 1971, where some 50 000 waders had lived in midwinter, 50–60% of them being Oystercatchers. Although the new lake remained marine-brackish, and its locally very shallow shorelines could still be exploited by some species, the entire wader population declined by 95%

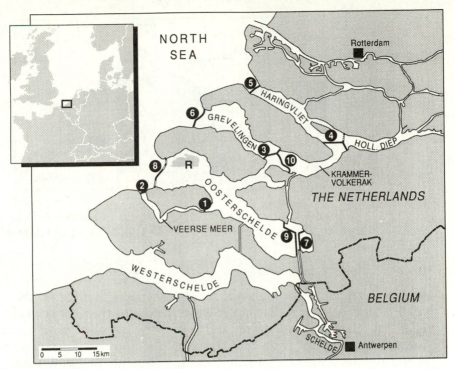

Fig. 11.4 Map of the Dutch Delta area, the estuarine complex of the rivers Rhine, Meuse and Schelde in the SW Netherlands, with various waterbodies that resulted from the Delta project engineering scheme. The particular engineering works, and the dates of their completions, are: 1 = Zandkreekdam, 1960; 2 = Veersegatdam, 1961; 3 = Grevelingendam, 1964; 4 = Volkerakdam, 1969; 5 = Haringvlietdam, 1970; 6 = Brouwersdam, 1971; 7 = Markiezaatskade, 1983; 8 = Oosterschelde storm-surge barrier, 1986; 9 = Oesterdam, 1986; 10 = Philipsdam, 1987. Also indicated, the location of the Roggenplaat tidal flat (R) in the western Oosterschelde, mentioned in the text.

(Wolff *et al.* 1976; Lambeck *et al.* 1989). Unfortunately, the wider consequences for waders within the Delta area could not reliably be judged because only two overall counts had been made before closure, both in 1966–1967 (Wolff 1967). But owing to its conspiciousness and limited year-to-year fluctuations in population size (Meire *et al.* 1994a), the Oystercatcher is perhaps the only species for which a tentative conclusion can be made. In the 1970s, hence after the loss of the Grevelingen estuary, midwinter numbers in the Delta area as a whole tended to be higher than in 1967 (Leewis *et al.* 1984) (Fig. 11.5). This does not suggest that the Oystercatchers displaced from the Grevelingen were unable to re-establish themselves elsewhere in the vicinity.

This result was confirmed by the only detailed study in that period which had been carried out since 1964 (Lambeck *et al.* 1989). A monthly

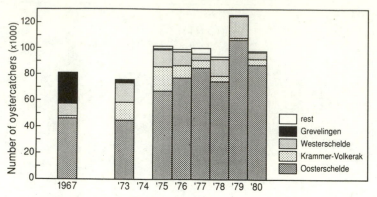

Fig. 11.5 Numbers of Oystercatchers in different areas of the Dutch Delta area in January 1967–1980. Compiled from data in Wolff (1967), Saeijs and Baptist (1977) and Meininger *et al.* (1984).

census was made of the high-tide roosts of waders that fed on the 16 km² sized Roggenplaat tidal flat in the western part of the Oosterschelde (Fig. 11.4). A comparison before and after the closure of the adjacent Grevelingen estuary showed Oystercatcher numbers increased significantly on the Roggenplaat after 1971. Although this could have been the result of a long-term population growth (Prater 1981; Becker and Erdelen 1987; Smit and Piersma 1989), a more sophisticated time-series analysis revealed unequivocally that a permanent change had occurred in May 1971, precisely the month of the enclosure of the Grevelingen. Furthermore, four out of 253 Oystercatchers cannon-netted in the Grevelingen in 1970 were re-trapped in the Oosterschelde some 15–16 years later (see below) and two more birds were found dead there, whereas there were no recoveries from other wintering areas. The Oystercatcher population of the Roggenplaat increased by 8.4% due to this Grevelingen effect (van Latestijn and Lambeck 1986). Apparently, there was enough capacity in the remaining tidal areas of the Dutch Delta to accommodate many, or all, of the birds displaced from the Grevelingen in 1971.

Nonetheless, there was evidence of some resistance to incoming birds at the local scale. The Roggenplaat had previously supported high densities of Oystercatchers, this undoubtedly being associated with the above-average biomass density of the prey (Lambeck *et al.* 1989; Coosen *et al.* 1994). In agreement with the expectations of Goss-Custard (1977b, 1985) and Goss-Custard *et al.* (1982b), the increase in density in this already densely-populated area after the closure of the Grevelingen—despite its proximity—was disproportionately low. Most interestingly, the increase was accompanied by a shift in the timing of the seasonal peak in numbers from midwinter to autumn (Fig. 11.6), even though this did not happen in the Oosterschelde as a whole (Lambeck *et al.* 1989).

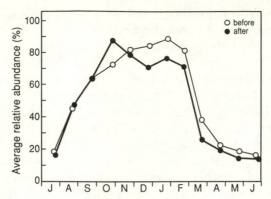

Fig. 11.6 Seasonal (1 July to 30 June) changes in the numbers of Oystercatchers on the Roggenplaat before (1965/66–1970/71) and after (1971/72–1978/79) the adjacent Grevelingen estuary was empounded. Monthly counts, as expressed in per cent of the seasonal peak value, were averaged for each of the two periods (modified, now showing original data, after Lambeck *et al.* 1989).

While the densities of birds in the remaining intertidal areas of the Ooster-schelde and Westerschelde estuaries increased after the Grevelingen was closed, the increase was apparently constrained in certain preferred areas where densities were already high. This implied that, by the early 1970s, the capacity of some of the remaining areas to support birds displaced by further habitat loss in the Delta was being approached.

According to the original Delta scheme, the Oosterschelde had been destined to become a freshwater lake also but, largely because of its great nature conservation and shellfishery value, this plan was dropped in 1976. The already partially-constructed 8 km dam at the estuary mouth became a storm-surge barrier with, in total, 65 piers spanning the com-bined 2850 m width of the three main tidal channels (Nienhuis and Smaal 1994). Although the gates between the piers would only be closed temporarily during storm floods to protect the hinterland, the planned barrier would reduce the cross-sectional area of the entrance by 77.5%, and thus severely restrict the incoming flow. To avoid the resulting drastic reduction in tidal amplitude, two compartment dams had also to be built in the eastern sector and in the Krammer–Volkerak, the northern branch of the Oosterschelde. These were required to decrease the surface area, and thus maintain a reasonably large tidal range, and to provide a non-tidal shipping lane between the ports of Rotterdam and Antwerp (Fig. 11.4). Nonetheless, the tidal amplitude ultimately decreased from 3.7 to 3.25 m. At the end of the engineering works in 1987, the combined effect of the two compartment dams and the decreased tidal amplitude resulted in a further loss of 36% of the original area of inter-tidal flats in this major European wader area (Nienhuis and Smaal 1994).

The construction in late 1981 of a small dam in the easternmost part of the Oosterschelde was the first step in this final engineering project. Probably a few thousand Oystercatchers lived in the area enclosed by this initially porous wall, with most of them feeding on the few local natural mussel banks (van Dessel 1983). Construction activities during November–December 1981 raised the average low tide level by 2.25 m and the low-lying mussel banks no longer emerged. Dense flocks fed at the moving tide edge, with as many as 200 Oystercatchers per ha. Upshore freezing during the second half of December further reduced the available feeding area and, on windy days, few birds fed there. After a week of mild weather, an even colder period began and the entire area froze over. Probably because their condition had already been weakened, at least 850 birds died (van Dessel 1983). It took a week before the survivors moved from their traditional feeding areas to those on the other side of the wall; this may reflect the normal value associated with site attachment in Oystercatchers (Ens and Cayford, Chapter 4, this volume). After the not yet completed wall burst during a stormy day in March, normal tides returned to the area for more than a year, but no information is available as to the response of the Oystercatchers before the final closure in spring 1983.

By 1980, some 100 000–120 000 Oystercatchers wintered in the Delta area, of which over 85% were in the Oosterschelde/Krammer-Volkerak. With the frequency of counts gradually increased to once a month from 1978 (Meininger *et al.* 1984), the effects of the large-scale changes during 1985–1987 could be monitored more closely than had been possible previously. Intensive studies were also begun in 1979 on the behaviour and ecology of Oystercatchers at a tidal flat in the central Oosterschelde (Meire and Kuyken 1984; Meire 1993) and a ringing project was started in 1984 (Lambeck 1991). From spring 1985, as more and more piers of the barrier were put in position, the tidal amplitude started to decline; in the central Oosterschelde, it had decreased from 3.7 m to 2.6 m by the autumn of 1986, when the storm-surge barrier was completed (Fig. 11.7). The Oesterdam, the first of the two compartment dams (Fig. 11.4), was closed in October 1986. The resulting increase in the current velocities in the northern branch of the Oosterschelde was so much greater than had been predicted that the barrier was used almost daily to check the tides, further reducing the tidal amplitude to an average of 2.2 m. It was not until the closure of the Philipsdam in mid-April 1987 that the tidal amplitude could be allowed to return to its new equilibrium average of 3.25 m (Fig. 11.7).

By chance, the winters of 1984–1985, 1985–1986 and 1986–1987 were all severe. During the second of these winters, the tidal amplitude was declining while, in the third, there was a relatively small tidal amplitude combined with frequent manipulations of the tides. As happens widely (Goss-Custard *et al.*, Chapter 6, this volume), mortality from

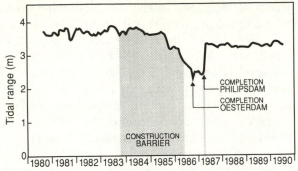

Fig. 11.7 Changes in the mean tidal range in the central part of the Oosterschelde during the 1980s, reflecting the direct effects of the engineering works in progress (data Rijkswaterstaat).

starvation was high in the first winter, before these tidal regime changes had started, and 1516 corpses were found. The proportion of juveniles amongst the victims (58%) was much higher than the 4.3% found in cannon-net samples taken from the live population during the preceding two months (Fig. 11.8); 20–25% of the juveniles may have died in marked contrast to an estimated 1.7% of immatures and 0.5% of adults. Nearly 50% of the starved adults had leg or bill deformities, injuries, visible tumours or an incomplete moult, as opposed to 3% in the samples of live birds. The characteristics of the frost victims were therefore similar to those found elsewhere (Goss-Custard *et al.*, Chapter 6, this volume).

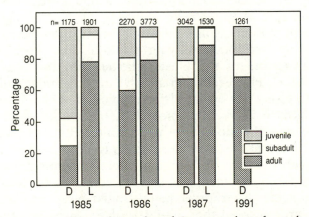

Fig. 11.8 The age composition of samples of Oystercatchers from the Delta area of SW Netherlands found dead during severe winter weather in January or February (D) compared with that of the whole live population at the time (L), as found in cannon-net catches from October–February. Data, and sample sizes, from four periods in 1985, 1986, 1987 and 1991 are shown; in the last case, no samples of live Oystercatchers were obtained (Lambeck 1991, and unpublished information).

The pattern of mortality in the two following winters, when the tidal conditions changed, was quite different. Despite a slightly less severe frost than in 1984–1985, no fewer than 5100 dead Oystercatchers were found in February and March 1986, representing about 5.7% of the Delta population (Lambeck 1991; Meininger *et al.* 1991). Compared to the previous winter, the number of juveniles that died was only slightly higher and represented a similar proportion of the live population of first-winter birds. Their proportional share of the total mortality, however, declined sharply due to a remarkable increase in adult mortality, most of it involving apparently healthy individuals. Nonetheless, conditions were not apparently bad enough to trigger an exodus southwards. Interestingly, the mortality was largely confined to the Oosterschelde and was insignificant in the adjacent Westerschelde estuary. The mortality in the Waddenzee during this one month of frost was also relatively small and predominantly involved juveniles, and hence followed the normal severe winter pattern (Stock *et al.* 1987b). All the evidence thus points to a local phenomenon restricted to the Oosterschelde itself.

The obvious cause of the high mortality was the, by then, 0.5 m reduction in tidal amplitude (Fig. 11.7). The highest densities of cockles and mussels, the main prey of Oystercatchers, are found at the lower levels of the flats (Meire and Kuyken 1984; Meire *et al.* 1994b; R.H.D. Lambeck, unpublished information). The reduced tidal amplitude meant that the best feeding areas were exposed for a very short time, if at all. Furthermore, the opportunities for compensatory feeding in the poor quality upshore areas in the intertidal zone (Meire 1993) or in fields would have been much reduced during severe frost (Goss-Custard *et al.*, Chapter 6, this volume). Under such conditions, a much greater part of the population than normal was at risk.

Since prey biomass as well as elevation differ between tidal flats, the chances of Oystercatchers dying would also be expected to have differed between sectors of the Oosterschelde, and this was indeed the case (Table 11.1). However, data simply on the numbers of frost victims per area can give a somewhat distorted picture of the local conditions when many birds have recently moved in from other feeding areas. This is because birds that do not normally winter in an area may be most at risk in severe conditions as, indeed, ringing data from that time illustrates. So, though the number of victims would suggest differently, the mortality in the second winter was below average in the western sector of the Oosterschelde. Re-traps of metal-ringed birds and sightings of colour-ringed birds showed that 80–90% of the survivors that bore rings had also been ringed in the same sector, strongly suggesting that they were residents there. In contrast, only 54% of the frost victims with rings had been originally trapped in that sector, so that the remaining 46% were probably newcomers. In addition, all 16 marked birds found dead in the Westerschelde originated from the Oosterschelde whereas, had mortality

Table 11.1 The winter mortality of Oystercatchers during February–March 1986 in the Westerschelde estuary and four sectors of the Oosterschelde estuary: (a) numbers of birds found dead as a per cent of the local population, and the number of ringed victims per sector, split according to their original ringing place; (b) the local death risk of Oystercatchers, winter 1986, as expressed by the ratio between the number of ringed victims found (F) and the number to be expected (E) from the totals ringed in a sector (Lambeck *et al.* 1991, and unpublished information).

(a)

Area	Victims (% of local population)	Ringed birds		
		Local	Other area	% Local
Westerschelde	2.1	0	16	0
Oosterschelde				
West	6.4	64	54	54
Centre	4.2	185	36	84
East	2.7	18	7	72
North	8.4	38	4	90

(b)

Area	Expected	Ringed birds	
		Found	Ratio F/E
Westerschelde	13	0	0
Oosterschelde			
West	141	79	0.56
Centre	144	231	1.60
East	68	31	0.46
North	57	81	1.42

been distributed without regard to origin, 13 should have been local birds (Table 11.1). The clear implication is that moving to another feeding area does not necessarily safeguard a bird from starvation, with lack of experience of the local feeding conditions and conflict with unknown residents perhaps comprising additional handicaps.

The severe weather in the third winter was preceded by a period during which the tides had been manipulated over six days in October 1986, to facilitate the closure of the Oesterdam. Feeding became impossible in the preferred low-lying areas over six consecutive low tide periods and, after one normal tide, over three more. Despite feeding more upshore, the daily food consumption of the average Oystercatcher declined by one third (Meire 1993). Surprisingly, the Oystercatchers did not increase their rate of food intake after the tides had been allowed to return to normal. The low prey biomass combined with a considerably reduced feeding area due to a 1.1 m smaller tidal range may have meant that Oystercatchers had fed at the maximum possible rate anyway before the tidal manipulations occurred (Meire 1993). After the subsequent closure of the Oesterdam, the average tidal amplitude decreased by an extra 0.4 m to 2.2 m and, moreover, as outlined earlier, became much more variable and thus unpredictable to the birds. December and early January were very windy and rather frequently the ebb tide did not recede further than the mid-tide level, sometimes for two or three tides in succession. This presumably explains the 30–40 g drop in the average weight of Oystercatchers in the western Oosterschelde in early January 1987. This provided an unfavourable starting point for the birds when, a few days afterwards, severe winter weather again arrived. With 3100 birds found dead, the mortality rate was again considerable and the composition of the victims was similar to that in the previous winter (Fig. 11.8). About 35% of the adults had arrested moult, suggesting that they had encountered difficulties the preceding autumn, perhaps especially during the tidal manipulations in October. Nonetheless, and despite some southward emigration, the size of the Delta population remained roughly at its previous level (Meininger *et al.* 1991); this is in contrast to the Wadden Sea where many Oystercatchers were forced southwards by even more extreme weather. That the conditions in the Delta as such were reasonable and that the problems there were only caused by the reduced accessibility of feeding areas, was illustrated by what happened following a change in tidal policy during the last part of the frost period. The incoming tide was partly held back while the ebb tide was allowed to flow freely, thus increasing access to the food. Catches revealed that this policy worked as, despite the frost, the Oystercatchers were quickly able to regain their normal winter weights (Lambeck 1991). In April 1987, the Philipsdam was finally closed and the associated habitat behind the Philipsdam lost.

The effects of the completion of all the engineering works on Oyster-catcher numbers in the Oosterschelde were explored by Schekkerman *et al.* (1994). These authors compared bird numbers in five seasons (1978–1982) in the pre-barrier period with the first three seasons (1987–1990) after it had been completed. The seasonal (July-June) number of Oyster-catcher-days hardly differed between the two periods which shows that, because of the reduction in the intertidal area, the usage of the area increased. However, there were significant changes in the seasonal pattern of occurrence of Oystercatchers. Mean peak numbers in 'autumn' (July–November) increased by 27%, although this difference was not significant. This change was largely due to the earlier arrival of Oystercatchers in the Delta in some years, whereas the October population was quite stable; 85–90 000 birds in 1978–1982 and 90–95 000 in 1987–1990 (calculated from Meininger *et al.* 1984, 1994, Meininger and van Haperen 1988). But because of the reduction in tidal area this still implies a considerable increase in the density of Oystercatchers, from 4.8 to 7.8 birds per ha, when averaged over the whole Oosterschelde. In contrast, mean peak numbers in winter fell significantly ($p < 0.05$) from

98 000 to 76 000 birds, a reduction of 22.5%, notwithstanding an increase in average density from 5.4 to 6.4 birds per ha. Changes in spring numbers were negligible. Summarizing, the overall result was that the seasonal peak numbers of 90 000–95 000 in the post-barrier period occurred earlier than it had done before, with numbers peaking in October or even September. In contrast to the pre-barrier period, when numbers increased slightly towards mid-winter, the numbers now decreased from late autumn to winter. Although higher densities were supported in autumn, the Oosterschelde seemed no longer able to accommodate the larger numbers, and densities, of Oystercatchers later in the winter. The forward shift in peak numbers that had originally been found on a pre-ferred tidal flat, the Roggenplaat, after the influx of displaced birds from the Grevelingen estuary in 1971, now applied to the entire Oosterschelde population.

Most interesting was the fate of the Krammer–Volkerak birds, the majority of which lost their feeding grounds. Colour ringing showed that some of them moved to other parts of the Oosterschelde. However, their post-barrier rate of disappearance from the Delta area as a whole was about twice that of birds ringed elsewhere in the Delta, the percentage disappearing in winter being higher than in autumn (Fig. 11.9). By the second winter after the completion of the engineering works, some two-thirds of the colour-ringed Krammer–Volkerak birds were missing from the Delta area (Brenninkmeijer *et al.* 1993). Since virtually all birds involved were adults and so presumably of comparable status, the data strongly suggest that the post-barrier redistribution did not simply involve a random mixing of newcomers and established birds. Rather, individuals that were already established in the feeding areas that remained after the engineering works had been completed seemed to

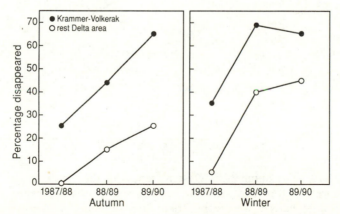

Fig. 11.9 The percentage rates of disappearance from the Delta area in autumn and in winter of Oystercatchers ringed in the Krammer–Volkerak and elsewhere in the Delta (Brenninkmeijer *et al.* 1993).

have had a better chance of maintaining their position in the reduced Oosterschelde than the immigrants that had been displaced from elsewhere; residents again have an advantage, it seems, over immigrants. This resembles the finding that displaced birds are disproportionally hit by severe weather, except that in this case, many of the immigrants were probably also subdominants. The implications of these findings from the Krammer–Volkerak birds are considerable because they imply that habitat loss can effect a large proportion of an existing population of adult birds, and not just the particularly vulnerable young ones.

Although Schekkerman *et al.* (1994) only found a negligible difference in the yearly number of Oystercatcher-days between 1978–1982 and 1987–1990, the real effect of the engineering works on the capacity of the estuary may have been somewhat larger. Despite the difficulties already encountered by the birds due to the changing tidal conditions, the number of Oystercatcher-days in the two last years before completion was *circa* 21.6 million as opposed to an average of 19.1 million, roughly 10% less, in the first three post-barrier years (Fig. 11.10). Apart from the statistical uncertainty due to the small sample size, the significance of this for investigating the capacity of the Oosterschelde to support Oystercatchers cannot yet be assessed without data on their food supply. A study of 14 sampling stations did not reveal any major post-barrier changes in the intertidal macrozoobenthos community structure; fluctuations in benthos seemed to be determined more by the occurrence of severe and mild winters than by the hydrodynamic changes in the Oosterschelde (Seys *et al.* 1994). Indeed, an extensive pre-barrier survey

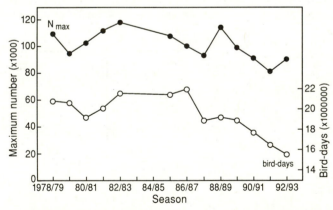

Fig. 11.10 The numbers of Oystercatchers in the Oosterschelde estuary over the whole year before and after the engineering works associated with building the storm-surge barrier were completed. Bird numbers are shown as the maximum count in the season or as the total number of Oystercatcher-days between 1 July and 30 June (Lambeck 1991, updated from Meininger *et al.* 1994 and P.L. Meininger, unpublished information).

of three major tidal flats in August 1985 and repeated, post-barrier, in August 1989 showed a more than doubling of the average biomass of all potential prey species from 43 g to 97 g AFDW per m^2, which was due almost entirely to an increase in the biomass of adult cockles (recalculated from Meire *et al.* 1994b). Although the harvestibility of the prey should also be taken into account (Zwarts *et al.*, Chapter 2, this volume), the feeding conditions must have been relatively favourable in August 1989. This presumably allowed greater Oystercatcher densities to persist in the reduced intertidal area that remained immediately after the engineering works had been completed, as was also clearly documented for another tidal flat by Meire (1991).

From the point of view of drawing general conclusions about the effect of winter habitat loss on Oystercatcher numbers, this was unfortunate; reductions in habitat area seemed to be confounded by changes in the food supply in the remaining areas. However, the late summer prey stocks do not reflect the feeding conditions in winter. The cockle fishery expanded after the barrier was built (Coosen *et al.* 1994). In autumn 1987, for example, the fleet removed over half the cockle stock of the Roggenplaat and the densities of cockles remaining approached the limits to profitability for Oystercatchers (Lambeck *et al.* 1988). Similarly, the high cockle biomass of 65 g AFDW per m^2 on the same flat in August 1989 had dropped to 25 g by November (R.H.D. Lambeck, unpublished information). Fisheries may therefore have contributed to the decrease in Oystercatcher numbers over the winter. Since any deleterious effect of habitat loss may have been further masked by the record mild winters of the first three post-barrier years, Oystercatcher numbers may now be tightly linked to the food supply and the Oosterschelde may have been at or near capacity since 1987 (see also Meire 1991).

These conclusions can only be tested by further monitoring the longer-term changes brought about by the massive change in the Delta system. Surveys show that cockle spatfall has been very meagre during the early 1990s. Since shellfishing continued in the better locations, the cockle biomass has been very low during 1990–1993. This reduction in the abundance of a key prey coincides with a decline in the numbers of Oystercatcher-days recorded in the Oosterschelde since 1989 (Fig. 11.10). Three processes involved in this reduction in numbers have been identified so far. First, there has been some dispersal to rather marginal areas near the Oosterschelde, such as North Sea beaches and a small Belgian estuary, and to areas in France. Although the numbers in the adjacent Westerschelde area also tended to increase (Meininger *et al.* 1994), this turbid estuary with its strong salinity gradient is of much less importance to Oystercatchers than the Oosterschelde, supporting peak numbers of 10 000–15 000 birds only. Second, there is a general impression that inland feeding by Oystercatchers has increased considerably over the same period, as might be expected if more birds found the intertidal food supplies to be inadequate (Goss-Custard et al., Chapter 6, this volume). Third, ringing recoveries suggest that some birds are now returning earlier to the breeding areas (R.H.D. Lambeck, unpublished information).

The greater use of fields and earlier return to the breeding areas are both quite risky for Oystercatchers should severe weather arrive late in the winter (Goss-Custard et al., Chapter 6, this volume) or early in spring (Hulscher et al., Chapter 7, this volume). Indeed, the cold spell in February 1991 did suggest an increased vulnerability in cold weather; the 1260 corpses collected comprised a higher proportion of the reduced winter population than was found dead in the more severe winter of 1984–1985. An unusually high proportion (69%) of the victims were adults also (Fig. 11.8; R.H.D. Lambeck, unpublished information), reminiscent of the winters with a reduced tidal amplitude. Although the experience of the Delta area is that extensive habitat loss will eventually lead to a reduction in local numbers through increased rates of emigration and winter mortality, especially in severe weather, the longer term effects on Oystercatcher survival of all the inter-connected environmental changes that have occurred, and continue to take place, are still in the process of being evaluated.

Any subsequent effects of winter habitat loss on reproductive performance were beyond the scope of the Delta studies. Understanding the essential contribution this makes to Oystercatcher population dynamics will require the long-term monitoring of breeding populations known to winter mainly in the Delta area.

Summary

This chapter reviews the large number of potential, often complex, interactions between man and Oystercatchers in the coastal zone. There is no firm evidence that potentially toxic compounds, such as heavy metals, pesticides and PCBs, have killed many Oystercatchers, not even when birds metabolize their energy reserves during severe winter weather. However, sub-lethal effects cannot be ruled out, especially on Oystercatcher reproductive success and prey abundance.

Oystercatchers mainly inhabit the well-mixed marine zones of estuaries where heavy eutrophication is least likely to depress the food supply; indeed, nutrient input may have increased prey biomass in several areas and contributed to the long-term growth of the Oystercatcher population. Although the laying of artificial intertidal mussel beds may have improved the food supply on occasions, people and Oystercatchers compete directly for shellfish and bait. Several consecutive years of poor spatfall in combination with continued shellfishing has been shown to reduce Oystercatchers numbers locally and to increase winter mortality. The effect of bait-digging on prey is insignificant yet. There is no convincing evidence either that the disturbance arising from, for example, bait-digging, leisure activities in the coastal zone and aircraft, have so reduced the time available for feeding that Oystercatchers have suffered. An increased protection has reduced shooting mortality over this century. In France, the only country where hunting remains legal, 2000 birds are killed annually, with up to ten times that number of immigrants being taken during exceptional spells of severe weather to the north and east.

The loss of coastal feeding areas due to various human activities has been a continuing process in many countries over the last few centuries. The most comprehensive studies of its effect on Oystercatchers have been made in the Delta region in the South-west Netherlands. When the feeding grounds of 30 000 wintering Oystercatchers were removed when the Grevelingen estuary was closed in 1971, the overall numbers in the region did not decrease, suggesting that there had been spare capacity. However, there was evidence of resistance to the resulting increase on Oystercatcher densities in some favoured feeding areas. A further loss of feeding area, and also of feeding time, occurred during 1985–1987 when the engineering works in the Oosterschelde estuary—which is crucial for Oystercatchers—were in their last phase. This resulted in high numbers of Oystercatchers—including unusually large numbers of adults—dying in severe winter weather. When the works had been completed, the feeding time returned to normal but 36% of the original feeding areas had been lost. Over the next three years, the numbers of Oystercatchers settling on the estuary at the end of the summer remained the same as before the habitat was lost. However, whereas in the past the peak numbers had been reached in midwinter, Oystercatcher numbers now declined over the winter, even though the weather was extremely mild and the feeding conditions were favourable. Individuals whose feeding areas remained were the least likely to disappear. When in subsequent winters the feeding conditions deteriorated again because of a lack of spatfall while cockle fishing continued, winter numbers of Oystercatchers declined still further and their vulnerability in severe weather increased. The next step is to predict how particular combinations of severe weather, habitat loss and shellfishing will affect Oystercatcher winter mortality rates and numbers in the future.

Acknowledgements

Many people have participated in the studies in the Delta area, but a vital role was played by Peter L. Meininger, Patrick M. Meire, Erik G. J. Wessel, Emiel B.M. Brummelhuis, Cor M. Berrevoets, Alex Brenninkmeijer, Eric C. L. Marteijn, Henk W. Spiekman and Rob C. W. Strucker. The help of all other people, and the support from colleagues elsewhere, the Dutch Ringing Centre, authorities and land-owners was highly appreciated. Additional funds for this study were made available by the former Dienst Getijdewateren of Rijkswaterstaat, now the National Institute for Coastal and Marine Management/RIKZ. The colour-ring studies were carried out in collaboration with the Research Institute for Nature Management/RIN, now Institute for Forestry and Nature Research/IBN-DLO, and supported by travel grants from the Beijerinck–Popping fund.

12 The carrying capacity of coastal habitats for Oystercatchers

John D. Goss-Custard, Andrew D. West, Ralph T. Clarke,
Richard W.G. Caldow and Sarah E.A. le V. dit Durell

Introduction

The need to predict the consequences for Oystercatcher numbers of the many anthropogenic influences to which the coastal zone is prone (Lambeck *et al.*, Chapter 11, this volume) has stimulated much of the research that has been done on the species. Broadly, predictions are required at two scales; locally, for the estuary or stretch of coast where environmental changes are expected to occur, and globally, for the wider population across a region, country or the whole of Europe. If a fishery, for example, was considering removing half the mussel (*Mytilus edulis*) beds within one estuary, the following questions would be asked. Will the food supply that remains be sufficient to maintain the present number of Oystercatchers? If not, and some starve or emigrate, how many would remain? Would any emigrants be able to establish themselves elsewhere and, if so, would they survive as well as at present? If more birds do die, either locally or as a knock-on consequence of the redistribution of birds to other areas, what would be the effect on the metapopulation, the greater population of which the local Oystercatchers are one part? The metapopulation consequences are discussed by Goss-Custard *et al.* (Chapter 13, this volume). This chapter deals only with predicting the effect of a change in the local environment on local numbers.

Throughout much of their range, the coastal feeding and breeding areas of Oystercatchers are frequently considered for recreational, industrial or agricultural purposes. Many of these activities would remove Oystercatcher habitat directly and permanently. Others, such as walking and picking shellfish by hand, leave the habitat essentially unchanged but disturb the birds and prevent them from using it; they also remove habitat, although only temporarily. Still other activities, such as large-scale shell-fishing on the flats, affect habitat quality by changing the abundance of a critical resource. The question arising in all cases can be phrased in two ways. The version that makes the fewest assumptions about the underlying biological processes is simply to ask how such habitat changes would affect Oystercatcher numbers locally. A more controversial approach is to ask how these activities might affect the ability of the area to support Oystercatchers or, in other words, alter the local carrying capacity.

Phrasing the question in terms of the carrying capacity is contentious because the term has been used in so many different ways that some authors have questioned its value in general (Dhondt 1988), as well as its appropriateness for migratory shorebirds in particular (Evans and Dugan 1984). Much of the confusion has arisen because the concept has sometimes been used as an equivalent to equilibrium population size (Dhondt 1988) and because implications have been read into it that are neither necessary nor justified (Evans 1984; Goss-Custard 1985). But its advantage is that it does encapsulate the useful commonsense notion that resources must ultimately set a limit to local numbers. The carrying capacity of an area can be unambiguously and operationally defined to be reached when, as a result of an intensifying density-related feedback process, such as competition for space or food, no more birds can establish themselves in the area, however large the numbers trying to do so. From then on, for every new bird that tries to settle, one bird either emigrates or dies (Goss-Custard 1985).

Recently, the usefulness of this essentially demographic definition has received support from empirical studies showing that the numbers of some birds, such as the Grey Plover (*Pluvialis squatarola*) and Brent Geese (*Branta bernicla*), do reach such a plateau in some localities (Moser 1988; Ebbinge 1993) as, indeed, had been suggested in small intertidal areas by counts of Redshank (*Tringa totanus*) and of Oystercatchers (Goss-Custard 1977b,c). These findings are significant because the mechanisms involved are likely to be very different between the carnivorous plovers, Redshank and Oystercatchers, which are dispersed while feeding, and the flocking and herbivorous geese (Goss-Custard and Charman 1976; Evans 1984). Nonetheless, many workers are still reluctant to use the term as it makes rather specific assumptions about the way in which local numbers are determined. The appropriateness of the concept therefore remains in doubt.

It is argued here that a demographic analysis of itself is not adequate for testing for carrying capacity and that the only effective way to evaluate the concept is to model the underlying biological processes. While investigating how habitat change might affect local Oystercatcher numbers, this chapter explores the concept of carrying capacity using a model of Oystercatchers feeding on the main mussel beds of the Exe estuary. This reveals that the key question does not concern the concept itself but the identification of the circumstances in which, in practice, the supply of potential recruits is large enough for the local carrying capacity to be reached.

The demographic approach

As bird densities increase and the resistance to further settlement intensifies, an increasing proportion of birds will fail to establish themselves,

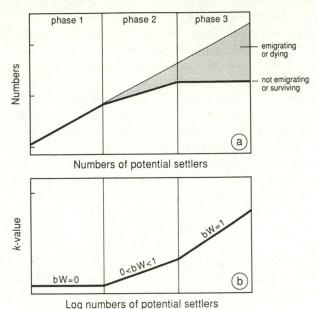

Fig. 12.1 How carrying capacity is defined. (a) The numbers surviving or not emigrating in relation to the numbers of potential settlers on the estuary. (b) The proportion emigrating or dying in relation to the number of potential settlers. In phase 1, competition is absent or so weak that all potential settlers on the estuary are able to establish themselves and survive the winter. As the number of birds (N) trying to occupy the area rises in phase 2 so that competition intensifies, an increasing proportion (P) of potential settlers is unable to become established and so moves elsewhere or dies. Nonetheless, the number settling continues to rise. But in phase 3, competition is so intense that, for every new bird that tries to settle, one either emigrates or dies. At this point, numbers reach a plateau and the capacity has been reached. From then on, the plot of the k-value ($k = -\log(1 - P)$) (Varley and Gradwell 1960) against log N has a slope (bW) of 1.

even though the absolute numbers that are able to do so continues to rise; this process is discussed in more detail, for birds establishing breeding territories, in Goss-Custard (1993). With mortality or emigration expressed as a *k*-value (Varley and Gradwell 1960), the rate of increase (*bW*) of *k* with population density is 1 when the carrying capacity has been reached (Fig. 12.1). At this point, the habitat is saturated and numbers cannot increase further without a change in either the birds' density-related adaptations, such as their social behaviour, or in the habitat itself.

In practice, it is difficult from surveys or censuses alone to establish that saturation has been reached. It is not enough simply to demonstrate stable numbers over many years in an area whose quality does not change and, by removal experiments, then show that potential settlers have been excluded (Goss-Custard 1993). With little annual variation in the supply of potential settlers, numbers may remain stable at a level well below the capacity of the habitat and, with $bW < 1$, birds may be excluded even before the habitat has become saturated. Rather, it needs to be shown that local numbers remain stable despite wide variations in the supply of potential recruits and that the resource level does not change. Both conditions are difficult to meet in field studies. The size of the total population from which a local population is drawn may change rather slowly. Furthermore, the boundaries of the region from which birds recruit to the local population may be difficult to define. For example, the fact that breeding numbers on Skokholm have fluctuated independently of the 40% increase in the size of the British population (Goss-Custard *et al.*, Chapter 13, this volume) may not mean that the capacity of the island has been reached but that the size of the recruiting population to this area has not increased in line with the national population. Alternatively, stable numbers on Skokholm might have arisen because a declining proportion of the British total, although the same number of birds, has been attracted to an area of diminishing quality. Showing that local numbers have remained level while national numbers have increased is thus not a critical demonstration that capacity has been reached. Some other approach is needed.

A model to predict carrying capacity of wintering areas

In wintering Oystercatchers, the most likely source of feedback that will limit density is competition for food, probably in combination with parasites, predators and accidents (Goss-Custard *et al.*, Chapter 6, this volume). Competition arises from prey depletion, both over one winter and in the long term, and from interference between foraging birds (Goss-Custard 1977b, 1980; Zwarts and Drent 1981; Goss-Custard *et al.*, Chapter 5, this volume). The idea underlying our model is that the distribution of competitive abilities across individuals determines the proportion that fail to compete effectively, and so die or emigrate, at different population sizes. This allows the form of the density-dependent emigration and mortality functions to be derived and, therefore, the carrying capacity to be deduced (Fig. 12.2). The proportion failing also depends on the spatial variation in the resource, with more individuals failing as the proportion of feeding areas that are of poor quality increases. As the overwinter depletion of the food supplies of Oystercatchers may occur at different rates in different places (Goss-Custard *et al.*, Chapter 5, this volume) and as the food value of individual prey items may change as they lose condition (Dare and Edwards 1975; Bayne and Worrall 1980; Zwarts and Wanink 1993), seasonal changes in food supply in different places must also be incorporated. The responses of individual Oystercatchers to each other and to their spatially and seasonally varying mussel food supply have therefore been synthesised in an individuals-based and physiologically-structured model. The model is game theoretic (Maynard Smith 1982) because the choices made by one competing individual as to where to feed are contingent on those made by all other individuals. The model tracks the body condition of each individual and, by running simulations over a range of initial population sizes, can be used to derive the function in Figs 12.1 and 12.2.

Details of the model are given in Goss-Custard *et al.* (1995a,b; Chapter 5, this volume), and Clarke and Goss-Custard (Appendix 1, this volume), and only its broad features are described here. The intake rates of an Oystercatcher feeding on mussels depend on several factors. They are affected by the density, size distribution, energy content and, in some cases, the average thickness of the mussel shells. All of these vary both spatially and seasonally (Goss-Custard *et al.* 1993) and affect the intake rates of all Oystercatchers. However, the intake rates of birds feeding in the same place at the same time differ because individuals vary in two aspects of competitiveness. One is their interference-free intake rate

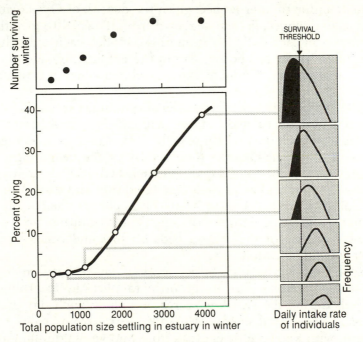

Fig. 12.2   Scheme showing how the density dependence in winter mortality or emigration can be derived from knowledge of the variation in the behaviour of individuals. With increasing numbers of birds settling on an estuary, the distribution of intake rates across individuals becomes increasingly skewed towards low values as the foraging opportunities of the inferior competitors are increasingly constrained by the intensifying competition. On the simplifying assumption that a certain threshold intake rate is needed for a bird to survive or to remain on the estuary, an increasing proportion of individuals will fall below the threshold and so either die or emigrate (Goss-Custard 1985, 1993; Goss-Custard and Durell 1990; Ens *et al.* 1994).

(*IFIR*), thought to reflect a bird's foraging efficiency. The other is their dominance-related susceptibility to interference (*STI*), the amount by which intake rate is reduced below *IFIR* as the density of Oystercatchers increases. An individual's intake rate at any one time and place thus depends on the food supply, on its own foraging efficiency and dominance and on the density of Oystercatchers, and thus its competitors, where it is feeding.

In the model, each individual Oystercatcher is given a unique combination of foraging efficiency and dominance, drawn from the empirically determined range of variation recorded among Exe Oystercatchers. Then, the *IFIR* of an Oystercatcher of average foraging efficiency is estimated on each day during winter for each mussel bed from the biomass density of mussels present, using a functional response derived from an empirical optimal foraging model (Goss-Custard *et al.* 1996b). An individual's potential maximum intake rate, or its *IFIR*, is then calculated from its own foraging efficiency relative to that of the average bird. The reduction due to interference, given the density of Oystercatchers present and its dominance, is then deducted from the *IFIR* to give the actual rate of intake. In each daily iteration of the model, each individual is selected in random order to choose, within the empirically-determined 3% limit of its ability to discriminate, the mussel bed where it achieves the highest gross intake rate at that time.

The model tracks the feeding location, intake rate and body condition of each bird on each day from September to mid-March. The mussel biomass on each of the 12 mussel beds in September, averaged over eight years of study, defines the food supply at the beginning of the winter. From then on, the food supply on each bed declines as it is depleted by Oystercatchers and as mussels lose condition and disappear from storms and other mortality agents. Many birds change their feeding site as the winter progresses; because of mussel deterioration and depletion, the relative quality of the mussel beds change and other birds also move their foraging location.

Each bird is given an initial level of fat reserves in September and subsequently either puts on fat, or metabolizes it, according to how well it feeds during the two low water periods of each day and the current temperature-related energy demands. Any surplus energy consumed is stored, with known efficiency, up to a known maximum daily rate of fat gain. Each bird attempts during autumn and winter to accumulate fat up to a specified maximum amount. The reserves are used to maintain the bird on days when it fails to meet its current requirements solely through foraging. The present version of the model only considers reserves of energy stored as fat and ignores the small, and only slowly metabolizable fraction, stored as muscle protein (Davidson and Evans 1982).

Model predictions for aspects of the behaviour and distribution of the birds at different times in the winter and for the rates that prey are depleted have been tested (Goss-Custard *et al.*, Chapter 5, this volume). In general, most trends were correctly predicted, but their magnitudes were sometimes under-predicted. In the present context, the critical test, of course, is how well the model predicts overwinter mortality. Compared with the observed rates, the model predicts the mortality rates of young birds quite well but considerably over-predicts those of adults because it under-estimates the ability of real Oystercatchers to lay down fat reserves during autumn and early winter. The most important explanation for this is that the model assumes that the birds can only obtain food from the mussel beds whereas, in fact, they can take other prey before and after the mussel beds have been exposed and covered by the tide (Goss-Custard *et al*, Chapter 6, this volume). That this explains the discrepancy is supported by two successful predictions; the model accurately predicts that the first Oystercatchers should start to collect supplementary food from fields over high tide in late October and that an increasing number need to do so as the winter progresses (Goss-Custard *et al.* 1995b). By this and other means, real Oystercatchers can extend their feeding period beyond the 12 hours in every 24 that the model currently assumes.

As the amount of extra food obtained from sources other than mussels has yet to be estimated, a re-scaling coefficient was used as a temporary measure to adjust the daily food consumption until the model predicted the observed mortality rates in each age-class. Winter weight gains and losses then coincided closely with the observed changes. These re-scaling coefficients, which varied only between 1.2 and 1.7, provide a measure of by how much the current model falls short of its objective and empha-size that more model development is required before definite predictions can be made. However, they also allow us to make preliminary predic-tions of the form and parameter values of the density-dependent winter emigration and mortality functions (Fig. 12.2), and thus to explore the local carrying capacity.

Carrying capacity of the Exe estuary mussel beds

Capacity within one winter

The respective roles of immigration and emigration and of mortality in determining Oystercatcher numbers on an estuary are unknown. In Oys-tercatchers, as in shorebirds in general, estuaries may be partly, perhaps mainly, colonized by juveniles prospecting for the first time for a place to spend the winter (Goss-Custard *et al.* 1977c; Sutherland 1982d). Whether all arriving juvenile Oystercatchers remain or whether those that cannot feed adequately move on to seek a more suitable site before the winter is unknown. Juveniles, immatures and perhaps small numbers of adults may change their wintering site in subsequent years (Goss-Custard *et al.* 1982c). However, the majority return year after year to the estuary where they spent their first winter, perhaps so as not to lose the benefits arising from an increasing familiarity with the area and increasing dominance (Ens and Cayford, Chapter 4, this volume; Lambeck *et al.*, Chapter 11, this volume). Many wintering Oystercatchers also die during their first two winters, especially in severe weather (Goss-Custard *et al.*, Chapter 6, this volume). While both emigration and mortality clearly play a role in reducing numbers in one estuary, no-one has yet worked out their relative importance.

Simulations were therefore run on the assumption that local numbers could be reduced either by emigration or by mortality at any time during the winter. As it was the simplest way in which to start, the first simulations were run for just one winter by varying the numbers of Oystercatchers introduced in September by up to ten times the numbers that have occurred on the mussel beds of the Exe up to now. The age and feeding method distributions were the same as those of the present Exe population. In simulations assuming that birds were lost only through mortality, birds died on the day their fat reserves were exhausted. In simulations assuming that birds emigrated from the estuary before they reached this point, birds left when their fat reserves reached 8% of their weight, as has been shown by Hulscher (1989, 1990).

In fact, there is only a slight difference between the predicted density-dependent mortality and emigration functions (Fig. 12.3a). In accordance with the concept of carrying capacity developed in Fig. 12.1, the simulations show that as the number of potential recruits in September increases, both the numbers surviving and the numbers remaining tend towards plateaux, although at a lower level when this is brought about by emigration (Fig. 12.3b). Uncertainty about whether bird numbers are determined by emigration or mortality is not, therefore, important for exploring the concept of carrying capacity itself, but does influence the precise value predicted for a particular site. The difference may be greater if birds emigrated on other decision rules. In particular, birds might move as soon as their fat reserves begin to fall, rather than wait until their reserves are dangerously low. Adults, for example, might leave if they did not expect to reach the heavy mass in spring that is required to fuel their migration back to the breeding areas. But simulations, not shown here, show that if they do this, the carrying capacity is rather little affected (A. D. West, unpublished information). Nonetheless, a precise prediction as to the actual carrying capacity of the Exe mussel beds must await further research into the basis on which prospecting juvenile Oystercatchers and established adults decide whether to move on or stay.

The model can be used to explore the effect of removing an increasing proportion of the mussel beds. By way of illustration, the capacities resulting from removing the worst or best 30%, 60% or 90% of the mussel bed area are shown in Fig. 12.4a; the slightly higher maximum capacity in this figure compared with Fig. 12.3 simply reflects small differences in the values of some model parameters. The proportionate

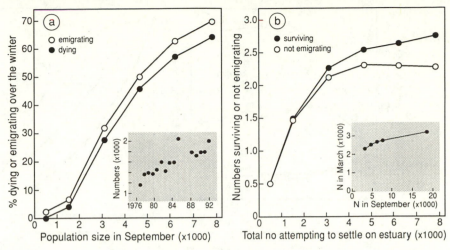

Fig. 12.3 The predicted carrying capacity of the mussel beds of the Exe estuary. (a) The density-dependent mortality and emigration functions. The inset shows the mean winter numbers actually recorded on the beds since the winter of 1976–77. (b) The numbers of Oystercatchers on the 12 main mussel beds of the Exe estuary that are predicted to survive or not to emigrate over the winter in relation to the numbers attempting to settle on the mussel beds in autumn. The inset shows how, even with 20 000 recruits in September, the numbers surviving still increases slightly as the numbers settling in autumn increases. Means of three simulation runs shown.

reduction in capacity following the loss of a given proportion of the mussel bed area of highest, average or lowest quality is summarized in Fig. 12.4b. Unsurprisingly, removing habitat of average quality leads to a directly proportionate reduction in capacity. But the figure shows just how big a difference it makes whether habitat of above or below average quality is lost. For example, whereas removing the best 50% reduces capacity over five-fold, removing the worst 50% barely reduces it by 20%. The magnitude of the difference depends on the precise distribution of bed qualities; the more uniform the bed qualities, the closer the extremes move towards the $y = x$ line. But, insofar as the Exe mussel beds are representative of spatial variation in the quality of the feeding areas of Oystercatchers in general, the results indicate the very different consequences of removing areas of different quality.

An important point arising from these simulations is that the current winter numbers on the Exe lie within the region of the mortality and emigration functions where both these rates become density-dependent. Since the study on the Exe began in 1976, numbers on the 12 mussel beds have steadily increased (inset in Fig. 12.3a), probably because other minor mussel beds have declined (J. D. Goss-Custard, unpublished information). As density dependence begins at *circa* 1500 birds, Oystercatcher density would have begun to affect emigration or mortality in the early

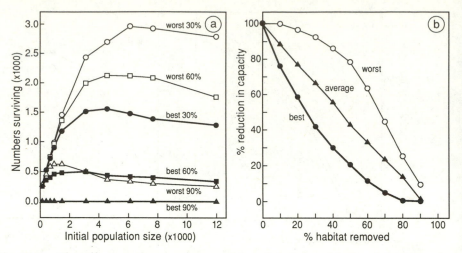

Fig. 12.4 The effect on the carrying capacity of the mussel beds of the Exe of removing habitat of above average, below average or average quality. (a) The numbers of birds surviving in relation to the numbers settling in September according to the quality and size of the area removed; 30% or 60% or 90% of the worst or best habitat was removed. (b) The proportionate reduction in capacity following a given proportionate reduction in the area of food supplies, starting with the best 10%, worst 10% or 10% of average quality.

1980s. Insofar as the present model mimics reality, competition seems already to influence the numbers of mussel-eating Oystercatchers on the Exe. But even though mortality or emigration became density-dependent a decade ago, numbers have nonetheless continued to increase. An increasing *number* of birds remain even though a gradually increasing *proportion* of potential settlers is presumably now dying or emigrating. That mortality or emigration is already density-dependent does not necessarily mean that no further increase in numbers can take place; it depends on the strength of the feed-back processes. According to the model, the point at which, within one winter, $bW = 1$ has not yet been reached in the mussel-feeding Oystercatchers of the Exe.

Indeed, the model suggests that capacity is not even reached when 20 000 birds, or ten times the present number, arrive on the estuary in autumn, even though only about 3000 of these 20 000 would survive the winter. Even with this number of recruits in September, the value of the slope, bW, is only 0.82 (SE = ± 0.01). Accordingly, the numbers surviving when 20 000 birds settle in September is rather higher than when 5000 do (inset to Fig. 12.3b). Strictly, therefore, the carrying capacity of the mussel beds has not yet quite been reached in these simulations. However, simulations not shown here show that even a slightly higher average *STI* leads to carrying capacity being very clearly reached at a much lower number of potential recruits. For investigating it as a concept, simulations over a ten-fold range of potential recruits were regarded as adequate, and the carrying capacity of the mussel beds is provisionally estimated to lie between 3000 and 3500.

How carrying capacity within one winter is achieved

The two feed-back processes involved in competition for food in Oyster-catchers, interference and depletion, operate over different time-scales; indeed, this is the main way of distinguishing between these processes in shorebirds generally (Goss-Custard 1980). Interference acts immediately; as bird density increases, the intake rate of subdominants declines at once. In contrast, depletion of most Oystercatcher prey takes place slowly. This difference in time-scale raises the possibility of determining the respective roles of these two feed-back processes in setting the winter carrying capacity by examining the stage in the winter when capacity is reached.

The simulations show that, for the Exe population, bird numbers are not much reduced in autumn (Fig. 12.5a). Very few die or emigrate during the first month, for example, unless very large numbers arrive (inset to Fig. 12.5a). Capacity is clearly not reached at the time of settlement in autumn. Rather, the reduction in numbers occurs throughout the winter, after recruitment has finished. As a result, the overwinter reduction in Oystercatcher densities over all 12 mussel beds is strongly density-dependent (Fig. 12.5b). That the carrying capacity is set late in the winter explains why, in Fig. 12.4a, there is a decline in the numbers surviving the winter when very large numbers arrive in autumn. This arises because, before birds die, they deplete the food supply, which therefore supports rather fewer birds in late winter.

The increasing mortality over the winter implies that it is depletion that primarily sets the overwinter survival rate, and thus the capacity of the mussel beds. This is further suggested by the steep overwinter decline in the standing crop biomass of the mussels (Fig. 12.5c), especially in the simulations in which recruitment is high (Fig. 12.5d). However, two other processes also reduce food abundance. First the flesh content of a mussel of given length decreases by 33–35% over the winter, and other mortality agents remove a further 10% of the mussel biomass. In comparison, the birds take only a small proportion of the initial standing crop biomass, being only *circa* 11% when 3500 birds settle in autumn (Fig. 12.5d). Second, the ambient temperature drops from autumn to spring, increasing the energetic, and hence food requirements, of the birds. The fact that the carrying capacity is reached late in the winter may therefore not reflect the depletion of prey by the birds themselves but the gradual deterioration in the feeding conditions, largely associated with other factors.

This was tested in simulations in which the overwinter declines in ambient temperature and prey condition were removed, either singly or

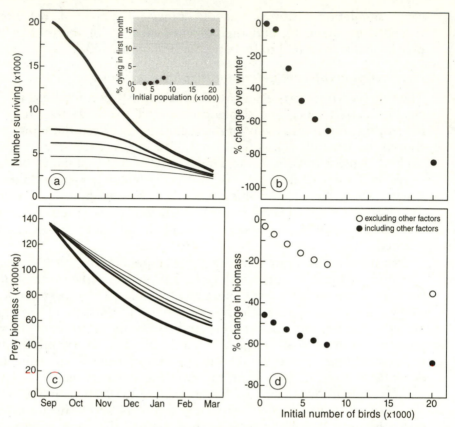

Fig. 12.5 The stage in winter when the carrying capacity is reached according to the size of the recruiting population in September. (a) The numbers surviving at different stages of the winter. The inset shows the percentage dying during the first month in relation to population size. (b) The overwinter percentage reduction in bird numbers in relation to the inital numbers settling. (c) The change in prey biomass over the winter; the thickness of the line identifies the equivalent simulation in (a). (d) The overwinter reduction in prey biomass in relation to the inital numbers settling, including or excluding the losses due to factors other than depletion by Oystercatchers.

in combination. As might be expected in an estuary with such mild winters, maintaining the ambient temperature at high autumn levels throughout the winter made little difference to the predicted capacity, whereas removing the overwinter loss of prey biomass increased it by at least 25% (Fig. 12.6). This is not surprising; it is another way of saying that food abundance affects the numbers of Oystercatchers that survive the winter. But it illustrates that carrying capacity is not only determined by the abundance of mussels in autumn but also by the rate at which other factors reduce the biomass as the winter progresses and the critical feeding conditions arrive.

Indeed, the determination of carrying capacity may be even more

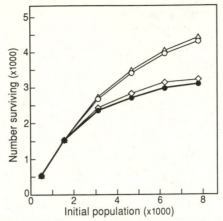

Fig. 12.6 The predicted carrying capacity of the mussel beds of the Exe estuary at present (●); with no overwinter decline in ambient temperature from their autumn value (◇); with no decrease in prey biomass due to loss of condition and non-Oyster-catcher mortality (○); and with no decrease in both temperature and prey biomass (△).

restricted in time than this. As well as changing seasonally, the intake rates of Oystercatchers on mussel beds vary through the neap–spring tidal cycle. On spring tides, the birds can feed on the best quality mussels found at the lower shore levels (Goss-Custard *et al.* 1993) and, for most of the exposure period, the density of birds, and so interference levels, are relatively low (Goss-Custard and Durell 1988). But on neap tides, only poorer quality mussels at the higher shore levels are accessible and bird densities, and so interference levels, are high because of the reduced feeding area available. The model does indeed predict that most birds die on neap tides (Fig. 12.7). The precise number dying depends, however, on the assumption made about the relative quality of the upshore and downshore feeding areas. More birds die when the quality of the upper-most parts of the beds are poor relative to the quality of the low-lying areas which are only exposed on spring tides. Clearly, it is now impor-tant to measure the downshore gradient in mussel bed quality as the numbers of Oystercatchers that survive the winter seems largely to be established on neap tides late in the winter, and thus by the food supply available on these restricted occasions.

That survival is set both late in the winter, when the food supply is low, and on neap tides, when interference competition is most intense and the food supply of poorest quality, suggests that both a bird's forag-ing efficiency and susceptibility to interference influence its chances of survival. This is confirmed by examination of the characteristics of the birds that died over the winter in a year when capacity was exceeded by the numbers attempting to settle in September (Fig. 12.8a,b). Both com-ponents of competitiveness affect the birds' survival chances. Conversely,

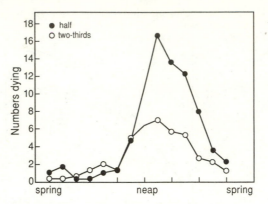

Fig. 12.7 Model predictions of the stage in the neaps–springs cycle that Oyster-catchers die in late winter. Two curves are shown according to the difference that is assumed in the quality of the food supply at the upper and lower shore levels; this has not yet been measured on all mussel beds of the Exe, hence two probable values are shown. The quality of the mussels at the top of the shore, which provide the only feeding areas accessible on neap tides, is either two-thirds or half that of those exposed on spring tides at the lower shore-levels. The average quality of the beds in the three conditions when the beds are all fully exposed on spring tides is the same.

the carrying capacity is in turn affected by the average foraging efficiency of the birds; at a given food abundance and level of interference competition, more birds will survive if the average efficiency is high (Fig. 12.8c). This does not apply to the birds susceptibility to interference, however, as this is determined in the model by the relative, and not absolute, dominance of the birds (Goss-Custard *et al.* 1995a).

The late-winter determination of capacity implies that numbers in September could increase considerably above present values, if a sufficiently large supply of birds were available. The simulations show, in fact, that as the total supply of birds increases well beyond the late-winter carrying capacity, bird numbers increase on mussel beds of all qualities, as birds move between them seeking the best place to settle (Fig. 12.9a). Even with over ten times the number that currently settle on the estuary—equivalent to six times the carrying capacity in these simulations—numbers do not level off on any of the feeding areas in autumn; they would only begin to do so when most birds die in September. Rather, numbers level off in late winter and do so on beds of all qualities at approximately the same initial population size (Fig. 12.9b). The numbers on each feeding area in autumn reflect the behavioural decisions of individuals as to whether to remain or to move on. The population consequences arise later in the winter when the feeding conditions deteriorate and the carrying capacity is set. In terms of predicting the effect of habitat loss, this implies that bird densities would be able to increase initially in autumn as birds whose feeding areas had been removed re-

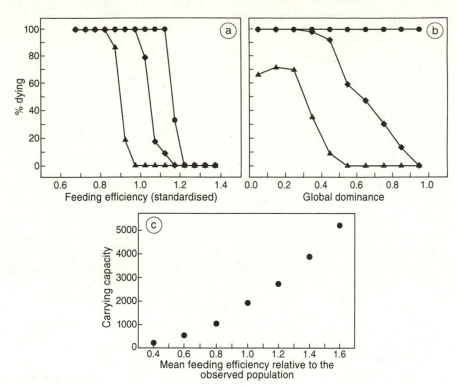

Fig. 12.8 Carrying capacity and the characteristics of the birds involved in achieving it. The mortality of hammering Oystercatchers according to their (a) feeding efficiency, and (b) dominance status, with a September population of 4650 birds. The three lines refer to categories of dominance in (a) and to categories of feeding efficiency in (b) at the extremes and at the midpoint of the range. These are: dominance—(●) 0–0.2; (■) 0.4–0.8; (▲) 0.8–1.0; feeding efficiency—(●) -3— -1 SD; (■) -1 SD—+1 SD; (▲) +1—+ 3 SD. (c) The carrying capacity of the mussel beds as a function of the average feeding efficiency of Oystercatchers.

distributed themselves among the areas that remained. The overall numbers would only be reduced below previous levels later in the winter when the feeding conditions deteriorated. Encouragingly, this is precisely the pattern that was observed in the Oosterschelde in the South-west Netherlands when the intertidal area was reduced by engineering works (Lambeck *et al*, Chapter 11, this volume).

Capacity over a series of winters: the effect of annual variability in bird numbers and environmental conditions

So far, the unrealistic assumption has been made that capacity is determined within only one winter—the within-winter capacity. In reality, both the standing crop prey biomass (Hancock 1971; McGrorty *et al.* 1990; Triplet and Etienne 1991; Desprez *et al.* 1992; Beukema *et al.*

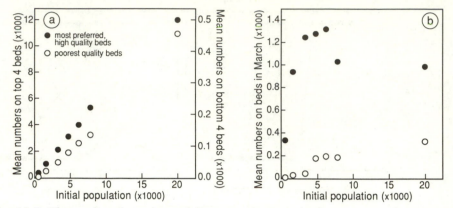

Fig. 12.9 The predicted numbers of Oystercatchers on the four most preferred, high quality, mussel beds combined and on the four poorest quality beds combined in relation to the number of birds settling on the Exe estuary in autumn. (a) Numbers in September, at the time of settlement; note the different scales. (b) Numbers surviving the winter.

1993; Zwarts and Wanink 1993; Coosen *et al.* 1994) and the supply of birds available to occupy the estuary (Goss-Custard *et al.*, Chapter 13, this volume) will vary between years. What effect do such annual fluctuations have on the carrying capacity predictions in the long term, across many winters?

The question was examined by running simulations over a period of years. At the beginning of the second and subsequent winters, the same number of birds that had survived the previous winter returned to the Exe, along with a variable number of juveniles per adult. In some simulations, the same individuals that had survived the previous winter returned to the Exe the following autumn; the average efficiency of the post-juvenile birds therefore increased year by year through selection of the most efficient individuals. In other simulations, selection for efficiency was not allowed; the distribution of foraging efficiencies amongst the birds returning in autumn was re-set to be the same each year. The initial food supply in autumn either remained the same year by year or was allowed to fluctuate at random within a uniform distribution with a range of plus or minus two standard deviations of the observed variation. The within-winter carrying capacity of the beds was also determined each year of the simulation to calculate the maximum number of birds the mussel beds could, in theory, support.

With a constant food supply but fluctuating juvenile recruitment, and without selection for efficiency, the population on the mussel beds by the end of the winter built up to a level so that the long-term capacity was similar to the within-winter value of *circa* 3500 birds (Fig. 12.10a). However, with selection, the numbers of overwinter survivors increased

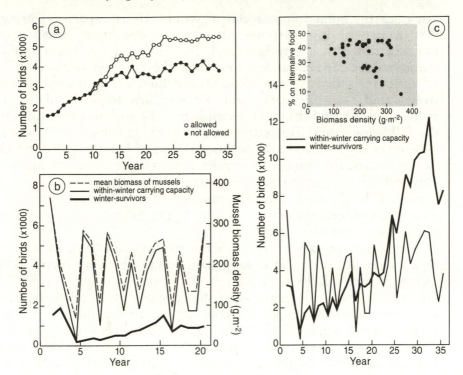

Fig. 12.10 The effect of including annual variations in the recruitment of juvenile birds and of the food supply on carrying capacity predictions for the Exe estuary mussel beds. (a) Constant food supply but fluctuating numbers of juvenile recruits, which varied at random between 0.1 and 0.5 juveniles per adult. The graph shows the numbers of Oystercatchers surviving each winter, with selection year by year for the most efficient individuals either not allowed or allowed. (b) The effect on the number of birds surviving the winter and the within-winter carrying capacity of introducing a fluctuating food supply. The mean biomass of mussels across the 12 beds is also shown. (c) As (b), but with an alternative food supply provided. The inset shows the proportion of birds using the alternative feeding area in late winter as a function of the initial biomass of mussels in September. No selection for efficiency was allowed.

year by year as more and more birds were able to feed fast enough to survive on the food supply remaining at the end of each winter. But numbers again levelled off when the reduction in mortality through selection for efficiency was matched, on average, by the extra mortality caused by the increased depletion and interference competition. This again illustrates how the capacity of the estuary to support Oystercatchers depends in part on the average efficiency of the birds.

When the food supply was allowed to fluctuate, the fluctuations in the within-winter capacity closely followed, of course, those of the food supply (Fig. 12.10b). However, in the long-term, Oystercatcher numbers did not track the annual fluctuations in within-winter capacity. During the first winter of low carrying capacity, Oystercatcher numbers were knocked down to a very low level. Numbers did not again attain the within-winter capacity at any time over the next ten years, even when selection for efficiency was allowed and numbers recovered at a relatively rapid rate. The next winter of very low food abundance again reduced Oystercatcher numbers well below the within-winter capacity of the following years. The within-winter carrying capacity of the mussel beds was only reached in the occasional year of low shellfish abundance. In most years, the number of recruits was never sufficient to enable bird numbers to increase to the point where the mussel beds were fully exploited. In the long term, the number of Oystercatchers in any one year is determined, not by the capacity of the habitat in that year, but by that in preceding poor years.

These two simulations represent the two extreme cases of a constant or a widely fluctuating food supply. In reality, the food supply of most Oystercatchers will fluctuate, but not so violently as shown in Fig. 12.10b. One reason for this is that Oystercatchers can turn to other foods, such as *Nereis diversicolor* and *Macoma balthica*, when shellfish are scarce (Triplet and Etienne 1991; Desprez *et al.* 1992; Beukema *et al.* 1993). The effect of having an alternative food source was explored by providing a thirteenth feeding area in which neither interference nor depletion competition were assumed to occur. The quality of this area was arbitrarily set at quite a high level; specifically, one at which, in the most vulnerable group (juvenile stabbers), the individuals of above average foraging efficiency only were able to achieve the intake rate necessary to survive the winter. The mussel food supply was made to fluctuate between years in exactly the same pattern as in the previous simulation. Selection for foraging efficiency was not allowed.

More birds now survived each winter because large numbers turned to the alternative food supplies in years of low mussel abundance (inset to Fig. 12.10c). Although it is not shown in the Figure, the local wintering population gradually increased as a result so that, after 25 years, it was fluctuating between 7000 and 9000 birds. This is an unrealistically large number, so the quality of the alternative feeding areas provided in this simulation may have been too high. Nonetheless, the simulation illustrates the important role that an alternative food source can play. Once the population had built up sufficiently, the numbers remaining on the shellfish beds themselves at the end of each winter now tracked the fluctuations in the within-winter capacity much more closely than they had done when no alternative food supply had been made available. But although the numbers of survivors did approach the within-winter carrying capacity on a number of occasions, in most winters the capacity was still not reached (Fig. 12.10c). This happened because many birds moved off the shellfish beds during the winter as the quality of the food supplies there deteriorated because of depletion by the Oystercatchers themselves, mortality from other causes and the reduction in the condition of individual mussels. Clearly, the extent to which, and the frequency with which, the within-winter capacity of the shellfish beds will be approached will depend on the quality, and therefore attractiveness, of the alternative food supply. This affects both the total numbers of Oystercatchers on the estuary each year that are available to exploit the shellfish beds and the proportion of them that actually do so.

These simulations, which were run for quite extreme conditions, allow the following conclusions to be drawn about the circumstances in which the within-winter carrying capacity is likely to be reached in Oystercatchers. The capacity of the shellfish beds themselves can be reached when the shellfish food supply is the same each year, even at the low rates of juvenile recruitment typical of this species (Goss-Custard *et al.*, Chapter 13, this volume). With fluctuating shellfish abundance, the proportion of years in which the within-winter capacity of the shellfish beds is achieved will be greater when the fluctuations in shellfish are small than when they are large and is likely to be increased if an alternative food supply is also available. However, the quality of the alternative feeding areas is also likely to have an important effect on the numbers of Oystercatchers that remain on the shellfish beds at the end of the winter. The alternative food supply also has an important influence on winter survival, and thus the size of the whole population on the estuary. Because of this, it is more appropriate to consider the carrying capacity of the combined food supplies of the whole estuary—including, of course, the supplementary food supplies in the adjacent terrestrial habitats—than to isolate that of the shellfish beds alone. It is clear from the simulations that the extent to which the combined food supplies fluctuate will determine how frequently the within-winter capacity of the estuary as a whole can be reached.

Discussion

Carrying capacity and food limitation

Some early attempts at predicting the carrying capacity of estuaries for shorebirds assumed that measuring depletion would give a good indication of how close an estuary was to capacity; see the review in Hale (1982). This is probably justified when such a high percentage, say 95%, of the food supply—including the alternative and supplementary sources—is removed over the winter that it is most unlikely that many more birds could be supported. However, in shorebirds, the rates of prey depletion are seldom more than 40%, even in the most preferred feeding areas, and are usually much less (Goss-Custard 1980). Such low rates tell us little about whether an area is at capacity. Our model shows that when food declines for reasons other than depletion, and when interference competition also has an important influence on bird density and survival, depletion over a wide range of values gives no indication as to whether capacity has been reached; as the simulations showed, at capacity Oystercatchers may only remove < 10% of the food supply. Numbers are not limited by the gross quantity of food available during one winter but by the rate at which the birds can consume it relative to their demands for energy. Their ability to do this is not related in a simple manner to the proportion of the food they remove over the winter.

The idea that numbers are limited and the carrying capacity achieved only when depletion is equivalent to the annual production of the food supply (Piersma 1987; Meire 1993) is also inappropriate. First of all, it confuses the concepts of limitation and carrying capacity. Unless the slope bW of the relationship in Fig. 12.1 switches instantaneously from 0 to 1, the phrase 'limited by the food supply' does not necessarily have the same meaning as 'the carrying capacity has been reached'. As the simulations show, local bird numbers can be limited by the food supply—in the sense that starvation contributes to the mortality rate (Goss-Custard 1993)—well before capacity is reached. Second, so many features of the birds and of their prey interact to determine carrying capacity that it would be highly fortuitous if depletion when the estuary is at capacity were equivalent to the annual production of the prey. For example, the average foraging efficiency of Oystercatchers has a direct effect on predicted carrying capacity. The average foraging efficiency, in turn, may reflect the trade-off between the risk of the bill being damaged, of the birds being taken unawares by a predator or of ingesting possibly damaging parasites (Hulscher, Chapter 1, this volume; Sutherland *et al*, Chapter 3, this volume); these factors probably provide the constraint on the continuing increase in average foraging efficiency. Since group selection seems most unlikely to arise in such mobile birds, there is no reason to believe that the trade-off decisions of fitness maximizing individuals should necessarily result in a communal efficiency in the exploitation of a local population of prey.

Is carrying capacity ever reached?

The results of the model simulations are reasonably consistent with the concept of carrying capacity as defined in Fig. 12.1. On the simplifying assumption that capacity is achieved only by mortality, the simulations show that, within a single winter, bird numbers can approach the level at which one bird dies before March, generally on neap tides, for every additional one that arrives the previous autumn. In the long term, this within-winter capacity can be reached every winter when the food supply is constant. But in reality, food abundance and weather conditions fluctuate, often very widely. With such annual variations in the feeding conditions represented in the model by fluctuations in shellfish abundance, bird numbers do not reach the within-winter capacity set by the shellfish beds themselves; their numbers are reduced too much in the occasional years of very low shellfish stocks. Without an unrealistically high level of immigration, capacity in years of very high shellfish abundance is unlikely to be reached because the supply of birds is far too low. However, the presence of an alternative food supply that allows most birds to survive in years when shellfish are scarce enables the within-winter capacity of the shellfish beds to be tracked more closely. For this reason, it is appropriate to consider the carrying capacity of all the food supplies on the estuary and adjacent terrestrial supplementary feeding areas combined and to investigate now the extent to which they, in aggregate, fluctuate between winters.

The results underline the obvious but easily overlooked point that the within-winter carrying capacity can only be attained when there is a sufficient supply of birds available to do so. Nonetheless, it is striking that it requires a very large number of potential recruits for the within-winter capacity to be realized. We must ask, therefore, whether in the real world, the supply of Oystercatchers is ever great enough for the capacity of coastal areas to be reached. If not, the concept of carrying capacity would represent a theoretical maximum whose only interest would be in providing a potential limit against which existing numbers might be compared, but it would have little practical value.

Capacity could be reached if the size of the greater population, or metapopulation, to which the birds in each estuary belong was so large that the supply of potential recruits to each estuary exceeded their capacity to support them. But this is most unlikely ever to occur in a species with such a low per capita reproductive rate (Goss-Custard 1993). Consider a metapopulation that fluctuates around an equilibrium level and assume that each bird is free to move between estuaries and chooses the one where it maximizes its intake rate, and thus chances of surviving the winter. An increase in the reproductive rate or decrease in the summer mortality rate would cause numbers to increase in parallel in all estuaries as the increasing number of individuals sought to locate the currently best feeding conditions. As the metapopulation increased, winter bird densities, and so mortality, would also begin to increase. With such a low reproductive rate, the increase in metapopulation size would be checked before the rate of increase, bW, in winter mortality reached the value of 1 (Fig. 12.1). As argued by Goss-Custard (1993) and Sutherland (1996), the metapopulation would move back towards the equilibrium position well before the capacity in all the individual estuaries had been reached.

This conclusion has been confirmed for Oystercatchers by simulations with a metapopulation model of the European population (Goss-Custard *et al.*, Chapter 13, this volume). To give the best chance of winter capacity being reached, the important density dependence operating through competition for territories in the summer was removed and the productivity

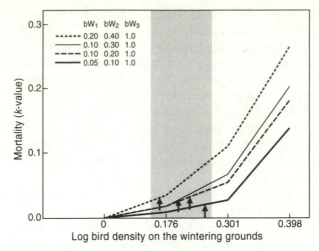

Fig. 12.11 The density on the wintering grounds, and thus the point along the winter density-dependent mortality function, at which the European population of Oystercatchers reaches equilibrium under different assumptions of the form of the mortality function. The shaded area shows the range within which equilibrium was reached across all simulations. The arrows show the point along the function, and thus the density, at which equilibrium was reached in each case. The rate of increase of mortality, expressed as a k-value (Varley and Gradwell 1960), with Oystercatcher density increases in three steps, the respective slopes being bW_1, bW_2 and bW_3. Despite removing all other sources of density dependence in the model, it proved impossible to force the population to reach equilibrium in the third segment of the function where $bW_3 = 1$ and the carrying capacity is reached over the entire wintering range.

of each breeding pair was increased eight-fold above the levels actually observed in the field. The population was now regulated only by density-dependent mortality in the winter. As Fig. 12.11 shows, this mortality was represented by a curve similar to that shown in Fig. 12.1b with, in all simulations, the final of three density-dependent segments having a slope of 1; this, of course, is when all the wintering area would be at capacity. The slopes and intercepts of the first two segments were varied over a range of values in an attempt to force the winter density into the third segment. In fact, it proved impossible to do this because the metapopulation always attained equilibrium well before this point was reached, at a level of nearly twice its present size. Even on the entirely unrealistic assumption of there being no constraints on production in the summer, the Oystercatcher metapopulation never reached the density at which capacity was reached throughout the wintering range.

This result suggests that the only way in which an estuary could be at capacity for Oystercatchers is that some of them attract a disproportionately large number of birds for fitness-related reasons other than the food supply. One real possibility is that estuaries close to the breeding areas may be preferred if they allow the birds that winter there an advantage when competing for breeding sites (Hulscher *et al.*, Chapter 7, this volume). Furthermore, if prospecting birds are able to assess in autumn their chances of overwinter survival, the local carrying capacity could be reached more by emigration of potential recruits in autumn than by late-winter mortality, as is assumed in the current model simulations. Since the simulations show that carrying capacity is lower when it is brought about by emigration, capacity could be reached at even lower numbers of potential recruits, further increasing the chances that it could be realized in some preferred estuaries.

The effect of habitat loss and change on local numbers

The removal or deterioration of feeding areas on estuaries already at capacity would, of course, reduce local bird numbers. Our model can then be used to predict the carrying capacity simply from the predicted feeding conditions in the estuary itself. However, so long as local densities exceed the critical value at which emigration or mortality becomes density-dependent, local bird numbers will be reduced by habitat loss even in estuaries that are not yet at capacity. This is illustrated in Fig. 12.12 which shows the numbers of birds that are, on average, predicted by the model to survive the winter after an increasing proportion of the habitat is lost. The number of birds arriving on the estuary each autumn was either 517, 1550 or 7750; the carrying capacity is only reached in the last case. Simulations were run by removing habitat of average quality or by removing the best or worst 10%, 20%, 30% and so on. The results show that the numbers of survivors is reduced by habitat loss, even when only 1550 birds—and so well below capacity—arrive on the estuary each autumn (Fig. 12.12b). Local habitat loss can thus reduce local numbers in an estuary that is far below capacity, with the numbers surviving depending a great deal on the numbers that settle there in the first place.

The last point means that the effect of habitat loss on local numbers depends in part on the effect that the loss has on the supply of potential recruits, and thus on the size of the metapopulation itself. If only a small fraction of the metapopulation spends the winter in the affected area, the numbers of potential recruits to that area may be little affected. But if the feeding grounds of a large number of Oystercatchers is affected, the size of the metapopulation itself would be reduced because of the generally increased winter mortality rate. The effect of habitat loss and change on local numbers has then to be predicted in three inter-related steps. The first is to predict how the size of the metapopulation would be affected. The second is to predict how the reduced number of birds would distribute themselves among the wintering areas that remain. The third is to predict how many of the recruits to an estuary would survive the winter. Modelling the second step—the redistribution of birds between estuaries —is quite feasible using a modification of the approach discussed in this

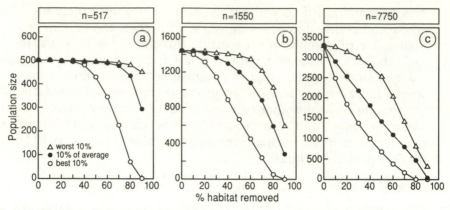

Fig. 12.12 The effect on local numbers of removing an increasing proportion of the feeding areas with different numbers of recruits arriving in autumn: (a) 517 recruits; (b) 1550 recruits, the average between 1976 and 1983 on the Exe estuary; and (c) 7750 recruits, which would allow the within-winter capacity of the mussel beds to be reached. The numbers of birds surviving the winter is shown as a function of the proportion of the mussel beds that are removed, starting with the best 10%, worst 10% or 10% of average quality.

chapter (Goss-Custard *et al.* 1995b), and is in progress (R.W.G. Caldow, unpublished information). Modelling the third step has been the subject of this chapter. The first step requires modelling the dynamics of the population at a regional, even global scale, and is discussed in the next chapter (Goss-Custard *et al.*, Chapter 13, this volume).

Summary

Carrying capacity in one estuary is reached when, as a result of a density-related feedback process such as competition for food, no more birds can establish themselves, however large the numbers of potential settlers available to do so. Simulations over a single winter with an individuals-based, game theoretic and physiologically-structured empirical model show that bird densities can indeed reach the level at which for every additional bird that arrives in autumn, one dies later in the winter. This within-winter carrying capacity is not reached until late in winter when a combination of declining prey biomass and, to a much lesser extent, increasing energy demands of the birds causes the mortality rate to increase, especially on neap tides when feeding conditions are particularly poor. However, in nature, both the feeding conditions and the number of potential recruits fluctuate annually. Capacity can be reached every winter when the shellfish food supply is the same each year, even at the low rates of juvenile recruitment typical of Oystercatchers. With fluctuating shellfish abundance, the proportion of years in which within-winter capacity is achieved will be greater when the fluctuations in shellfish are small than when they are large. The presence of an alternative food supply allows the late-winter bird numbers on the shellfish beds to track fluctuations in the carrying capacity of the beds more closely. It is thus more appropriate to consider the carrying capacity of the combined food supplies of the whole estuary and those of the supplementary sources in adjacent terrestrial habitats than to isolate those provided by the shellfish beds alone. The degree to which this combined food supply fluctuates will determine how frequently the within-winter capacity of the estuary as a whole can be reached.

The results raise the question of how often, in nature, estuaries are at capacity. On the assumption that Oystercatchers move freely between estuaries and decide where to settle only on the basis of the feeding conditions, it is unlikely that carrying capacity is reached anywhere in the wintering range of the European Oystercatcher metapopulation. Simulations with a metapopulation model show, indeed, that this would still be the case even if all the present constraints on the reproductive rate of Oystercatchers were to be removed. The reason is that, as the metapopulation increases, the mortality rate increases and prevents numbers reaching the point at which, across the entire winter range, one bird dies for every new recruit that arrives in autumn. In a species with such a low per capita reproductive rate, capacity seems only likely to be reached if disproportionately large numbers of birds are attracted to particular estuaries because a factor which affects fitness, other than the food supply itself, varies between them; for example, the birds may prefer estuaries near to their breeding areas.

If the estuary is already at capacity, our model can be used to predict the carrying capacity following habitat loss only from the predicted feeding conditions in the estuary itself. But habitat loss would also reduce numbers in estuaries that are not yet at capacity because the local mortality or emigration rates would still increase. The exact number of survivors would depend critically on the number of recruits, and thus on the size of the metapopulation itself. Therefore, the effect of habitat loss and change on local numbers has to be predicted in three inter-related steps: (i) predicting how the size of the metapopulation would be affected; (ii) predicting how the reduced number of birds would re-distribute themselves among the wintering areas that remain; and (iii) predicting how many of the recruits to a particular estuary would survive the winter.

13 Population dynamics: predicting the consequences of habitat change at the continental scale

John D. Goss-Custard, Sarah le V. dit Durell, Ralph T. Clarke, Albert J. Beintema, Richard W.G. Caldow, Peter L. Meininger and Cor J. Smit

Introduction

As the previous chapter discussed, Oystercatcher numbers breeding or wintering in one area may depend not only on factors operating in that area but also on the supply of birds available to live there. This, in turn, depends on the attractiveness of the area relative to others and on processes occurring on the, often distant, wintering or breeding grounds and on migration. In such a species, population processes must be investigated at large spatial scales. But how large must the scale be before, in some meaningful sense, we can say that a population is being studied? Oystercatcher numbers on one mussel bed, for example, are determined almost entirely by the behavioural decisions of individuals and few would regard investigating this as a population study. But as the scale increases relative to the scale of movement of individuals, numbers become increasingly dominated by the balance between reproductive and mortality rates until all of the important changes in numbers are caused this way. Some ecologists argue that population studies can only be done satisfactorily at the scale at which immigration and emigration are for all practical purposes non-existent. In other words, population dynamics should deal only with the balance between reproductive and mortality rates; studies of movement should be done only to understand better the processes that determine these rates. This approach is adopted in this chapter which explores the population dynamics of Oystercatchers at the continental scale required. We also model how numbers might be affected by changes in the breeding and wintering habitats that could take place over the next few decades (Lambeck *et al.*, Chapter 11, this volume).

Overall approach

Since Oystercatcher numbers do not increase indefinitely or go extinct, the rates of reproduction and/or mortality can be assumed to be density-dependent and to regulate numbers. In a regulated population, numbers

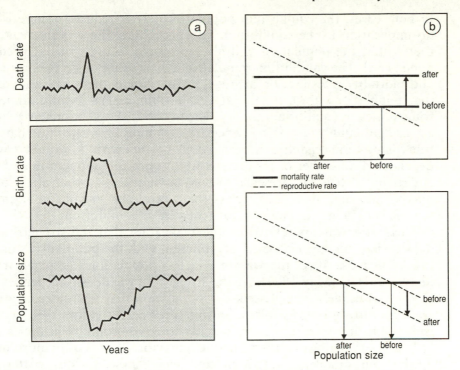

Fig. 13.1 Characteristics of a regulated population, as defined in this chapter. (a) Following a sudden increase in the death rate, the hypothetical population slowly returns to its previous level because of a density-dependent compensatory increase in the reproductive rate. (b) How changing either a density-independent mortality rate (top) or density-dependent reproductive rate (bottom) causes the equilibrium population size to change as indicated by the arrow on the horizontal axis.

tend to return to some typical level if, for some reason, a large increase or decrease has occurred because of a temporary imbalance between reproductive and mortality rates; for example, when many die in severe weather (Fig. 13.1a). Such a response to a 'perturbation' provides, in fact, the most convincing evidence for regulation (Sinclair 1989).

The importance of such feed-back processes should not, however, obscure the point that the level at which reproductive and mortality rates are, on average, equal can be greatly affected by a change in either the density-dependent or density-independent rates. Figure 13.1b (top) illustrates what happens when the density-dependent reproductive rate remains the same but the density-independent mortality rate increases. This could arise, for example, because the average winter weather conditions deteriorate so that more birds die, on average, each year. Conversely, Fig. 13.1b (bottom) illustrates what happens when the density-independent mortality rate remains the same but the density-dependent reproductive rate is lowered because, for example, a change in the breeding habitat makes a greater proportion of chicks vulnerable to predators.

In both cases, the equilibrium population size is reduced. There is thus no implication in the equilibrium approach that, like a permanently set thermostat, the population will always return to a particular predetermined level. The equilibrium population size is simply where reproductive and mortality rates are, on average, equal. Nor is there any implication that populations are generally at equilibrium, as has been assumed by some critics of equilibrium theory (Krebs 1991). Populations may fluctuate around equlibrium as reproductive and mortality rates fluctuate and track long-term changes in the environment with a rapidity which depends on a number of characteristics of both the population and their environment. It is the presence of feed-back processes that affect reproductive and mortality rates that is the central element of the approach, not the belief that populations are generally at equilibrium.

This point was underlined by the realization that the strength of the density-dependent processes in conjunction with the population's natural rate of increase affect the way in which a population fluctuates around its typical level (May 1981). As the strength of the density dependence in theoretical models is increased, the population first fluctuates erratically and only slowly returns to equilibrium after perturbation, then becomes very stable and so returns very quickly to equilibrium, then cycles regularly and, ultimately, exhibits chaotic behaviour that is difficult to distinguish from random changes in numbers. These various patterns of population behaviour result simply from the deterministic properties of the density-dependent factors acting in conjunction with the average rate of increase. Real populations also have the additional source of variability caused by environmental fluctuations. In combination, properties of the population itself and environmental factors can cause a very wide range of patterns in the regularity and extent of population fluctuations, the average level around which a population occurs and the speed with which a perturbed population returns to its previous level or, indeed, its tendency to instability and extinction. Taken together, these complex properties define the dynamics of the population.

Clearly, it will be impossible to understand fully such population behaviour without a quantitative approach. A model constructed from words simply cannot trace through the effects on population size of a change in one or more interacting, especially non-linear, parameters; for example, an increased density-dependent reproductive rate, a decreased summer adult mortality but an increased winter juvenile mortality rate. But unless we can do this, can we really claim to understand the population to any meaningful extent?

Distribution and population trends

Total population size

The most recent estimate of the total numbers of Oystercatchers in Europe is 876 500, with 214 000–291 000 pairs breeding (Goss-Custard

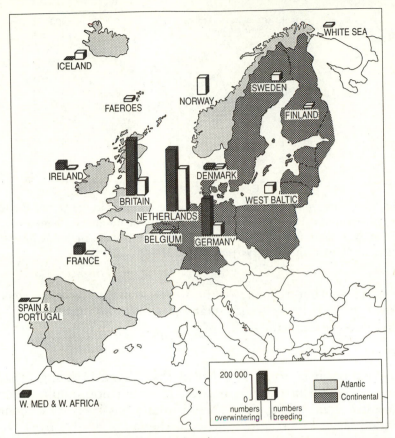

Fig. 13.2 The distribution of Oystercatchers during the breeding and nonbreeding seasons during the 1980s and early 1990s. The shading shows the division into the Atlantic and Continental regional populations used in the analysis and in the model.

et al. 1995d). For both data analysis and modelling, the European population was divided into the Atlantic and Continental regions (Fig. 13.2), largely because of the difference in winter climate and because there is rather little interchange of birds between them (Hulscher *et al.*, Chapter 7, this volume). Trends in total population size are best examined in midwinter when all Oystercatchers congregate in the coastal zone. British winter numbers have increased steadily, at least until recent years whereas, in The Netherlands, numbers have fluctuated more widely but without obvious trend (Fig. 13.3). Given the importance of these two areas to Oystercatchers (Fig. 13.2), the results indicate no general decline in Oystercatcher numbers over the last two decades.

Coastal breeders

Regular censuses of breeding pairs in German, Dutch and British coastal areas suggest numbers have either increased or fluctuated without trend over much of this century (Fig. 13.4a,b) and over the last three decades

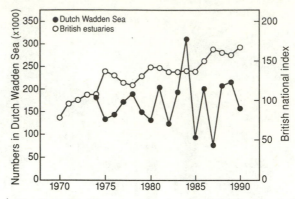

Fig. 13.3 Trends in wintering Oystercatcher numbers in the Dutch Wadden Sea and British estuaries. For The Netherlands, the total numbers counted in January are shown (C.J. Smit, unpublished information). For Britain, the midwinter (November–January) index is shown (Prys-Jones *et al.* 1994).

(Fig. 13.4c). But as many of the sites are nature reserves, it is not clear how typical are these trends; many populations in unprotected areas may have become extinct.

Inland breeders

In contrast, it is firmly established that inland breeding Oystercatchers have increased in many countries, although not simultaneously (Goss-Custard *et al.* 1995d). The spread inland has been well documented in Britain (Fig. 13.5a) and in The Netherlands where, by 1990, Oystercatchers were breeding throughout the country (Fig. 13.5b). However, local increases in numbers have only rarely been monitored. At most, a few well-spaced counts are available. For example, in Germany, Oystercatchers began moving up the river Rhine from The Netherlands in 1950. By 1962, there were 25–30 pairs; in 1971, 50–70; and by 1976–77, 75–85 pairs (Mildenberger 1982). In another study, the relative abundances of several species, including the Oystercatcher, were monitored in the Dutch province of Friesland. Figure 13.6 shows the numbers of Oystercatcher per Lapwing (*Vanellus vanellus*) and, for comparison, the numbers of Black-tailed Godwit (*Limosa lapponica*) per Lapwing. Had the Lapwing population remained constant, these figures would suggest a 100% increase of Oystercatchers in just over 20 years, contrasting with 50% decline in the Black-tailed Godwit. In reality, Lapwing populations probably increased (van Dijk *et al.* 1989), making the increase in Oystercatchers even more spectacular.

The cause of the spread inland

Oystercatchers are not, in fact, the only wader species that have adapted to breeding on agricultural grasslands. The inland spread of waders has

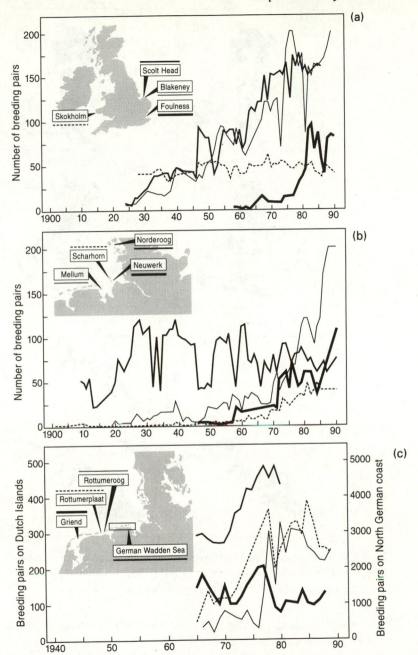

Fig. 13.4 Trends in the numbers of coastal breeding pairs. (a) In the British sites of Skokholm, Scolt Head, Blakeney and Foulness since the 1920s. (b) The islands of Norderoog, Scharhoern, Neuwerk and Mellum along the north German coast since 1900. (c) The whole German Wadden Sea and the islands of Griend, Rottumerplatt and Rottumeroog in the Dutch Wadden Sea since 1965. The line adjacent to each site name identifies the trends shown in the graphs.

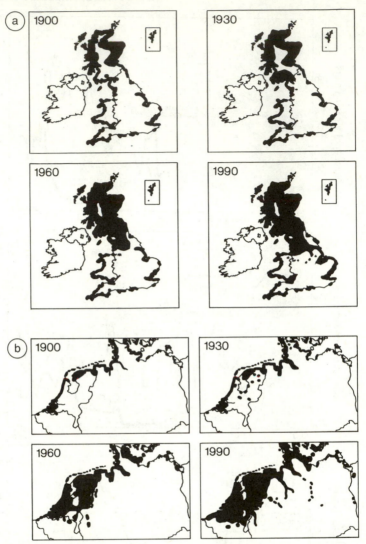

Fig. 13.5 The spread inland of breeding Oystercatchers: (a) Britain (Buxton 1962; Dare 1966; Sharrock 1976; Gibbons *et al.* 1993). (b) In The Netherlands and neighbouring countries (P.L. Meininger, unpublished information).

been most studied in The Netherlands, where the following are collectively known as meadow birds: Oystercatchers, Lapwing, Curlew (*Numenius arquata*), Black-tailed Godwit, Redshank (*Tringa totanus*), Snipe (*Gallinago gallinago*), and Ruff (*Philomachus pugnax*). All these species use the agricultural grasslands for nesting, but originate from natural open habitats. Meadow birds have only colonized farmlands where moist conditions prevail throughout the nesting season (Beintema *et al.* 1985). These special conditions are set by climate, hydrology, and soil.

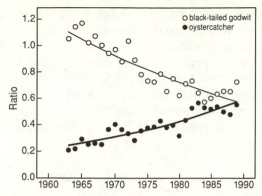

Fig. 13.6 The ratio between the numbers of breeding Oystercatchers and Black-tailed Godwits and the numbers of breeding pairs of Lapwings since 1963 in the Dutch province of Friesland (A.J. Beintema, unpublished information).

As a result, within Europe, meadow birds have arisen only in a limited area, ranging from Britain through The Netherlands, North-west Germany, Denmark, southern Sweden, Poland, and the Baltic States, with the centre of gravity in The Netherlands. The best Dutch meadow bird habitats occur where waterlogged peat deposits are covered up by clay sediments. The fertile top layer gives high productivity, while the wet peat underneath prolongs wet conditions into summer. This seems to guarantee favourable feeding conditions until late in spring and prevents early grazing and mowing, enabling birds to hatch safely; although many nests are destroyed by predators, losses due to trampling by cattle and mowing can also be considerable (Beintema and Muskens 1987). In general, inland breeding waders have no defence against cattle, and trampling is a chance process. However, Oystercatchers do vigorously attack cattle, although usually to little observable effect. Nonetheless, it may explain why Oystercatcher nests survive at a slightly higher rate than do those of other inland-breeding wader species. Oystercatchers thus seem to have spread inland because the conditions there became highly suitable for breeding, although it may have required a preceding change in behaviour allowing birds to feed on terrestrial prey (Heppleston 1972).

There is, however, an upper as well as a lower limit to the intensification of agriculture that allows meadow birds to flourish on grassland. The intensification of agriculture has had two opposing effects on meadow birds (Beintema 1983). On the one hand, increased fertility yields higher biomass densities of invertebrates. On the other, the probabilities of hatching and fledging decrease because of intensified agricultural activity, such as increased cattle densities and earlier mowing dates. As agriculture intensifies, meadow birds will colonize the area as soon as the minimum level of intensification needed has been reached. Populations and densities will then build up, until the upper limit has been reached as intensification proceeds further. After that, the entire system collapses, and populations will eventually die out, due to poor reproduction.

The points corresponding to the lower and upper limits of intensification differ between species. Little is known about lower limits, but upper limits have been well studied in The Netherlands where meadow bird populations have begun to decrease following further intensification. The different species can be ranked according to their vulnerability, with the species that will disappear first having the lowest upper limit of tolerance. From the least to the most vulnerable, the species are ranked Curlew, Oystercatcher, Lapwing, Black-tailed Godwit, Redshank, Snipe and Ruff. Vulnerability depends on characteristics such as the mortality rate, clutch size, timing of nesting, length of nesting season, and relaying capability (Beintema 1983). If we order the species' lower limits—which may be set by the need for nutrient rich earthworms to produce eggs—according to body weight, and the upper limits according to vulnerability, the relative ranges of tolerance of the different species to agricultural intensification can be ranked (Beintema 1983). This ranking implies that the heavier-bodied species would have colonized meadows later than the lighter ones. The records are insufficient to test this prediction precisely. However, Black-tailed Godwits probably were still increasing in The Netherlands as recently as the 1940s and 1950s whereas the heavier Oystercatcher had its most marked increase in the 1960s to early 1980s. The Curlew, the heaviest of all, only started to increase markedly during the last decade.

Demographic consequences of population increase

The increases in the British wintering population and in many inland and coastal breeding populations throughout Europe allow the possibility that reproductive and mortality rates are density-dependent to be tested.

Reproductive rate

For the reproductive rate to be density-dependent across the entire population, either the proportion of adults that breed or the fledgling production per breeding pair, or both, must decrease as the total population increases.

Number of pairs breeding

Both observation and experiment suggest there is intense competition for breeding territories (Ens et al., Chapter 8, this volume). It would therefore be expected that a decreasing proportion of Oystercatchers have bred as numbers have risen. In accordance with this, the numbers of pairs breeding in some British coastal sites has fluctuated without trend, despite a substantial increase in the total British population (Goss-Custard et al. 1995d). But this may be weak evidence because the local

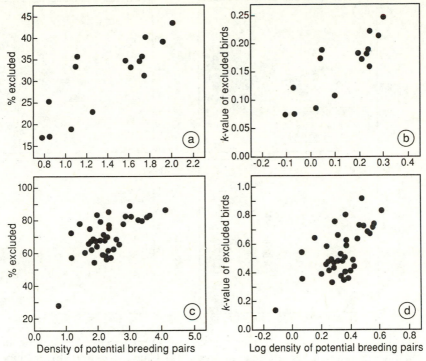

Fig. 13.7 The proportion of Oystercatchers unable to obtain a breeding territory in relation to the numbers available to do so in May; (a and b) the Lune Valley, in North-west England; (c and d) Mellum Island, North Germany. Proportions are expressed as percentages in (a) and (c) and as k-values (Varley and Gradwell 1960) in (b) and (d).

numbers of competitors may not have risen in parallel with the national population. However, in the Lune Valley in North-west England (K. B. Briggs, unpublished information) and on Mellum Island in North Germany (K.-M. Exo, unpublished information), the proportion of adults that are nonbreeders increased with the total number of potential breeders (Fig. 13.7). As a result of increasing competition for territories, it therefore seems likely that the per capita production of young across the whole population, including both breeders and nonbreeders, has decreased in many places across Europe over recent decades.

Fledgling production

In the plots in Figure 13.7, in which the proportion of birds excluded is expressed as a k-value, the slopes in both areas are less than one. Therefore, the numbers of pairs breeding continued to increase even though the proportion of the potential breeders failing to do so also increased. The fledgling production per breeding pair may therefore also have decreased over the decades, either because more pairs bred in poorer

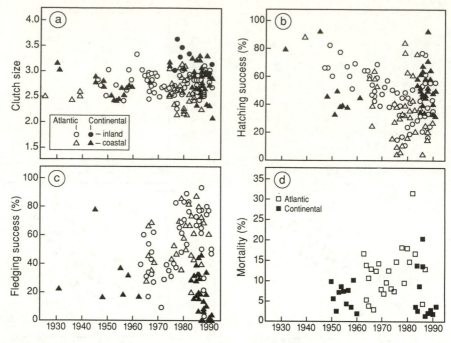

Fig. 13.8 The main demographic variables in Oystercatchers in Atlantic and Continental sites plotted against year of study. (a) Clutch size and (b) hatching success and (c) fledging success in inland and coastal sites. (d) Between breeding season disappearance rates of adults.

habitats or because of an increasing density-dependent mortality during the egg or chick phase.

The data on the three stages of fledgling production were obtained from the published and many unpublished sources detailed in the Appendix to this chapter. Step-down multiple regression analyses were carried out with each site and year record as a datum (Goss-Custard *et al.* 1995d). Along with the year of study, the region (Atlantic or Continent) and the habitat (inland or coastal) from which most of the chick prey were known or suspected of coming were included as dummy (0/1) independent variables.

The year of study had no significant effect on variations in clutch size when all three variables were considered together (Fig. 13.8a). Clutch size was significantly larger on inland sites and in the Continental region ($P < 0.001$). The percentage of eggs hatching declined significantly ($P < 0.002$) over the study period at a rate of 0.37% per year (Fig. 13.8b). Hatching success was higher in inland sites ($P < 0.011$) and on the Continent ($P < 0.001$). Neither the year of study (Fig. 13.8c) nor the habitat had a significant effect ($P = 0.961$ and 0.903 respectively) on fledging success, but it was significantly lower ($P < 0.001$) in the Continental region. The only evidence that fledgling production per breeding pair has decreased is thus provided by the trend in hatching success.

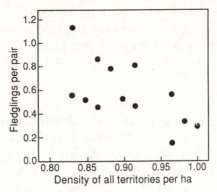

Fig. 13.9 Fledglings produced per pair of territorial birds on Skokholm in relation to the total numbers of territories, including those where laying may not have occurred.

Direct tests for density-dependence in fledgling production could only be made for the Lune Valley and Skokholm. Loss from egg stage to fledging (expressed as k-values) was unrelated ($P > 0.05$) to the total number of eggs laid on Skokholm ($n = 13$) and the Lune ($n = 14$). However, on Skokholm, the fledglings produced per breeding pair decreased sharply as the numbers, and thus density, of all territorial pairs increased (Fig. 13.9). This provides further evidence that production per breeding pair may have decreased in some areas where numbers have increased.

Mortality rate

Most studies have measured mortality as the probability of return from one breeding season to the next of marked breeding birds. Annual return rates do not differ between the sexes (Safriel *et al.* 1984; Briggs 1985), which were therefore averaged. High rates associated with severe winter weather on the Continent were excluded from a regression analysis of the percentage annual disappearance. Neither the year of study (Fig. 13.8d) or habitat was significantly related to disappearance rate and only the region was important, with more adults disappearing in the Atlantic region.

Conclusions

The strongest evidence that per capita reproductive rate has decreased in the many areas where Oystercatcher numbers have increased is that an increasing proportion of potential breeders have failed to acquire a territory. There is also evidence that fledgling production per breeding pair has also decreased. Improvements in methodology mean that the decrease in hatching success is only weak evidence; early clutches that are quickly predated (Ens 1991) may have been missed in the early years, thus increasing the apparent hatching success at that time. By the same argument, any density-dependent increase in recent years in winter mortality may have gone undetected as searches for returning ringed birds have become more efficient or as food supplies in many areas increased (Beukema *et al.* 1993). Sources of density dependence other than competition for territories have thus yet to be convincingly demonstrated.

Population model

The basic model for exploring the dynamics of the European Oyster-catcher population assumes that the only density dependence arises from competition for breeding space. Within each region, birds were divided into coastal and inland breeders, in the proportions indicated by the review in Goss-Custard *et al.* (1995d), but, within each region, inland and coastal birds all wintered together on the coast.

Parameter values

Proportion breeding

The proportion (P) of potential breeding pairs (N) unable to gain a territory is expressed as a k-value ($k = -\log_{10}(1 - P)$). When plotted against $\log_{10}N$, the increasing resistence of territory owners, or the declining compress-ibility of territories, is measured by the slope (bT). With $bT < 1$, territo-ries can still be compressed further so more pairs breed as population size increases, even though an increasingly large proportion is excluded. With $bT = 1$, territories cannot be compressed any further so that a con-stant number of pairs breed, irrespective of any further increase in num-bers attempting to do so. The shape and parameter values of this relationship are important but difficult to estimate. The values of bT for the Lune and Mellum were 0.31 and 0.74 respectively but, for the model, an estimate of bT is required for an entire subpopulation consist-ing of many local populations with their own values of bT. Furthermore, the data in Fig. 13.7 were collected over a narrow range of N. Over the much wider range of values required in model simulations, the relation-ship is likely to be curvilinear, with bT gradually increasing as the num-bers of potential breeders increases (Klomp 1980; Goss-Custard 1993). In the absence of data for an entire subpopulation, the effect of varying the slope and function shape, from the linear to markedly curvilinear, was explored.

Fledglings produced per breeding pair

The mean parameter values for clutch size, hatching and fledging success used in the model simulations were obtained from the above data analyses. The normally distributed annual variations in the values of each parame-ter were calculated as deviations from site means, enabling the typical standard deviation between years in the average site to be estimated (Goss-Custard *et al.* 1995d). These annual within-site variations would not be reproduced across the whole breeding range because annual varia-

Table 13.1 Parameter values for each of the four subpopulations used in the basic population model (Goss-Custard *et al.* 1995d). Mean annual values for clutch size and all subsequent mortalities are shown. Standard deviation (SD) are the same across all four sub-populations and refer to annual variations.

	Atlantic		Continental		S.D.
	Coast	Inland	Coast	Inland	
Clutch size	2.604	2.750	2.840	2.986	0.0062
Egg mortality	0.682	0.598	0.518	0.435	0.0040
Chick mortality	0.445	0.445	0.796	0.796	0.0046
Mortality in 1st and 2nd Winter	0.200 above adult rate				
Mortality in 3rd and 4th Winter	Same as adult rate				
Adult mortality in Winter	0.09126	0.09126	0.03855	0.03855	0.00247
Additional severe winter mortality at all ages	0	0	0.07–0.15	0.07–0.15	0
Adult mortality in summer	0.010	0.010	0.010	0.010	0
Mortality in 2nd and 4th Summer	0	0	0	0	

tions are not correlated across sites within regions, or even between nearby sites. Therefore, the statistical central limit theorem was used to estimate the annual variation across a region, the number of sites being notionally set at 1000; varying this value over the probable range made little difference to the estimate. The subpopulation means and standard deviations used in the model are detailed in Table 13.1. The model assumes that all birds are capable of breeding four years after hatching (Ens *et al.*, Chapter 8, this volume).

Mortality rates

With little mortality occurring in summer or, it is believed, on migration, most post-fledging mortality was assumed to occur in winter (Goss-Custard *et al.* 1995d). Annual variations were estimated in the same way as the production parameters, on the assumption of 200 independent over-wintering estuaries in each region. The extra severe-weather mortality occurring in 15% of winters in the Continental region was modelled by simply adding an amount to the mortality that would have occurred had it been a typical winter. The additional mortality was varied at random in the range 1–15%, 15% being the highest additional adult disappearance rate recorded in severe winters (Goss-Custard *et al.*, Chapter 6, this volume). In view of the milder climate, the model assumes that no large-scale severe-winter mortalities occur across the entire Atlantic region. Mortality in third- and fourth-winter birds was assumed to equal that in adults while that in juvenile and second-winter birds was assumed to be 20% higher (Goss-Custard *et al.* 1982c; Kersten and Brenninkmeijer 1995).

Test of the basic model

In exploratory runs, the coastal subpopulations in both regions quickly declined to extinction. Either fledgling production had been under-estimated or the nonbreeding season mortality had been over-estimated. Both explanations are likely. Fledgling production may sometimes increase at low breeding densities (Fig. 13.9), and so prevent extinction, and the adult mortality is probably over-estimated because some dis-appearing adults may not actually have died (Goss-Custard *et al.* 1995d). Extinction was avoided in the model by gradually reducing winter mortality rates until the coastal populations in each region stabilized; this only required an absolute reduction of 2% in the mortality rate.

Typical time-series for the Atlantic and Continental populations are shown in Fig. 13.10. The effects of a single massive habitat loss in winter are also shown, but are not discussed here. As has been observed in nature (Fig. 13.3), larger fluctuations occurred in the Continental than in the Atlantic region because of chance runs of severe-winter mortality. The fluctuations were unrealistically large when fledging success was density-independent. Perhaps density-dependent fledgling production is widespread in the Continental region or the effect of severe winters on the mortality across the Continent as a whole has been overestimated.

The only direct model test available was to compare the model's pre-dictions for the proportion of birds of all ages that do not breed with observed values. Using data in Goss-Custard *et al.* (1995d), the total numbers of individuals breeding in each region was compared with the total numbers of birds of all ages. On most assumptions (Goss-Custard *et al.* 1995e), the model predictions fell within the observed range of 32–44%. Bearing in mind that the model deals with an entire European population, the comparisons are encouraging.

Predicting the effects of habitat change and loss

Some anthropogenic factors that might affect Oystercatchers, such as certain kinds of coastal pollution, would act independently of bird density. Others, particularly habitat loss, would not; Oystercatcher density will inevitably increase, at least temporarily, unless the species range can ex-pand. The effect on population size of changing the density-independent and density-dependent rates were therefore both explored.

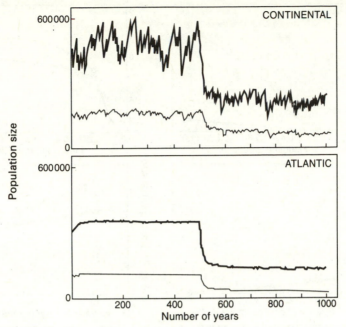

Fig. 13.10 Typical simulation trajectory of the total population in autumn (heavy line), and the number of breeding pairs (light line) for the Atlantic and Continental regions before and after the removal of 60% of the winter habitat in year 500. The parameters for the density-dependent functions in both populations were $bT = 0.5$, $bW = 0.01$, $cW = 1.0$. Mortality rates were as detailed in Table 13.1.

Effect of changing density-independent rates

It was assumed for simplicity that the intensity of competition for territories, bT, shows a linear increase across the range of $\log_{10}N$. Two values were used, representing possible region-wide extremes; 0.3 and 0.7. Simulations were run for almost 1000 years after equilibrium had been reached and the outputs averaged over the final 100.

Although shown only for the Continental population in Fig. 13.11, the sizes of both subpopulations considered were very sensitive to the levels of all the major density-independent rates: (i) the production rate of fledglings (Fig. 13.11a); (ii) the survival of young birds during their first and second winters (Fig. 13.11b); and (iii) the adult mortality rate (Fig. 13.11c). As would be expected with weaker regulation, population size was particularly sensitive to the density-independent rates at the lower value of bT. These simulations illustrate the important point that it not just a change in density-dependent rates that affect population size.

Density dependence and habitat loss

No direct test has yet been made for density dependence in winter; mortality is technically very difficult to measure, even in one estuary, and

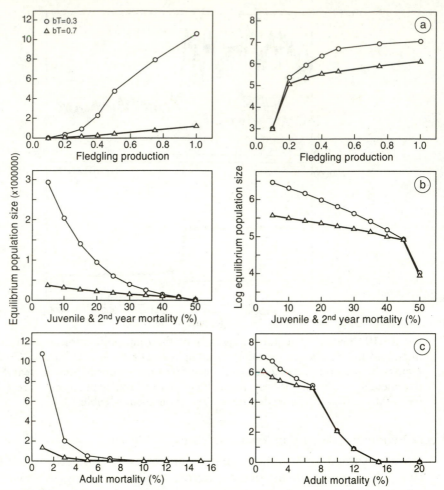

Fig. 13. 11 Equilibrium population size in the Contintental population in relation to (a) the production rate of fledglings, (b) the survival of birds during their first two winters, and (c) the adult mortality rate. To show effects at low population sizes, population size is transformed to logarithms in the right-hand graphs. The population is regulated only by competition for breeding territories at two levels of bT. The values of the rates not varied in these simulations are shown in Table 13.1.

the population varies annually by so little that any dependence of mortality on density could take decades to emerge (Goss-Custard *et al.* 1995b,e), especially if the food supply also varies (Beukema *et al.* 1993). Accordingly, the densities at which mortality is likely to become density-dependent has been examined indirectly from studies of individual variation in competitiveness, using a physiologically-structured game theoretic model as detailed in Goss-Custard *et al.* (Chapters 5 and 12, this volume) and Clarke and Goss-Custard (Appendix 1, this volume). Briefly, for the

Oystercatchers that feed on the 12 main mussel beds of the Exe estuary, the proportion of birds whose energy reserves fell to zero, and so died, were plotted against the initial numbers settling in September. Mortality became density-dependent at 1–1.5 birds per ha of intertidal flats and then increased rapidly, although at a decelerating rate. As current observed densities exceed these values, the population may already be experiencing density-dependent mortality and/or emigration.

In winter at the regional scale

The basic population model has no density dependence in the winter mortality as its real strength is difficult to quantify from field observations. However, without density dependence in the winter mortality rate, a loss of winter habitat would have no effect on winter survival and hence long-term population levels, which is obviously unrealistic in many cases. Winter density dependence was represented in the model simulations in the rest of this chapter as follows: for winter bird densities below a threshold value of cW, mortality was density-independent and equal to the values given in Table 13 for each region and age class (mW). When density exceeded this threshold, mortality increased at a rate of bW for every bird per hectare increase in bird density.

As Figure 13.10 illustrates, the total population breeding in summer and arriving on the wintering areas in autumn reaches a new equilibrium quite rapidly after winter habitat loss. To explore the effect of habitat loss, the area available in winter was reduced in another series of simulations by steps of 10%, until 90% had been removed. The same coefficients of the winter density-dependent mortality function were kept throughout so it was assumed that habitat of average quality was removed at each step. The population density at which the winter mortality became density-dependent (cW) simply determined the point at which habitat loss started to effect population size (Goss-Custard *et al*. 1995e). Importantly, habitat loss began to have an affect earlier in the more fluctuating Continental population because winter densities began periodically to exceed cW earlier in the process of habitat removal. Winter habitat loss would therefore be expected to affect population size sooner in populations that fluctuate widely than in ones that do not.

Once enough habitat had been lost to increase density above cW, the magnitude of the impact then depended on the slope of the density-dependent winter mortality function (bW). The effect on equilibrium population size is shown for each regional population in three ways in Figure 13.12: (i) the population size following a particular percentage reduction in winter area at a given value of bW (Fig. 13.12a); (ii) the population size at different values of bW resulting from a given amount of habitat loss (Fig. 13.12b); and (iii) the percentage reduction in population size following different percentage reductions in habitat at a given value of bW (Fig. 13.12c). In its early stages, habitat loss has a large

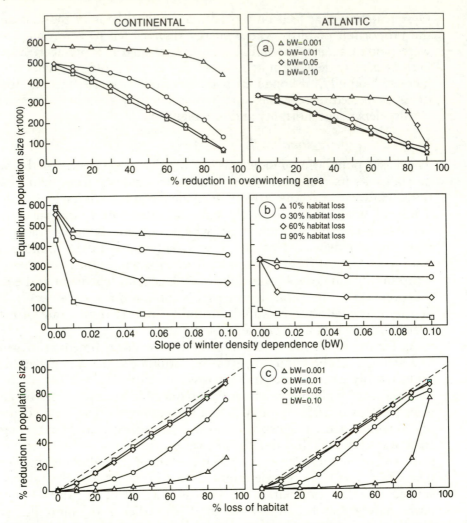

Fig. 13.12 The effect of various amounts of winter habitat loss on the Atlantic and Continental Oystercatcher populations as predicted by the model. (a) Population size following a particular percentage reduction in winter area at a given value of bW. (b) Population size at different values of bW resulting from a given amount of habitat loss. (c) Percentage reduction in population size following different percentage reductions in winter area at a given value of bW. The diagonal dashed line represents proportionality at which a given percentage reduction in habitat results in the same percentage reduction in equilibrium population size. In all cases, $bT = 0.5$ and $cW = 1.0$. The values of mW were mean observed mortality rates detailed in Table 13.1.

effect on population size in both regions only at the higher values of bW, above 0.01. However, above 0.05, further increases in the strength of the density dependence make little difference to the predicted effects of habitat loss—at least over the range explored here. The percentage reductions in population size never exceeded the percentage reductions in habitat loss, but they were roughly equivalent at the higher values of bW (Fig. 13.12c). Except in these circumstances, the decrease in population size was always subproportional to the decrease in the wintering area itself.

Further simulations were run to test how sensitive these conclusions were to the values of other model parameters. The strength of the competition for breeding territories (bT) in linear functions, and whether bT took the same or different values inland and on the coast, made a difference but it was rather small in these simulations (Goss-Custard *et al.* 1995e). A concave curvilinear dependence of bT on $\log_{10}N$ generally made little difference, although the reduction in population size was slightly greater with a convex function at intermediate (0.01–0.05) values of bW (S.E.A. le V. dit Durell, unpublished information). As might be expected, changing the winter density-dependent function from linear to curvilinear also made rather little difference, whether in the convex or concave, when the curvilinearity was slight. But a strong curvilinearity made a large difference, but only in the concave (Fig. 13.13a). The population was rather little affected until large areas had been removed because the increase in density initially only slightly increased mortality. With a convex function,

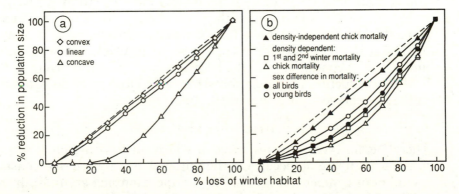

Fig. 13.13 The effect of making various assumptions on the model predictions of the effect of winter habitat loss on equilibrium population size. The percentage reductions in population size associated with increasing percentage reductions in the area of winter habitat are shown. (a) Comparison between a concave, linear and convex winter density-dependent mortality function. bW in the linear was 0.037; in the concave it was successively over three equal density segments, 0.001, 0.01 and 0.1; in the convex it was 0.1, 0.01 and 0.001. (b) The effect of restricting winter density-dependent mortality to just first- and second-winter birds; changing chick mortality from density-independent to density-dependent; and introducing a sex difference in mortality in all birds or just in young birds. In both diagrams, the diagonal dashed line represents proportionality at which a given percentage reduction in habitat results in the same percentage reduction in equilibrium population size. In all cases, $bT = 0.5$. All graphs refer to the whole European population, with the density-independent parameters as given in Table 13.1.

of course, mortality increased rapidly from the beginning of habitat loss and the population reductions were almost proportional to the amount of the habitat lost.

The consequences of habitat loss were also sensitive to assumptions about the basic population biology of Oystercatchers on which there is some doubt. First- and second-year birds may be the most vulnerable to competition (Goss-Custard and Durell 1984) and the population size was initially much less affected by habitat loss if only these birds were assumed to be subject to density-dependent mortality (Fig. 13.13b), even though they comprise all the birds of future generations. Introducing density-dependent fledging success greatly reduced the predicted effects because the elevated winter mortalities resulting from habitat loss were partly compensated by the increased fledgling production at low breeding densities. As there may be sex differences in mortality, particularly in young birds (Durell and Goss-Custard 1996), mortality rate in females was increased to 2% above that in males in either first- and second-winter birds or in birds of all ages. Without also including density dependence in fledging success, all populations quickly went extinct because there were too few females to produce young. With density dependence included, introducing a sex difference in mortality in either young birds or birds of all ages made little difference to the consequences of habitat loss. Taken together, these simulations strengthen the emerging general conclusion as to the effect of removing winter habitat of average quality at the regional and European scale; in Oystercatchers, only in a limited set of conditions will the loss of winter habitat of average quality lead to a reduction in population size that is close to being proportional to the percentage reduction in the habitat itself.

In winter in inland and coastal breeding populations

There are, however, circumstances in which supraproportional reductions in population size do occur. On the untested assumption that coastal and inland birds compete on equal terms on the wintering grounds, the model predicts that the faster breeding inland populations will survive winter habitat loss much more successfully than the slower breeding coastal birds, which are simply outproduced as winter competition intensifies (Fig. 13.14). Whereas the inland breeding population may be little affected until much of the winter habitat has been lost, the reduction in the coastal population may be strongly supraproportional during the early stages of habitat loss.

Effect of removing the best or worst habitat first in winter

The simulations so far have used the same density-dependent winter mortality function throughout the various stages of habitat loss, as if habitat of average quality was removed at every stage. In reality, habitat of above or below average quality may be removed first so that the shape

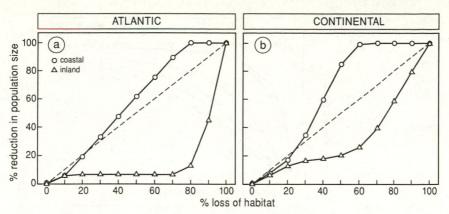

Fig. 13.14 The effect of removing winter habitat on the size of ⸱⸱ɪe inland and coastal breeding populations that share it: (a) Atlantic population; and (b) Continental population. The diagonal dashed line represents proportionality at which a given percentage reduction in habitat results in the same percentage reduction in equilibrium population size. In all cases, $bT = 0.5$. The density-independent parameters are as detailed in Table 13.1.

and parameter values of the density-dependent function will change as the habitat losses proceed. Given the sensitivity of the predicted effects to the parameter values of this function, this is likely to make a large difference.

The first step in this investigation was to establish how the density-dependent mortality function would change if habitat of above or below average quality was removed. This was done through simulations with the individuals-based, physiologically-structured game theoretic model (Goss-Custard *et al.*, Chapter 12, this volume). The total area of the the mussel beds was reduced in steps of 10%, starting with the best (or worst) 10%, and then the best (or worst) 20%, and so on. This made a large difference to the predicted density-dependent mortality functions (Fig. 13.15a) with, for example, the function (unsurprisingly) becoming much steeper as an increasingly large proportion of the best habitat was removed.

The consequences of these changes for equilibrium population size following habitat loss were explored with the population model parameterized for the Exe estuary (Goss-Custard and Durell 1984). Figure 13.15b illustrates the output from simulations in which 10% portions of the feeding area were successively removed and the appropriate density-dependent winter mortality functions, as derived from Figure 13.15a, inserted at each stage. After every additional 10% loss, the population quickly reduced to a new, and lower, equilibrium level. As would be expected, compared with removing habitat of average quality, the population was much more reduced when the best quality feeding areas were removed first, especially in the early stages of habitat removal. Indeed, the effects of habitat loss were now supraproportional (Fig. 13.15c). In

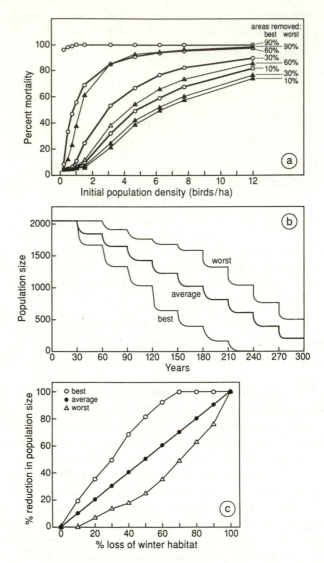

Fig. 13.15 The predicted effect of removing either the best or worst quality feeding areas on the population size of Oystercatchers of the Exe estuary. (a) The density-dependent winter mortality functions associated with removing 10% or 30% or 60% or 90% of the best or worst mussel beds; derived from an individuals-based game theoretic model of the Exe estuary (Goss-Custard *et al.*, Chapter 12, this volume). (b) The predicted effect of successively removing 10% of the best, worst or average mussel beds on population size. (c) The predicted effect of successively removing 10% of the best, worst or average mussel beds on population size expressed as the percentage reductions in population size associated with increasing percentage reductions in the area of the mussel beds. Chick mortality was density-independent and $bW = 0.104$ in the region of equilibrium at the start, which is why the effect of removing average quality habitat approached proportionality in this case (Fig. 13.13b). Density-independent parameters as detailed in Goss-Custard and Durell (1984).

contrast, of course, removing habitat of below average quality in the early stages had very little effect on population size and was subproportional throughout.

In summer

The population model can also be used to predict the effects of the loss of breeding habitat on the total population size and on both the number and density of breeding pairs. For the purposes of illustration, the dependence of bT on $\log_{10}N$, where N is the density of pairs competing for territories, was assumed to be linear and fledgling production to be density-independent. Without winter density dependence, summer habitat loss led, of course, to exactly proportional reductions in total numbers and breeding numbers, while the density of breeding pairs remained constant, at both high and low values of bT (Fig. 13.16a). The results were rather different when winter density dependence was introduced. Summer habitat loss had almost a proportional effect on population size with bT set at 0.7 but a clearly subproportional effect when bT was reduced to 0.3 (Fig. 13.16b). Varying bW, with bT set at 0.5 in all cases, also had an effect, with the reduction in population ranging from being approximately proportional to clearly subproportional (Fig. 13.16c). These examples illustrate the general point that, in a migratory species such as the Oystercatcher, the effect of habitat loss in one season depends not only on the density dependence in that season but also, to varying degrees, on the density dependence operating in the other season.

Simulations were also run on the effect of increasing, rather than decreasing, the amount of breeding space available by up to a factor of two simultaneously in inland and coastal habitats. Without winter density dependence, habitat gain simply resulted in a proportional increase in breeding numbers and population size, with breeding density remaining constant. But when winter density dependence was introduced, the increase in both total and breeding numbers was subproportional, and rather similar, with bT set at either 0.3 and 0.7 (Fig. 13.17a). At a given value of bT, the population size increased hardly at all with strong winter density dependence (Fig. 13.17b); for example, with bW set at 0.35, an 80% increase in summer habitat area led only to a 12% increase in the total population size in autumn and in the number of breeding pairs. The increase in population size was much greater when competition on the wintering areas was weak but, even then, the increase in population size was subproportional. Whether increasing the summer habitat makes a material difference to population size therefore depends a great deal on the strength of the density dependence in the coastal areas in winter.

One reason for this is that, as the areas of both the inland and coastal breeding areas are increased, the inland breeding population increases at the expense of the coastal breeding population (Fig. 13.18). The more slowly reproducing coastal population is unable to take advantage of the

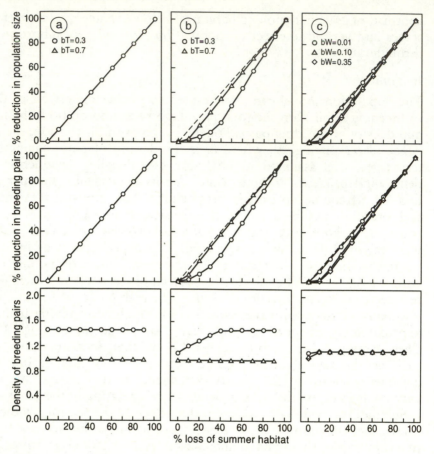

Fig. 13.16 The predicted effect of removing summer breeding habitat on the total population size in autumn (top), the numbers (middle) and density (bottom) of breeding pairs. (a) No winter density dependence. Linear dependence of bT on the total numbers of pairs attempting to breed, with bT = 0.3 or 0.7. (b) With winter density dependence; bW = 0.05 and bT = 0.3 or 0.7. (c) Linear dependence of bW on winter density, with bW = 0.01 or 0.1 or 0.35; bT = 0.5. In (a) and (b) the symbols lie on top of each other and so cannot be distinguished. In the top and middle graphs, the diagonal dashed line (where visible) represents proportionality at which a given percentage reduction in habitat results in the same percentage reduction in equilibrium population size.

increased space because it is outcompeted on the wintering grounds by the more rapidly reproducing inland breeding population. This conclusion depends heavily, however, on the untested assumptions that inland and coastal breeding birds compete on equal terms on the coastal wintering grounds and that, as densities on the inland breeding grounds change, the strength of the density dependence is still adequately described by the value of bT used. But on these assumptions, there is not

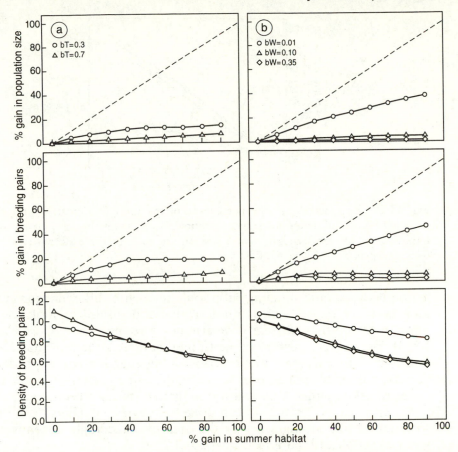

Fig. 13.17 The predicted effect of adding summer breeding habitat on the total population size in autumn (top), the numbers (middle) and density (bottom) of breeding pairs. Inland and coastal breeding areas were increased by the same percentage. All density-dependent relationships were linear. (a) $bT = 0.3$ or 0.7; $bW = 0.05$. (b) $bW = 0.01$ or 0.1 or 0.35; $bT = 0.5$. In the top and middle graphs, the diagonal dashed line represents proportionality at which a given percentage reduction in habitat results in the same percentage reduction in equilibrium population size.

a proportional increase in population size when additional breeding areas are provided if birds compete strongly on the wintering grounds.

Discussion

The simulations with the Oystercatcher population model illustrated how the year-round population size is affected by processes occurring on both the breeding and wintering grounds. Globally, population size depends on the combined effect of many factors, including the inter-

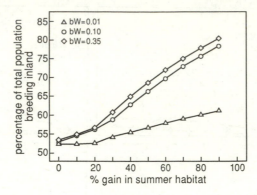

Fig. 13.18 The increase in the percentage of the total Oystercatcher population breeding inland following an equal percentage increase in the amount of summer habitat available for breeding inland and on the coast at three strengths of winter density dependence, $bW: bT = 0.5$.

action between intermingling subpopulations with different reproductive and mortality rates. In this situation, it is not useful to look for 'the' limiting factor that currently determines bird numbers (Goss-Custard 1993). In most circumstances, anything that affects the reproductive and mortality rates, whether their action is independent or dependent of density, will affect population size. For the same reason, it is not useful either to ask whether a population is limited on its breeding or its wintering grounds, as often has been done (see review by Hale 1982); to do so is rather like asking whether the area of a rectangle depends on its width or its breadth (Goss-Custard 1993).

The simulations also showed that the effects of habitat loss at one time of year on population size could depend a great deal on the shape of the density-dependent function and, within a certain range, the precise parameter values of the density dependence in both the summer and winter. This means that it is important to describe these functions. Needless to say, this is difficult at the large scale required, as the continent-wide investigations of duck quarry species also illustrate (Nichols 1991). There are bound to be many errors in estimation, even in such an amenable study species as the Oystercatcher.

The errors can be divided into errors of representativeness and of methodology. Most investigations have been made in areas with many birds, where densities may be particularly high. Little is known about the more dispersed populations, although they are likely to comprise only a small proportion of the whole. Perhaps more importantly, we do not know how the proportion of pairs breeding in different habitats with different levels of productivity changes with population size, as would be expected to occur on theoretical (Fretwell 1972) and empirical (O'Connor 1980) grounds. It is therefore not yet possible realistically to include such spatial variations in performance into the estimates of subpopulation parameter values. For a global population model, these are important omissions which may become more easily addressed as remote-sensing techniques develop.

There are a number of methodological problems in parameter estimation. Most estimates of adult annual mortality rates, for example, are based on the assumption that birds return to their previous breeding site and that they will be located. In fact, this is not always so (Goss-Custard *et al.* 1995d) and, with such low annual mortality rates, estimates may be quite seriously affected if only a few adults move outside the immediate study area or are simply overlooked. As the model showed, an error in annual mortality of only 2% in such a long-lived bird can make the difference between population persistence and extinction.

Both sources of error are likely to affect the critically important density-dependent functions. Most studies have neither been in progress for long enough nor made the necessary measurements to even test for density dependence in fledgling production rates or summer and winter mortality rates, let alone to estimate function parameter values. Only two studies have counted the numbers of nonbreeding adults in May in relation to the numbers attempting to breed. Even then, it was only possible to estimate how the proportion failing to breed changed within a limited range of population size. Yet the importance of being able to define and parameterize these functions was highlighted by several simulation results. For example, changes in density-independent rates most affected population size when the density dependence arising from competition for territories was weak. Similarly, increasing the area of breeding habitat available increased the population most when the winter density dependence was weak.

Another general conclusion is that a given percentage reduction in winter habitat of average quality did not ever result in an even greater percentage reduction in population size. Reductions even approaching proportionality to the area lost only occurred when the winter density-dependent function was steep. This important conclusion was robust across the probable range of values of many of the other parameters and clearly implies that, at the European scale, the Oystercatcher population is unlikely to be greatly reduced in the early phases of the removal of typical winter habitat.

This conclusion was, however, very sensitive to the condition that the lost habitat was of average quality. When habitat of above average quality was removed first, the reduction in population size became markedly supraproportional. Habitat loss in certain areas will therefore have a much greater affect than it will in others. The conclusion is also conditional on the scale at which the predictions are made. When winter habitat of only average quality was removed, the coastal breeding populations were initially most affected, and often, supraproportionally to the

amount of habitat lost. In fact, in many simulations, the inland populations virtually maintained their initial size until a high proportion of the
habitat had been lost. Even though they intermingled on the wintering
grounds and were subject to the same competitive pressures, habitat
loss initially most affected the slower breeding coastal subpopulations.
Therefore, the two conditions in which the model predicts supraproportional reductions in population size are both associated with spatial variation. While in a species such as the Oystercatcher it is necessary to
model the population at a regional scale, it is nonetheless important to
investigate demographic differences between sites within a region to
establish how changes in one site affect numbers in another.

Summary

The Oystercatcher population has increased in recent decades, perhaps
partly as a result of the continuing increase in inland breeding. This
occurs where agricultural intensification has improved the food supply,
and so the parents' ability to produce eggs and to provision young, without increasing by too much the risk that eggs and young will be killed by
cattle trampling or agricultural machinery. There is no indication that, in
response to increased population size, the clutch size, hatching success,
fledgling success and breeding adult mortality rate have declined in
recent decades; however, sampling and methodological difficulties may
have prevented any trends from emerging. The only detectable density-
dependent consequence of the population increase is that a greater proportion of potential breeding pairs may now fail to acquire a territory in
some areas. Fledgling production per breeding pair may also decrease in
some areas as the numbers of competing pairs increases.

Based on a literature review, a model is developed for the Atlantic and
Continental regional populations. Within each region, birds either breed
inland or on the coast but the birds from both habitats winter together
on the coast. Simulations show that anthropogenic factors which change
density-independent rates, even by quite a small amount, could greatly
affect population size, especially when the summer density dependence is
weak. Habitat loss in winter would cause density to increase, at least
temporarily, so that whether or not winter mortality is density-dependent
is critically important. Up to a certain point, the strength of the winter
density dependence has a large influence on the effect that habitat loss
has on population size. Most simulations show that, in comparison to
the proportion lost, removing habitat of average quality leads only to
subproportional reductions in population size. Two circumstances lead
to supraproportional reductions: (i) where two subpopulations with different reproductive rates use and lose the same winter habitat, the poorer-
breeding population decreases at a disproportionately high rate, and (ii)
when the best quality, rather than average quality, habitat is removed.

Removing summer habitat has an almost proportional or clearly subproportional effect on population size, depending on the density dependence in both summer and winter. Providing extra breeding habitat leads only to subproportional increases in population size, the increases being smallest when the winter density dependence is strong. One reason for this is that the more slowly reproducing coastal birds are unable to take advantage of the increased breeding space because they are out-competed on the wintering grounds by the more rapidly reproducing inland breeders. It is concluded that investigating spatial variation in the parameters of the critical density-dependent functions is required if more reliable predictions are to be made at both the local and global scales. This needs to be done in both summer and winter because, in a migratory species like this, the effect of habitat loss in one season depends not only on the density dependence in that season but also on the density dependence in the other season.

APPENDIX to Chapter 13: Production parameters of Oystercatchers. *N* refers to number of years studied. SD refers to the SD of the annual variations in each measure. C = Coast; I = Inland; na = not available.

Country	Habitat Nesting	Feeding	Period	Clutch size N	$\bar{x}$	SD	Hatching success N	$\bar{x}$	SD	Fledging success N	$\bar{x}$	SD	Fledglings per breeding pair N	$\bar{x}$	SD	Source*
Germany	Sand dunes	C	1948-59	10	2.57	0.12	6	40.5	5.9	4	26.1	10.0	12	0.32	0.25	1
Germany	Sand dunes	C	1986-88	3	3.03	0.15	3	44.3	9.4	3	6.4	0.4	3	0.09	0.02	2
Scotland	Sand dune	C	1966-68	3	2.56	0.13	3	46.4	21.8	3	25.7	8.4	3	0.36	0.20	3
Spain	Sandy beach	C	1979-80	2	3.09	0.02	1	84.0	–	1	48.0	–	1	1.20		4
England	Saltmarsh	C	1974-88	15	2.46	0.17	15	23.1	15.4	10	50.6	19.9	15	0.29	0.24	5
England	Saltmarsh/seawalls	C	1974-90	15	2.65	0.26	16	38.4	18.2	10	69.8	12.0	13	0.78	0.37	5
Netherlands	Saltmarsh	C	1984-90	7	2.79	0.18	7	58.3	8.5	7	22.3	13.2	7	0.40	0.24	6
Netherlands	Saltmarsh/dunes	C	1985-91	7	2.56	0.35	7	51.8	22.1	7	16.2	18.4	7	0.26	0.34	7
Netherlands	Saltmarsh/fields	C(I)	1983-91	9	2.98	0.29	9	53.3	9.0	9	13.1	11.0	9	0.23	0.21	7
Wales	Fields	I(C)	1948-73	26	2.70	0.20	23	58.0	11.3	13	37.6	13.3	13	0.58	0.26	8
Scotland	Fields	I	1966-67	2	2.68	0.14	2	44.5	3.1	2	50.8	4.0	2	0.75	0.14	3
Scotland	Fields	I	1966-68	3	2.92	0.01	3	47.3	20.1	3	40.6	5.1	3	0.60	0.14	3
England	Gravel	I	1976-91	16	2.72	0.19	16	24.2	7.5	14	70.3	10.3	16	0.47	0.18	5
England	Fields	I	1978-91	14	2.87	0.18	14	50.1	11.3	9	71.3	13.8	14	1.05	0.23	5
Denmark	Saltmarsh	C	1971-76	6	3.16	na		na			na			na		9
S France	Sand/gravel	C	1980	1	2.80			na			na			na		10
Germany	Sand dunes	C	1948	1	2.80		1	46.0	–	1	16.2	–	1	0.67	–	11
Germany	Sand dunes	C	1931	1	3.01		1	79.1	–	1	20.0	–	1	0.48		12
Finland	Gravel	C	1945	1	2.88		1	92.5	–	1	78.3	–	1	2.09		13
Netherlands	Sand dunes	C	1977		na			na			na		1	0.30	–	14
Netherlands	Various	I/C	1940-45	1	2.77	–		na			na			na		15
Wales	Sand dunes	C	1965		na		1	59.0	–	1	28.0	–		na		16

Country	Habitat	I/C	Year	n			n			n			n			Ref
England	Riversides	I	1967-68	2	2.71	na	2	58.3	na	2	69.1	na	2	1.11	na	17
England	Saltmarshes	C	1967-68	2	2.81	na		na			na			na	na	18
Scotland	Not given	C	1921-29	1	2.49	na		na			na			na	na	19
Scotland	Saltmarsh	C	1925-46	1	2.42	na		na			na			na	na	20
France	Sand dunes	C	1974-79	3	2.98	0.11	1	77.0	–	1	36.0	–	1	0.79		21
Scotland	Saltmarsh	C	1974-78	2	2.42	0.38	2	34.3	5.7	2	49.8	4.31	2	0.42	0.17	22
Germany	Saltmarsh	C	1986	1	3.21	–	1	67.8		1	30.2		1	0.66		22
Germany	Saltmarsh	C	1986	1	3.29	–	1	39.7		1	44.8		1	0.59		23
Germany	Fields	I	1990-91	2	3.0	0.14		na			na		2	0.65	0.35	24
Wales	Cliffs	C	1939	1	2.50	–		na			na			na		24
	Fields	I	1939	1	2.57	–	1	88.2			na			na		25
Netherlands	Fields	I	1974-80	6	3.24	0.24		na			na			na		25
Netherlands	Fields	I	1980-82	2	3.16	0.22		na			na			na		26, 27
Netherlands	Fields	I	1976-90		na		1	52.0	–	1	32.5	–		na		28
Scotland	Fields	I	1986-88	3	2.52	0.09	3	56.6	21.8	3	45.9	37.2		na		28
Scotland	Fields	I	1986-88	3	2.75	0.09	3	49.0	9.7	3	64.4	32.0		na		28
Scotland	Fields	I	1986-88	3	2.61	0.05	3	34.2	11.1	3	44.5	20.7		na		28

* 1, Schnakenwinkel 1970; 2, K.-M. Exo, unpublished information; Pleines 1990; 3, Heppleston 1972; 4, Martinez et al. 1983; 5, K. B. Briggs, unpublished information; 6, B. J. Ens and J. B. Hulscher, unpublished information; 7, C. J. Smit, unpublished information; 8, Harris 1967,1969; Nedderman 1953; Keighley 1949; 9, Møller 1978; 10, Blondel and Isenman 1981; 11, Rittinghaus 1950; 12, Dircksen 1932; 13, Nordberg 1950; 14, Swennen and de Bruijn 1980; 15, Haverschmidt 1946; 16, A. J. Mercer in Heppleston 1972; 17, Greenhalgh 1969,1973; 18, Campbell 1947; 19, Brown 1948b; 20, P. Triplet, unpublished information; 21, Rankin 1979; 22, Becker 1987; 23, P.L. Meininger, unpublished information; 24, Buxton 1939; 25, A. J. Beintema, unpublished information; 26, Beintema and Müskens 1987; 27, A. J. Beintema, unpublished information; 28, N. Picozzi and D. Catt, unpublished information.

Conclusions:
From individuals to populations:
progress in Oystercatchers

John D. Goss-Custard

The aim of this book has been to illustrate, with a particularly suitable study species, how theoretically-based empirical studies of the behavioural, morphological and physiological adaptations of individuals can be used to understand and predict at the population level. The conceptual framework used to address issues of individual adaptation has been the fitness cost–benefit analysis that has developed in behavioural ecology over the last thirty years (Mitchell and Valone 1990; Krebs and Davies 1991). The conceptual framework used at the population level has been the dynamic equilibrium approach that has developed over the same period (May 1981). The potential interdependence of the two paradigms has been recognized throughout, as is illustrated by the publication in the mid-1980s of a Symposium on this theme by the British Ecological Society (Sibly and Smith 1985). In fact, much of the impetus that began behavioural ecology in the first place was the desire of evolutionary ecologists, notably Robert MacArthur, to make sense of bewildering present-day ecological complexity. The belief was that establishing basic principles of species adaptation would enable ecologists to define and predict species niches and interactions, population size and community structure. Now that the demand on ecologists to predict to new circumstances is becoming even stronger, many of us involved in this book believe that the research paradigm of predicting events at the population level from behavioural ecological studies will be increasingly employed (Ens *et al.* 1994).

One reason for this belief is the truism that the response of populations to new circumstances depends entirely on the adaptations of the organisms themselves. More precisely, it depends on the responses of the individuals because it is individuals, not populations, that must adapt to new circumstances. Indeed, developing in parallel with behavioural ecology and population dynamical theory has been a growing awareness that ecological processes can be better understood by investigating and modelling interactions among individual organisms (Fretwell 1972; Łomnicki 1978, 1980; Goss-Custard 1985; Hassel and May 1985; Koehl 1989; Ens *et al.* 1994; Sutherland 1996). To quote Huston *et al.* (1988), ecological 'models often combine many individual organisms and assume that they can be described by a single variable, such as population size. This procedure violates the biological principle that each individual is different, with behaviour and physiology that result from a unique combination of genetic and environmental influences. Second, most models do not distinguish among organisms' locations. Each individual is assumed to have an equal effect on every other individual. This assumption violates the biological principle that interactions are inherently local. An organism is affected primarily by the other organisms with which it comes into contact.'

This is not to say that earlier workers failed to appreciate the importance of individual variation. The enormously influential empirical studies of Jenkins *et al.* (1963) on territorial behaviour and the limitation of breeding density dealt explicitly with differences in competitiveness between birds. Again, the distinction between contest and scramble competition rests on how unequally individuals are able to capture resources (Varley *et al.* 1973). The shift in emphasis that has arisen is that studies of individual variation are now being used to derive, or to predict, population functions and responses to environmental change rather than to explain patterns that have already been observed.

One essential precondition for this change of emphasis was the enormous computational power of modern PCs which enables population models to be based on the differing responses of individual organisms to each other and to their common environment. The factor limiting progress now is not computational capacity but acquiring the data on individual animals; field studies of individuals are very time-consuming. The individuals-based approach is certainly not a short-cut. But, in ecology, prolonged research is not unusual. Indeed, as this book has illustrated, there are circumstances in which the individuals-based approach may be quicker than traditional paradigms; it could take decades to parameterize the density-dependent overwinter mortality functions in the Oystercatchers, even for one tractable estuary such as the Exe. Studies of individuals therefore allow population functions to be derived that would otherwise elude the researcher, permitting population models to be parameterized that might otherwise be impossible to construct.

The other major strength of the individuals-based approach is its potential for providing reliable predictions in new circumstances. The need for such predictions is likely to grow because of the increasing weight given by policy makers to environmental matters. Whereas much research in the past looked for explanations of why a decline in a species population had taken place, we need increasingly to forecast what would happen *if* a particular policy were to be implemented. If we are to predict, for example, what would happen to equilibrium population size if habitat of above or below average quality were to be removed, we must be able to estimate what the parameters of the density-dependent functions would be after the habitat had been removed. The existing function—even if it could be described—could not be used with confidence because, in effect, the new circumstances lie beyond the existing empirical range. The parameter values of the function would change as a result of the very environmental changes whose consequences we wish to predict. The predictive strength of individuals-based models founded on the cost-benefit analysis of fitness is that they are based on fundamental properties of the organisms that should remain the same in the new environmental circumstances, so allowing population functions to be derived reliably for the new conditions. This, of course, is the widely recognized advantage of mechanistic models over the phenomenological models that use only empirical population functions, such as regression equations; we have all been trained to be aware of the dangers of predicting beyond the range when using such expressions (Koehl 1989). Unless we can predict from sound basic principles how the animals will react to environmental change, our predictions may be unreliable. Although much remains to be done, sufficient progress has been made with Oystercatchers to provide an example of how the individuals-based approach can be used to predict the responses of populations to environmental changes that have not yet been observed directly.

This point can be illustrated even more forcefully when the quality of the food supply changes at the same time that part of it is removed by habitat loss. For example, simulations with a physiologically-structured mussel model show how the growth rates, body quality and density of size classes of mussel prey would change in response to the hypothetical circumstances of an increase in the levels of contaminants, such as hydrocarbons, within the estuary (Goss-Custard and Willows 1996). These changes would reduce the intake rates of the average Oystercatcher because these depend so much on mussel size and flesh content (Zwarts *et al.* 1996a). In addition, certain contaminants, for example tributyltin, can also increase the thickness of the mussel shells (D. S. Page, unpublished information) which, in turn, would make it more difficult for those Oystercatchers that open mussels by hammering to break into their prey (Meire 1996a). Any effect that this has on the rate at which the birds can feed, and thus their chances of obtaining their daily energy requirements, can be considered in the present version of the Oystercatcher model because the effect of shell thickness on intake rate can be included (Goss-Custard *et al.* 1996b). Other inputs can also be considered. Nutrient inputs may increase overall productivity, particularly of the mussels and other benthic animals which feed on phytoplankton, and so improve the food supply of the birds. The reduction of inputs to coastal waters may thus have opposing effects on the food supplies of the birds, as may many of the anthropogenic changes to which estuaries are prone (Goss-Custard 1995). Physiologically-structured models of mussel populations, in combination with individuals-based models of Oystercatcher populations, can be combined to predict the long-term consequences for both the birds and the shellfish of different combinations of habitat loss, pollution control and estuary management, the precise consequences of which might otherwise be very difficult to foresee (Goss-Custard and Willows 1996).

In the case of the birds, the key to the reliablity of such predictions is the soundness of the model assumptions about the way in which individuals will respond in the new circumstances. If the approach is to succeed, a sound understanding of the natural history of the species concerned is essential; it is no use making sophisticated model predictions if they are based on a false premise about the animal's repertoire of adaptations. Another vital aspect is that the ways in which individuals vary and how they make their decisions must be well known. This is where behavioural ecology is so important. By establishing the principles upon which different individuals make their decisions, it provides a reliable basis for forecasting how they as individuals will respond to the new circumstances brought about by environmental change. This illustrates how basic research that may initially seem to be rather esoteric may actually provide the fundamentals for making forecasts and planning the effective management of wild populations.

In addition to detailing important aspects of Oystercatcher natural history, all the chapters in this book on individual adaptation took behavioural ecological theory on animal decision-making as their starting point. In doing so, they have highlighted areas of theory that need to be strengthened. For example, optimal diet models have served as guidelines for much empirical work, which in turn have given new insights but also raised questions. To give two examples: (i) profitability must be defined for all prey of the same type, thus including the time 'wasted' in prey that are not taken (Meire and Ervynck 1986); (ii) Oystercatchers 'manipulate' their encounter rate with prey by adjusting their search behaviour, such as the probing depth, and are in this way able to increase their gross intake rate (Wanink and Zwarts 1996). Assumptions of constant search speed in existing models are thus clearly inadequate, and attention needs to be given to deriving a theoretical basis for predicting the rates at which animals should forage.

The model that explores the conditions for self-feeding versus parent-feeding in precocial birds is another example of the interplay between theoretical and empirical research that results in interesting predictions of how populations will be affected by new circumstances. The theory suggests that differences in bill size between parent and chick, the density of the food items, and their size, determine the circumstances under which parent-feeding is replaced by self-feeding. The model predicts that, as a result of man-induced changes in their food supply, inland-breeding Oystercatchers will become increasingly productive as more of their chicks feed themselves, with a consequent further increase in population size.

Finally, most progress in using the individuals-based approach to deriving population functions has been made in density-dependent mortality and emigration using a game theoretic approach. Game theory is highly suitable for this because it recognizes that, in a competitive situation, the decisions made by one competitor influence, and are influenced by, those made by others. The need now is to apply this methodology to another major source of density dependence, competition for breeding territories. This will require that much is learned about dominance in breeding birds. The research on Oystercatchers suggests that an individual's dominance is determined by a combination of fighting ability and prior residence, compared to the same attributes of the opponent. Because of the latter, dominance is site dependent. The challenge for empiricists is to find ways to estimate the contribution of each component to dominance. Also, due to the effect of residency on dominance, the decision to settle and build up local dominance in a given location has life-time fitness consequences (Ens *et al.* 1996e). Before resources can be allocated to different physiological processes, access must first be gained and then maintained. Present life history theory (e.g. Stearns 1992) pays insufficient attention to such career decisions and the underlying process of competition and associated trade-offs to the individual.

But, in return, current theory tells the empiricists just how much remains to be done. There are many aspects of Oystercatcher decision-making biology that remain to be investigated and included in the models. Of particular importance is the absence of state dependency in the present versions of the model. Real animals change their priorities according to their current state and anticipated prospects. A hungry animal with low body reserves, for example, will take more risks with predators and disease than will an animal whose reserves are complete. State-dependent models can make quite different predictions to the classical optimization models (Houston 1993) which have so far been used in Oystercatcher models. The absence of state dependency may not always matter; for much of the winter, most Oystercatchers are easily able to obtain their energy requirements so the basis of their decision-making may remain sufficiently stable to allow simple models to predict their behaviour quite adequately. But when an animal is in difficulty, so that its priorities change, state-dependent models are likely to give better predictions as to how it will respond. As the Oystercatcher model aims to predict mortality rates, this omission is clearly important and undoubtedly contributes to the model's current inability to predict accurately winter mortality in many sections of the Exe population.

The reviews of the empirical studies in this book have revealed many other areas of Oystercatcher biology which will have to be investigated if the individuals-based models are to provide a sound basis for prediction. Most studies have focused on local groups, whether wintering on one estuary or breeding on one island or marsh. But there are exchanges of birds between these groups, yet this is only included in existing models in a very simplistic way. The number of Oystercatchers wintering on one estuary depends not only on processes occurring in the many breeding sites from which birds are drawn but also on events going on in other estuaries. The next generation of individuals-based models will need to include the consequences of individual decision-making at a much larger geographic scale than has so far been attempted (Goss-Custard *et al.* 1995b). These models will have to include such life-history decisions as those made by prospecting juveniles seeking an estuary to spend the following and, quite probably, future winters. They will also have to include decisions made on migration and on whether to emigrate when local conditions deteriorate. And as the models look increasingly to long-term predictions, the possibility that certain individual characteristics will be favoured over others in new environmental circumstances, so that the animals evolve, will need to included. For example, the foraging efficiency of individuals affects the carrying capacity of an estuary and is a behavioural trait that might be subject to natural selection (Ritchie 1988). These are whole areas of avian biology which theoretical and empirical studies are only just starting to address.

Appendix 1:
The Exe estuary Oystercatcher-mussel model

Ralph T. Clarke and John D. Goss-Custard

Version 2 of the model extends the much simpler first version by including a time-base, by monitoring the daily consumption of mussel flesh by each individual Oystercatcher, along with its location and current level of fat reserves, and by tracking the changing food supply on each mussel bed from autumn to early spring. Version 1 is not detailed here as the computations involved, which are described in Goss-Custard *et al.* (1995a), are the same as those carried out by Version 2.

Version 2 is run in daily time steps from September 1 ($t = 1$) through to the end of March. Each day each bird chooses the mussel bed (food patch) where it currently achieves the highest intake rate, given its feeding efficiency and the food supply present and its dominance and the density of competitors present. Each bird is assumed to be both 'ideal' and 'free'; i.e. to have perfect knowledge of the daily intake rate it could currently achieve on each patch and to incur no costs in moving between patches.

Apart from its current instantaneous intake rate, an individual's energy consumption per 24 h is limited by the mussel bed exposure time, the limited capacity of the Oystercatcher gut to process mussel flesh and the efficiency with which it is assimilated. Daily energy requirements are temperature-related, with any surplus being stored as fat with known efficiency, up to a maximum of 5% per day. The model tracks the daily feeding location and body condition of each bird. Fat reserves maintain the bird on days when it fails through foraging to meet its current energy requirements. The model updates the mussel biomass on each bed daily by deducting the amount removed by Oystercatchers over the preceding 24 h, together with the (assumed) density-independent losses due to factors other than Oystercatchers. Birds may change feeding patch as relative patch qualities change through depletion and as competitors move or die.

Mathematical definition of version 2 of the model

m = number of food patches (mussel beds) on the estuary = 12
A_j = spring tide low water area of patch j; $j = 1, \ldots, m$ (range over patches: 3773–23528 m^2).

Q_{jt} = mussel biomass density of patch j on day t.
   (initial (i.e. 1 September) range: 80–405 g ash-free dry weight m^{-2}).
N = total number of birds arriving on the estuary in the Autumn.
E_i = fixed relative foraging efficiency of bird i; $i = 1, \ldots, N$.
   E_i are random deviates from a Normal distribution with mean $\mu = 1$ and standard deviation $\sigma = 0.12$; denoted by $E_i \sim N(\mu = 1, \sigma = 0.12)$.

Birds are classified by age class g (1 = juvenile, 2 = immature (2–4years), 3 = adult) and feeding method f (1 = stab, 2 = hammer mussels).

Interference-free intake rate (*IFIR*)

The *IFIR* of a bird of average foraging efficiency ($E_i = 1$) of feeding method f on patch j, denoted by I_{jtf}, is related to Q_{jt} by

$$I_{jtf} = \lambda_{1f}Q_{jt} - \lambda_{2f}Q_{jt}^2, \text{ subject to } I_{jtf} = \lambda_{1f}^2/4\lambda_{2f} \text{ for } Q_{jt} > \lambda_{1f}/2\lambda_{2f}; \quad (1)$$

where $\lambda_{11} = 3.2367$ and $\lambda_{21} = 0.00422$, $\lambda_{12} = 4.9041$ and $\lambda_{22} = 0.00639$.
   The *IFIR* of bird i (of feeding method f) on patch j on day t is therefore

$$R_{ij0} = E_i I_{jtf}. \quad (2)$$

Dominance

D_i = fixed global dominance of bird i, distributed uniformly over range (0–1) for adult and (0–0.7) for other birds.
L_{ij} = current 'local' dominance of bird i on patch j
   = proportion of birds on patch j with a lower D_i.

Bed exposure

The hours, H_t, available for feeding during each day t is modelled as a 14-day lunar cycle:

$$H_t = H_0 + (H_1 - H_0)(1 + \cos(2\pi(t-1)/14))/2 \quad (3)$$

where $H_1 = 10.6$ and $H_0 = 12.1$ denote hours exposed on spring and neap tides respectively.
   The proportion, P_{jt}, of each patch j exposed on day t is modelled as a 14-day lunar cycle:

$$P_{jt} = P_{j0} + (1-P_{j0})(1 + \cos(2\pi(t - 1)/14))/2 \quad (4)$$

where P_{j0} = fixed proportion of patch j exposed on a neap tide. P_{j0} ranges from 0.0 to 0.9, because some mussel beds remain completely covered on days of full neap tides.
   Stabbers, which feed throughout the patch exposure period, consequently feed, on average, at field-estimated densities of $k_f = 1.4$ times that of hammerers ($k_f = 1.0$) which predominantly feed at low water.

Intake rate

On day t, the food intake rate, R_{ij}, of bird i, using feeding method f, on patch j with N_j other birds is modelled as: either

$$R_{ij} = R_{ij0} - b_{ij} \log(S_{jft}) \tag{5a}$$

or

$$\log(R_{ij}) = \log(R_{ij0}) - b_{ij} \log(1 + S_{jft}) \tag{5b}$$

where $S_{jft} = k_f N_j / (A_j P_{jt})$ = average density of other birds currently on the patch, and b_{ij}, the susceptibility to interference (STI), is related to L_{ij} by:

$$b_{ij} = (\beta_{0f} - \beta_{1f} L_{ij}) s_j (1 - c_1)^t, \text{ subject to } b_{ij} \geq 0 \tag{6}$$

where β_{0f} and β_{1f} are calibrated for size-specific mussel flesh content on patch 1 on 1 September, s_j is the size-specific flesh content of mussels on patch j relative to patch 1 (s_j range 1–1.41), and $c_1 = 0.002163$ = the daily proportional decrease in the flesh-content of individual mussels over the winter. In the model simulations, which used interference form (5a), $\beta_{01} = 165$, $\beta_{11} = 520$, $\beta_{02} = 274$, $\beta_{12} = 85$.

Daily optimization procedures

On each day, each bird maximizes its daily intake rate, as $I_{it} = H_t \max_j \{R_{ij}\}$. At each iteration of the optimization for a day, the birds are considered to make their optimal decision of where to feed in an independent random order, each assuming they are the last bird to move. Birds are assumed to move patches only if it would achieve more than a 3% improvement in daily intake. The consequences of each movement are updated immediately. Iterations continue until the bird spatial distribution stabilizes, or more usually, until the proportion of birds moving between patches stabilizes.

Changes in body weight

Body condition of bird i on day t is measured by its body weight W_{it} (g), starting at $t = 1$ from an initial age-specific normal weight distribution, $N(\mu_{wg}, \sigma_{wg} = 52)$; where $\mu_{w1} = 463$, $\mu_{w2} = 487$, $\mu_{w3} = 507$. The daily energy requirements (F_t) of each bird is assumed to be size-independent and equivalent to a food intake of:

$$F_t = 26.75 + 0.19(10 - T_t) \text{ when } T_t < 10°C), \tag{7}$$

and $F_t = 26.75$ otherwise, where T_t = mean air temperature °C on day t.

(a) If daily assimilated intake $\alpha_1 I_{it} > F_t$, then bird i gains weight and:

$$W_{i(t+1)} = W_{it} + \alpha_2(\alpha_1 I_{it} - F_t), \tag{8a}$$

($\alpha_1 = 0.854$ = food assimilation efficiency, $\alpha_2 = 0.884$ = efficiency of converting between (to/from) assimilated food and body fat); subject to a maximum possible increase in body weight of 5% per day, and maximum type-specific target body weight of W_{1g} on 1 September, rising linearly to W_{2g} on day t_{2g}.; (W_{1g} = 465, 483, 503 g; W_{2g} = 494, 550, 598 g; t_{2g} = 106, 106, 196 for g = 1–3 respectively).

(b) If daily assimilated intake $\alpha_1 I_{it} < F_t$, then bird i loses weight and

$$W_{i(t+1)} = W_{it} - (F_t - \alpha_1 I_{it})/\alpha_2. \tag{8b}$$

If the W_{it} of a bird of type g falls below a starvation body weight of W_{0g}, it is assumed to die, or at least to leave the estuary (W_{01} = 300, W_{02} = 340, W_{03} = 350 g).

Daily changes to mussel biomass density

Finally, the total amount of food consumed (T_{jt}) by all birds on each patch j on day t, together with daily loss ($c_1 + c_2$) in biomass through non-Oystercatcher factors, is deducted from the patch total mussel biomass to give:

$$Q_{j(t+1)} = ((1 - c_1 - c_2)Q_{jt}A_j - T_{jt})/A_j. \tag{9}$$

where c_2 = 0.000837 = the daily proportional decrease in mussel density due to mortality agents other than Oystercatchers.

References

Alerstam, T. and Högsted, G. (1982). Bird migration and reproduction in relation to habitats for survival and breeding. *Ornis Scandinavica*, **13**, 25–37.

Allison, F.R. (1979). Life cycle of *Curtuteria australis* n. sp. (*Digenea: Echinostomatidae: Himasthlinae*), intestinal parasite of the South Island Pied Oystercatcher. *New Zealand Journal of Zoology*, **6**, 13–20.

Alting, D. (1990). IJking van hapgrootte schattingen van op *Macoma* fouragerende Scholeksters en *Macoma balthica* liften of in situ eten? Unpublished student thesis, University of Groningen.

Amadon, D. (1949). The seventy-five percent rule for subspecies. *Condor*, **51**, 250–258.

Anderson, K.R. and Minton, C.D.T. (1978). Origins and movements of Oystercatchers on the Wash. *British Birds*, **71**, 439–447.

Ankney, C.D. and MacInnes, C.D. (1978). Nutrient reserves and reproductive performance of female Lesser Snow Geese. *Auk*, **95**, 459–471.

Anonymous, (1987). Effecten van de kokkelvisserij in de Waddenzee. *Report Rijksinstituut voor Natuurbeheer* (RIN) **87/18**, Texel.

Arntz, W.E., Brey, T., Tarazona, J. and Robles, A. (1987). Changes in the structure of a shallow sandy-beach community in Peru during an El Niño event. *South African Journal of Marine Science*, **5**, 645–658.

Ashkenazie, S. and Safriel, U.N. (1979a). Breeding cycle and behaviour of the Semipalmated Sandpiper at Barrow, Alaska. *Auk*, **96**, 56–67.

Ashkenazie, S. and Safriel, U.N. (1979b). Time-energy budget of the Semipalmated Sandpiper *Calidris pusilla* at Barrow, Alaska. *Ecology*, **60**, 783–799.

Ashmole, N.P. (1971). Seabird ecology and the marine environment. In *Avian Biology*, Vol. I, (ed. D.S. Farner and J.R. King), pp. 224–286. Academic Press, London.

Baillie, S. (1980). The effect of the hard winter of 1978/79 on the wader populations of the Ythan Estuary. *Wader Study Group Bulletin*, **28**, 16–17.

Baird, D. and Milne, H. (1981). Energy flow in the Ythan Estuary, Aberdeenshire, Scotland. *Estuarine Coastal Shelf Science*, **13**, 455–472.

Baird, D., Evans, P.R., Milne, H. and Pienkowski, M.W. (1985). Utilization by shorebirds of benthic invertebrate production in intertidal areas. *Oceanographic Marine Biological Annual Review*, **23**, 573–597.

Baker, A.J. (1973). Distribution and numbers of New Zealand Oystercatchers. *Notornis*, **20**, 128–144.

Baker, A.J. (1974a). Ecological and behavioural evidence for the systematic status of New Zealand Oystercatchers (*Charadriiformes: Haematopodidae*). *Life Science Contributions of the Royal Ontario Museum*, **96**, 1–34.

Baker, A. J. (1974b). Criteria for ageing and sexing New Zealand Oystercatchers. *New Zealand Journal of Marine and Freshwater Research*, **8**, 11–21.

Baker, A.J. (1974c). Prey-specific feeding methods of New Zealand Oystercatchers. *Notornis*, **21**, 219–233.

Baker, A.J. (1975a). Lipid levels in the South Island Pied Oystercatcher (*Haematopus ostralegus finschi*). *New Zealand Journal of Zoology*, 2, 425–434.

Baker, A.J. (1975b). Morphological variation, hybridization and systematics of New Zealand Oystercatchers (*Charadriiformes: Haematopodidae*). *Journal of Zoology, London*, 175, 357–390.

Baker, A.J. (1977). Multivariate assessment of the phenetic affinities of Australasian Oystercatchers (*Aves: Charadriiformes*). *Bijdragen tot Dierkunde*, 47, 156–164.

Baker, A.J. and Cadman, M. (1980). Breeding schedule, clutch size and egg size of American Oystercatchers (*Haematopus palliatus*) in Virginia. *Wader Study Group Bulletin*, 30, 32–33.

Baker, A.J. and Hockey, P.A.R. (1984). Behavioural and vocal affinities of the African Black Oystercatcher (*Haematopus moquini*). *Wilson Bulletin*, 96, 656–671.

Barnard, C.J. and Sibly, R.M. (1981). Producers and scroungers: a general model and its application to captive flocks of House Sparrows. *Animal Behaviour*, 29, 543–550.

Bayne, B.L. and Worrall, C.M. (1980). Growth and production of mussels *Mytilus edulis* from two populations. *Marine Ecology Progress Series*, 3, 317–328.

Beaumont, A.R., Newman, P.B., Mills, D.K., Waldock, M.J., Miller, D. and Waite, M.E. (1989). Sandy-substrate microcosm studies in tributyltin (TBT) toxicity to marine organisms. *Topics in Marine Biology, Scientia Marina*, 53, 737–743.

Becker, M. (1987). Der Bruterfolg des Austernfischers, *Haematopus ostralegus*, auf Hallig Langeness (Schleswig-Holstein) in Abhängigkeit von Art und Lage des Nahrungsreviers. Unpublished Diploma thesis. Universität Berlin.

Becker, P.H. (1989). Seabirds as monitor organisms of contaminants along the German North Sea coast. *Helgoländer Meeresuntersuchungen*, 43, 395–403.

Becker, P.H. and Erdelen, M. (1987). Die Bestandsentwicklung von Brutvögeln der deutschen Nordseeküste 1950–1979. *Journal für Ornithologie*, 128, 1–32.

Becker, P.H., Heidmann, W.A., Büthe, A., Frank, D. and Koepff, C. (1992). Umweltchemikalien in Eiern von Brutvögeln der deutschen Nordseeküste, Trends 1981–1990. *Journal für Ornithologie*, 133, 109–124.

Beintema, A.J. (1983). Meadow birds as indicators. *Environmental Monitoring and Assessment*, 3, 391–398.

Beintema, A.J. and Müskens, G.J.D.M. (1987). Nesting success of birds breeding in Dutch agricultural grasslands. *Journal of Applied Ecology*, 24, 743–758.

Beintema, A.J. and Visser, G.H. (1989). Growth parameters in chicks of charadriiform birds. *Ardea*, 77, 169–180.

Beintema, A.J., Beintema-Hietbrink, R.J. and Müskens, G.J.D.M. (1985). A shift in the timing of breeding in meadow birds. *Ardea*, 73, 83–89.

Beintema, A.J., Thissen, J.B., Tensen, D. and Visser, G.H. (1990). Breeding ecology of charadriiform chicks in agricultural grassland. *Ardea*, 79, 31–43.

Bent, A.C. (1927). *Life histories of North American shorebirds*. Dover Publications, New York.

Berkhoudt, H. (1985). Structure and function of avian taste receptors. In *Form *and function in birds*, Vol. III, (ed. A.S. King and J. McLelland), pp. 463–496. Academic Press, London.

Berthold, P. (1990). *Vogelzug – eine kurze, aktuelle Gesamtübersicht*. Wissenschaftliche Buchgesellschaft, Darmstadt.

Beukema, J.J. (1974). Seasonal changes in the biomass of the macro-benthos of a tidal flat area in the Dutch Wadden Sea. *Netherlands Journal of Sea Research*, 8, 94–107.

Beukema, J.J. (1976). Biomass and species richness of the macro-benthic animals living on tidal flats of the Dutch Wadden Sea. *Netherlands Journal of Sea Research*, 10, 236–261.

Beukema, J.J. (1979). Biomass and species richness of the macrobenthic animals living on a tidal flat area in the Dutch Wadden Sea: effects of a severe winter. *Netherlands Journal of Sea Research*, 13, 203–223.

Beukema, J.J. (1982). Annual variation in reproductive success and biomass of the major macrozoobenthic species living in a tidal flat area of the Wadden Sea. *Netherlands Journal of Sea Research*, 16, 37–45.

Beukema, J.J. (1989). Long-term changes in macrozoobenthic abundance on the tidal flats of the western part of the Dutch Wadden Sea. *Helgoländer Meeresuntersuchungen*, 43, 405–415.

Beukema, J.J. (1991). Changes in composition of bottom fauna of a tidal flat area during a period of eutrophication. *Marine Biology*, 111, 293–301.

Beukema, J.J. (1993). Increased mortality in alternative bivalve prey during a period when the tidal flats of the Dutch Wadden Sea were devoid of mussels. *Netherlands Journal of Sea Research*, 31, 395–406.

Beukema, J.J. and Bruin, W. de (1977). Seasonal changes in dry weight and chemical composition of the soft parts of the tellinid bivalve *Macoma balthica* in the Dutch Wadden Sea. *Netherlands Journal of Sea Research*, 11, 42–55.

Beukema, J.J. and Cadée, G.C. (1986). Zoobenthos responses to eutrophication of the Dutch Wadden Sea. *Ophelia*, 26, 55–64.

Beukema, J.J. and Cadée, G.C. (1991). Growth rates of the bivalve *Macoma balthica* in the Wadden Sea during a period of eutrophication: relationships with concentrations of pelagic diatoms and flagellates. *Marine Ecology Progress Series*, 68, 249–256.

Beukema, J.J. and Essink, K. (1986). Common patterns in the fluctuations of macrozoobenthic species living at different places on the tidal flats in the Wadden Sea. *Hydrobiologia*, 142, 199–207.

Beukema, J.J., Essink, K., Michaelis, H. and Zwarts, L. (1993). Year-to-year variability in the biomass of macrobenthic animals on tidal flats of the Wadden Sea: how predictable is this food source for birds? *Netherlands Journal of Sea Research*, 31, 319–330.

Bianki, V.V. (1977). *Gulls, shorebirds and alcids of Kandalaksha Bay*. Israel Program for Scientific Translations, Jerusalem.

Bijlsma, R.G. (1990). Predation by large falcon species on wintering waders on the Banc d'Arguin, Mauritania. *Ardea*, 78, 75–82.

Birkhead, T. and Møller, A.P. (1992). *Sperm competition in birds: evolutionary causes and consequences*. Academic Press, London.

Blake, E.R. (1977). *Manual of Neotropical birds*, Vol. I. University of Chicago Press, Chicago.

Blake, R.W. (1979a). Exploitation of a natural population of *Arenicola marina*

(L.) from the north-east coast of England. *Journal of Applied Ecology*, **16**, 663–670.

Blake, R.W. (1979b). On the exploitation of a natural population of *Nereis virens* Sars from the north-east coast of England. *Estuarine Coastal Marine Science*, **8**, 141–148.

Blakers, M., Davies, S.J.J.F. and Reilly, P.N. (1984). *The atlas of Australian birds*. Melbourne University Press, Carlton.

Blomert, A.-M., Engelmoer, M. and Logemann, D. (1983). Voedseloecologie van de Scholekster op het Friese wad. *Report Rijksdienst voor de IJsselmeerpolders*, Lelystad. Unpublished student thesis, University of Groningen.

Blomqvist, S., Frank, A. and Pettersson, L.R. (1987). Metals in liver and kidney tissues of autumn-migrating Dunlin *Calidris alpina* and Curlew Sandpiper *Calidris ferruginea* staging at the Baltic Sea. *Marine Ecology Progress Series*, **35**, 1–13.

Blondel, J. and Isenmann, P. (1981). *Guide des Oiseaux de Camargue*. Delachaux and Niestle, Neuchâtel.

Boates, J.S. (1988). Foraging and social behaviour of the Oystercatcher *Haematopus ostralegus* in relation to diet specialization. Unpublished Ph.D. thesis. University of Exeter.

Boates, J.S. and Goss-Custard, J.D. (1989). Foraging behaviour of Oystercatchers *Haematopus ostralegus* during a diet switch from worms *Nereis diversicolor* to clams *Scrobicularia plana*. *Canadian Journal of Zoology*, **67**, 2225–2231.

Boates, J.S. and Goss-Custard, J.D. (1992). Foraging behaviour of Oystercatchers *Haematopus ostralegus* specializing on different species of prey. *Canadian Journal of Zoology*, **70**, 2398–2404.

Boere, G.C. (1976a). The significance of the Dutch Waddenzee in the annual life cycle of arctic, subarctic and boreal waders. Part I. The function as a moulting area. *Ardea*, **64**, 210–291.

Boere, G.C. (1976b). Enkele gegevens over de sterfte onder wadvogels op Vlieland tijdens de vorstperiode van 30 januari–7 februari 1976. *Rapport Staatsbosbeheer* **1976-7**.

Bolze, G. (1969). Anordnung und Bau der Herbstschen Körperchen in Limicolenschnäbeln in Zusammenhang mit der Nahrungsfindung. *Zoologischer Anzeiger*, **181**, 313–355.

Bonner, J.T. (1980). *The evolution of culture in animals*. Princeton University Press, New Jersey.

Boon, J.T., Eijgenraam, F. and Everaarts, J.M. (1989). A structure-activity relationship (SAR) approach towards metabolism of PCBs in marine animals from different trophic levels. *Marine Environmental Research*, **27**, 159–76.

Borgsteede, F.H.M., Broek, E. van den and Swennen, C. (1988). Helminth parasites of the digestive tract of the Oystercatcher, *Haematopus ostralegus*, in the Wadden Sea, The Netherlands. *Netherlands Journal of Sea Research*, **22**, 171–174.

Braithwaite, D.H. (1950). Notes on the breeding of Variable Oystercatchers. *Notornis*, **4**, 22–24.

Branch, G.M., Eekhout, S. and Bosman, A.L. (1990). Short-term effects of the 1988 Orange River floods on the rocky-shore communities of the open coast. *Transactions of the Royal Society of South Africa*, **47**, 331–354.

Branson, N.J.B.A. (1989). Oystercatcher mortality. *Wash Wader Ringing Group Report* **1987/88**, 53–56.

Brenninkmeijer, A., Lambeck, R.H.D. and Strucker, R.C.W. (1993). De invloed van de afsluiting van het Krammer-Volkerak op de verplaatsingen en mortaliteit van de scholekster, *Haematopus ostralegus*. Rapporten en Verslagen 1993–05. Nederlands Instituut voor Oecologisch Onderzoek-Centrum voor Estuariene en Mariene Oecologie, Yerseke.

Brey, T., Rumohr, H. and Ankar, S. (1988). Energy content of macrobenthic invertebrates: general conversion factors from weights to energy. *Journal of Experimental Marine Biology and Ecology*, **117**, 271–278.

Briggs, K.B. (1984a). Repeated polygyny by Oystercatchers. *Wader Study Group Bulletin*, **40**, 42–44.

Briggs, K.B. (1984b). The breeding ecology of coastal and inland Oystercatchers in north Lancashire. *Bird Study*, **31**, 141–147.

Briggs, K.B. (1985). The ecology of an inland breeding population of Oystercatchers (*Haematopus ostralegus*). Unpublished Ph.D. thesis. CNAA.

Brockmann, H.J. and Barnard, C.J. (1979). Kleptoparasitism in birds. *Animal Behaviour*, **27**, 487–514.

Brown, J.L. (1963). Aggressiveness, dominance and social organisation in the Stellar's Jay. *Condor*, **65**, 460–484.

Brown, R.A. and O'Connor, R.J. (1974). Some observations on the relationships between Oystercatchers *Haematopus ostralegus* L. and cockles *Cardium edule* L. in Strangford Lough. *Irish Naturalist*, **18**, 73–80.

Brown, R.H. (1948). Clutch size of the Oystercatcher. *British Birds*, **41**, 157–158.

Bryant, D.M. and Newton, A.V. (1994). Metabolic costs of dominance in Dippers, *Cinclus cinclus*. *Animal Behaviour*, **48**, 447–455.

Bull, K.R., Every, W.J., Freestone, P., Hall, J.R., Osborn, D., Cooke, A.S. and Stowe, T. (1983). Alkyl lead pollution and bird mortalities on the Mersey estuary, UK, 1979–1981. *Environmental Pollution* (Ser. A.), **31**, 239–259.

Bundy, G., Connor, R.J. and Harrison, C.J.O. (1989). *Birds of the Eastern Province of Saudi Arabia*. Witherby, London.

Bunskoeke, E.J. (1988). Over de fluctuaties van de prooikeus van de Scholekster (*Haematopus ostralegus*) in het broedseizoen 1986 op Schiermonnikoog. Unpublished student thesis. University of Groningen.

Bunskoeke, E.J., Ens, B.J., Hulscher, J.B. and Vlas, S. de (1996). Prey choice of the Oystercatcher *Haematopus ostralegus* during the breeding season. *Ardea*, **83**. (In press.)

Burger, J. (1984). Shorebirds as marine animals. In *Shorebirds: breeding behavior and populations*, (ed. J. Burger and B.L. Olla), pp. 17–81. Plenum Press, New York.

Busche, G. (1980). *Vogelbestände des Wattenmeeres von Schleswig-Holstein*. Kilda-Verlag, Greven.

Bush, A.P., Aho, J.M. and Kennedy, C.R. (1990). A comparison of ecological and phylogenetic factors as determinants of helminth parasite community richness. *Evolutionary Ecology*, **4**, 1–20.

Buskirk, J. van and Smith, D.C. (1989). Individual variation in winter foraging of Black-capped Chickadees. *Behavior of the Ecological Society*, **24**, 257–263.

Busse, P. (1986). Theoretical models in an interpretation of recovery patterns. *Ring*, **11**, 211–228.

Buxton, E.J.M. (1939). The breeding of the Oystercatcher. *British Birds*, 33, 184–193.

Buxton, E.J.M. (1962). The inland breeding of the Oystercatcher in Great Britain, 1958–59. *Bird Study*, 8, 194–209

Cadée, G.C. (1986). Recurrent and changing seasonal patterns in phytoplankton of the western most inlet of the Dutch Wadden Sea from 1969 to 1985. *Marine Biology*, 93, 281–289

Caldow, R.W.G. and Goss-Custard, J.D. (1996). Temporal variation in the social rank of adult Oystercatchers *Haematopus ostralegus*. *Ardea*, 83. (In press.)

Caldwell, H.R. and Caldwell, J.C. (1931). *South China birds*. Hester May Vandenbergh, Shanghai.

Campbell, B. (1947). Clutch-size in the Oystercatcher. *British Birds*, 40, 126.

Camphuysen, C.J. (1989). Beached bird surveys in The Netherlands 1915–1988: Seabird mortality in the Southern North Sea since the early days of oil pollution. *Technical Report 'Vogelbescherming'* no. 1. Stichting Werkgroep Noordzee, Amsterdam.

Camphuysen, C.J. and Dijk, J. van (1983). Zee- en kust-vogels langs de Nederlandse kust, 1974–1979. *Limosa*, 56, 81–230.

Cayford, J.T. (1988a). The foraging behaviour of Oystercatchers (Haematopus ostralegus) feeding on mussels (*Mytilus edulis*). Unpublished Ph.D. thesis. University of Exeter.

Cayford, J.T. (1988b). A field test of the accuracy of estimating prey size-selection in Oystercatchers from recovered mussel shells. *Wader Study Group Bulletin*, 54, 29–32.

Cayford, J.T. and Goss-Custard, J.D. (1990). Seasonal changes in the size selection of mussels, *Mytilus edulis*, by Oystercatchers, *Haematopus ostralegus*: an optimality approach. *Animal Behaviour*, 40, 609–624.

Chambers, M.R. and Milne, H. (1979). Seasonal variation in the condition of some intertidal invertebrates of the Ythan estuary, Scotland. *Estuarine Coastal Marine Science*, 3, 443–455.

Charnov, E.L. (1976). Optimal foraging: attack strategy of a mantid. *American Naturalist*, 110, 141–151.

Christian, P.D., Christidis, L. and Schodde, R. (1992). Biochemical systematics of the *Charadriiformes* (shorebirds): relationships between the *Charadrii*, *Scolopaci* and *Lari*. *Australian Journal of Zoology*, 40, 291–302.

Clark, N.A. (1982). The effects of the severe weather in December 1981 and January 1982 on waders in Britain. *Wader Study Group Bulletin*, 34, 5–7.

Clark, N.A. (1993). Wash Oystercatchers starving. *BTO News*, 185, 1 and 24.

Clark, N.A. and Davidson, N.C. (1986). WSG project on the effects of severe weather on waders: plans for 1986/87. *Wader Study Group Bulletin*, 47, 4.

Clutton-Brock, T.H. (ed.) (1988). *Reproductive success: studies of individual variation in contrasting breeding systems*. University Chicago Press, Chicago.

Coles, H.A. (1956). A preliminary study of growth-rate in cockles (*Cardium edule* L.) in relation to commercial exploitation. *Conseil Permanent International pour l'Exploration de la Mer*. 22, 77–90.

Collar, N.J. and Stuart, S.N. (1985). *Threatened birds of Africa and related islands*. International Council for Bird Preservation, Cambridge.

Cooper, J., Hockey, P.A.R. and Brooke, R.K. (1985). Introduced mammals on

South and South West African islands: history, effects on birds and control. In *Proceedings of the symposium in birds and man*, (ed. L.J. Bunning), pp. 179–203. Southern African Ornithological Society, Johannesburg.

Coosen, J., Twisk, F., van der Tol, M.W.M, Lambeck, R.H.D., van Stralen, M.R. and Meire, P.M. (1994). Variability in stock assessment of cockles (*Cerastoderma edule* L.) in the Oosterschelde (in 1980–1990), in relation to environmental factors. *Hydrobiologia*, **282/283**, 381–395.

Coulson, J.C. (1972). The significance of the pair-bond in the Kittiwake. *Acta XV Congressus Internationalis Ornithologici*, 424–433.

Coulson, J.C. and Thomas, C.S. (1985). Changes in the biology of the Kittiwake *Rissa tridactyla*: a 31 year study of a breeding colony. *Journal of Animal Ecology*, **54**, 9–17.

Cracraft, J. (1981). Toward a phylogenetic classification of the recent birds of the world (Class Aves). *Auk*, **98**, 681–714.

Craeymeersch, J.A., Herman, P.M.J. and Meire, P.M. (1986). Secondary production of an intertidal mussel (*Mytilus edulis* L.) population in the Eastern Scheldt (S.W. Netherlands). *Hydrobiologia*, **133**, 107–115.

Cramp, S. and Simmons, K.E.L. (1983). *Handbook of the birds of Europe, Middle East and North Africa: the birds of the Western Palearctic*, Vol. III, *Waders to Gulls*. Oxford University Press, Oxford.

Cresswell, W. and Whitfield, D.P. (1994). The effects of raptor predation on wintering wader populations at the Tyninghame estuary, southeast Scotland. *Ibis*, **136**, 223–232.

Curry-Lindahl, K. (1981). *Bird migration in Africa*. Vol. I. Academic Press, London.

Daan, S. and Koene, P. (1981). On the timing of foraging flights by Oystercatchers, *Haematopus ostralegus*, on tidal flats. *Netherlands Journal of Sea Research*, **15**, 1–22.

Daan, S., Dijkstra, C., Drent, R. and Meijer, T. (1989). Food supply and the annual timing of avian reproduction. *Acta XIX Congressus Internationalis Ornithologici*, 392–407.

Daan, S., Dijkstra, C. and Tinbergen, J.M. (1990). Family planning in the Kestrel (*Falco tinnunculus*): the ultimate control of laying date and clutch size. *Behaviour*, **114**, 83–116.

Dankers, N. and Beukema, J.J. (1983). Distributional patterns of macrozoobenthic species in relation to some environmental factors. In *Invertebrates of the Wadden Sea*, (ed. N. Dankers, H. Kühl and W.J. Wolff). pp. 69–103. Balkema, Rotterdam.

Dare, P.J. (1966). The breeding and wintering populations of the Oystercatcher (*Haematopus ostralegus* Linnaeus) in the British Isles. *Fishery Investigations*, Series II, Vol. 25, Nr. 5, 1–69.

Dare, P.J. (1970). The movements of Oystercatchers (*Haematopus ostralegus* Linnaeus) in the British Isles. *Fishery Investigations*, Series II, Vol. 25, Nr. 9, 1–137.

Dare, P. J. (1977). Seasonal changes in body-weight of Oystercatchers *Haematopus ostralegus*. *Ibis*, **119**, 494–506.

Dare, P.J. and Edwards, D.B. (1975). Seasonal changes in flesh weight and biochemical composition of mussels (*Mytilus edulis* L.) in the Conway estuary, North Wales. *Journal of Experimental Marine Biology and Ecology*, **18**, 89–97.

Dare, P.J. and Mercer, A.J. (1968). Wool causing injuries to legs and feet of Oystercatchers. *British Birds*, **61**, 257–263.

Dare, P.J. and Mercer, A.J. (1973). Foods of the Oystercatcher in Morecambe Bay, Lancashire. *Bird Study*, **20**, 173–184.

Dare, P.J. and Mercer, A.J. (1974a). The white collar of the Oystercatcher. *Bird Study*, **21**, 180–184.

Dare, P.J. and Mercer, A.J. (1974b). The timing of wing-moult in the Oyster-catcher *Haematopus ostralegus* in Wales. *Ibis*, 116, 211–214.

Davidson, N.C. (1981). Survival of shorebirds (*Charadrii*) during severe weather: the role of nutritional reserves. In *Feeding and survival strategies of estuarine organisms*, (ed. N.V. Jones and W.J. Wolff), pp. 231–249. Plenum Press, New York.

Davidson, N.C. and Clark, N.A. (1985). The effects of severe weather in January and February 1985 on waders in Britain. *Wader Study Group Bulletin*, **44**, 10–16.

Davidson, N.C. and Evans, P.R. (1982). Mortality of Redshanks and Oyster-catchers from starvation during severe weather. *Bird Study*, **29**, 183–188.

Davidson, N.C., Laffoley, D. d'A., Doody, J.P., Way, L.S., Gordon, J., Key, R., et al. (1991). *Nature conservation and estuaries in Great Britain*. Nature Con-servancy Council, Peterborough.

Davidson, P.E. (1967). A study of the Oystercatcher (*Haematopus ostralegus* L.) in relation to the fishery for cockles (*Cardium edule* L.) in the Burry Inlet, South Wales. *Fishery Investigations*, Series II, Vol.25, Nr. 7, 1–28.

Davidson, P.E. (1968). The Oystercatcher—a pest of shellfisheries. In *The problems of birds as pests*, (ed. R.K. Murton and E.N. Wright), pp. 141–155. Academic Press, London.

Davies, N.B. (1991). Mating systems. In *Behavioural ecology, an evolutionary approach*, (3rd edn), (ed. J.R. Krebs and N.B. Davies), pp. 263–294. Black-well Scientific Publications, Oxford.

Davies, N.B. (1992). *Dunnock behaviour and social evolution*. Oxford Univer-sity Press.

Dawkins, R. and Carlisle, T.R. (1976). Parental investment and mate desertion: a fallacy. *Nature*, **262**, 131–133.

Debout, G. (1990). L'huîtrier pie en Normandie: hivernage 1986/1987. *Le Cormoran*, **6**, 263–266.

Dement'ev, G.P. and Gladkov, N.A. (1969). *Birds of the Soviet Union*, Vol. III. Israel Program for Scientific Translation, Jerusalem.

Desprez, M. and Dupont, J.-P. (1985). Impact biosédimentaire d'amenagements portuaires en estuaire de Seine. In *La Baie de Seine: fonctionnement, con-séquences en matière d'environnement et d'utilisation de la mer*, Vol. 2, (ed. L. Cabioch), pp. 273–280. CNRS, Caen.

Desprez, M., Rybarczyk, H., Wilson. J.G., Ducrotoy, J.P., Sueur, F., Olivesi, R. and Elkaim, B. (1992). Biological impact of eutrophication in the bay of Somme and the induction and impact of anoxia. *Netherlands Journal of Sea Research*, **30**, 149–159.

Dessel, B. van (1983). Onderzoek naar de relaties tussen macrozoobenthos en enkele steltlopersoorten in het intergetijdegebied van het Verdronken Land van het Markiezaat van Bergen op Zoom. Student thesis project Zachstab, Delta Instituut voor Hydrobiologisch Onderzoek/Deltadienst Rijkswaterstaat, Yerseke.

Dewar, J.M. (1908). Notes on the Oystercatcher (*Haematopus ostralegus*), with reference to its habit of feeding upon the mussel (*Mytilus edulis*). *The Zoologist*, **12**, 201–212.

Dewar, J.M. (1910). A preliminary note on the manner in which the Oystercatcher (*Haematopus ostralegus*) attacks the Purple-Shell (*Purpura lapillus*). *The Zoologist*, **14**, 109–112.

Dewar, J.M. (1913). Further observations on the feeding habits of the Oystercatcher (*Haematopus ostralegus*). *The Zoologist*, **17**, 41–56.

Dhondt, A.A. (1988). Carrying capacity: a confusing concept. *Acta Oecologia*, **9**, 337–346.

Dijk, A.J. van, Dijk, G. van, Piersma, T. and SOVON. (1989). Weidevogelpopulaties in Nederland. De jongste aantalsschattingen in internationaal perspectief. *Vogeljaar*, **37**, 60–68.

Dircksen, R. (1932). Die Biologie des Austernfischers, der Brandseeschwalbe und der Küstenseeschwalbe nach Beobachtungen und Untersuchungen auf Norderoog. *Journal für Ornithologie*, **80**, 427–521.

Dobinson, H.M. and Richards, A.J. (1964). The effects of the severe winter of 1962/63 on birds in Britain. *British Birds*, **57**, 373–434.

Drent, R.H. (1970). Functional aspects of incubation in the Herring Gull. *Behaviour*, Suppl., **17**, 1–132.

Drent, R.H. and Daan, S. (1980). The prudent parent: energetic adjustments in avian breeding. *Ardea*, **68**, 225–252.

Drent, R.H. and Klaassen, M. (1989). Energetics of avian growth: the causal link with BMR and metabolic scope. In *Physiology of cold adaptation in birds*, (ed. C. Bech and R.E. Reinertsen), pp. 349–359. Plenum Press, New York.

Drinnan, R.E. (1957). The winter feeding of the Oystercatcher (*Haematopus ostralegus*) on the Edible Cockle (*Cardium edule*). *Journal of Animal Ecology*, **26**, 441–469.

Drinnan, R.E. (1958a). The winter feeding of the Oystercatcher (*Haematopus ostralegus*) on the Edible Mussel (*Mytilus edulis*) in the Conway estuary, North Wales. *Fishery Investigations*, **22**, 1–15.

Drinnan, R.E. (1958b). Observations on the feeding of the Oystercatcher in captivity. *British Birds*, **51**, 139–149.

Dugan, P.J. (1981). The importance of nocturnal foraging in shorebirds—a consequence of increased invertebrate prey activity. In *Feeding and survival strategies of estuarine organisms*, (ed. N.V. Jones and W.J. Wolff), pp. 251–260. Plenum Press, New York.

Dugan, P.J. (1982). Seasonal changes in patch use by a territorial Grey Plover: weather-dependent adjustments in foraging behaviour. *Journal of Animal Ecology*, **51**, 849–857.

Duinker, J.C., Hillebrand, M.T.J. and Boon, J.P. (1983). Organochlorines in benthic invertebrates and sediments from the Dutch Wadden Sea; identification of individual PCB components. *Netherlands Journal of Sea Research*, **17**, 19–38.

Dunbrack, R.L. (1979). A re-examination of robbing behaviour in foraging egrets. *Ecology*, **60**, 644–645.

Dunning, J.B. Jr. (1993). *CRC handbook of avian body masses*. CRC Press, Boca Raton.

Durell, S.E.A. le V. dit and Goss-Custard, J.D. (1984). Prey selection within a size-class of mussels, *Mytilus edulis*, by Oystercatchers, *Haematopus ostralegus*. *Animal Behaviour*, **32**, 1197–1203.

Durell, S.E.A. le V. dit and Goss-Custard, J.D. (1996). Sex ratios in overwintering shorebirds: Oystercatcher *Haematopus ostralegus* sex ratios on the Exe estuary. *Ardea*, **84**. (In press.)

Durell, S.E.A. le V. dit , Goss-Custard, J.D. and Caldow, R.W.G. (1993). Sex-related differences in diet and feeding method in the Oystercatcher *Haematopus ostralegus*. *Journal of Animal Ecology*, **62**, 205–215.

Durell, S.E.A. le V. dit, Goss-Custard, J.D. and Perez-Hurtada, A. (1996a). The efficiency of juvenile Oystercatchers feeding on Ragworms. *Ardea*, **84**. (In press.)

Durell, S.E.A. le V. dit, Ormerod, S.J. and Dare, P.J. (1996b). Differences in population structure between two Oystercatcher roosts on the Burry Inlet, South Wales. *Ardea*, **83**. (In press.)

Ebbinge, B.S. (1989). A multifactorial explanation for variation in breeding performance of Brent Geese *Branta bernicla*. *Ibis*, **131**, 196–204.

Ebbinge, B.S. (1993). Regulation of numbers of Dark-bellied Brent Geese *Branta bernicla bernicla* on spring staging sites. *Ardea*, **80**, 203–228.

Edelstam, C. (1972). The visible migration of birds at Ottenby, Sweden. *Vår Fågelvärld*, Suppl. 7, 1–360.

Edwards, C.A. and Lofty, J.R. (1977). *Biology of earthworms*. Chapman and Hall, London.

Eerden, M. van (1977). Vorstvlucht van watervogels door het oostelijk deel van de Nederlandse Waddenzee op 30 december 1976. *Watervogels*, **2**, 11–14.

Eilers, P.H. and Peeters, J.C.H. (1988). A model for the relationship between light intensity and the rate of photosynthesis in phytoplankton. *Ecological Modelling*, **42**, 199–215.

Emlen, J.M. (1966). The role of time and energy in food preference. *American Naturalist*, **109**, 427–435.

Ens, B. (1982). Size selection in mussel-feeding Oystercatchers. *Wader Study Group Bulletin*, **34**, 16–20.

Ens, B.J. (1991). Guarding your mate and losing the egg: an Oystercatcher's dilemma. *Wader Study Group Bulletin*, **61**, 69–70.

Ens, B.J. (1992). The social prisoner: causes of natural variation in reproductive success of the Oystercatcher. Ph.D. thesis. University of Groningen.

Ens, B.J. and Goss-Custard, J.D. (1984). Interference among Oystercatchers, *Haematopus ostralegus*, feeding on mussels, *Mytilus edulis*, on the Exe estuary. *Journal of Animal Ecology*, **53**, 217–231.

Ens, B.J. and Goss-Custard, J.D. (1986). Piping as a display of dominance by wintering Oystercatchers *Haematopus ostralegus*. *Ibis*, **128**, 382–391.

Ens, B.J. and Zwarts, L. (1980). Wulpen op het wad van Moddergat. *Watervogels*, **5**, 108–120.

Ens, B.J., Esselink, P. and Zwarts, L. (1990). Kleptoparasitism as a problem of prey choice: a study of mudflat-feeding Curlews, *Numenius arquata*. *Animal Behaviour*, **39**, 219–230.

Ens, B.J., Kersten, M., Brenninkmijer, A. and Hulscher, J.B. (1992). Territory quality, parental effort and reproductive success of Oystercatcher (*Haematopus ostralegus*). *Journal of Animal Ecology*, **61**, 703–715.

Ens, B.J., Klaassen, M. and Zwarts, L. (1993a). Flocking and feeding in the Fiddler Crab (*Uca tangeri*): prey availability as risk-taking behaviour. *Netherlands Journal of Sea Research*, 31, 477–494.

Ens, B.J., Safriel, U.N. and Harris, M.P. (1993b). Divorce in the long-lived and monogamous Oystercatcher *Haematopus ostralegus*: incompatibility or choosing the better option? *Animal Behaviour*, 45, 1199–1217.

Ens, B.J., Piersma, T. and Drent, R.H. (1994). The dependence of waders and waterfowl migrating along the East Atlantic flyway on their coastal food supplies: what is the most profitable research programme. *Ophelia*, Suppl. 6, 127–151.

Ens, B.J., Weissing, F.J. and Drent, R.H. (1995). The despotic distribution and deferred maturity in the Oystercatcher: two sides of the same coin? *American Naturalist*. (In press.)

Ens, B.J., D. Alting and Polman, G. (1996a). The effect of an experimentally created musselbed on intake rate, diet and feeding density of Oystercatchers *Haematopus ostralegus*. *Ardea*, 84. (In press.)

Ens, B.J., Dirksen, S. and Smit, C.J. (1996b). Changes in size selection and intake rate of Oystercatchers *Haematopus ostralegus* feeding on Cockles *Cerastoderma edule* and Mussels *Mytilus edulis* in the course of spring. *Ardea*, 84. (In press.)

Ens, B.J., Merck, T. and Smit, C.J. (1996c). Functional and numerical response of Oystercatchers Haematopus ostralegus on shellfish populations. *Ardea*, 84. (In press.)

Ens, B.J., Bunskoeke, A.J., Hulscher, J.B. and Vlas, S.J. de (1996d). The optimal diet model cannot explain the prey choice of Oystercatchers feeding on *Macoma* and *Nereis*. *Ardea* 84. (In press.)

Ens, B.J., Jeugd, H. v.d. and Heg, D. (1996e). Territory acquisition in the Oystercatcher: vital preparations for a breeding career. *Behaviour*. (In press.)

Es, F.B. van, Arkel, M.A. van, Bouwman, L.A. and Schröder, H.G.J. (1980). Influence of organic pollution on bacterial, macrobenthic and meiobenthic populations in intertidal flats of the Dollard. *Netherlands Journal of Sea Research*, 14, 288–304.

Esselink, P. and Zwarts, L. (1989). Seasonal trend in burrow depth and tidal variation in feeding activity of *Nereis diversicolor*. *Marine Ecology Progress Series*, 56, 243–254.

Essink, K. (1978). The effects of pollution by organic waste on macrofauna in the eastern Dutch Wadden Sea. *Netherlands Institute for Sea Resesarch*, Publication Series, No. 1, 1–135.

Etchécopar, R.D. and Hüe, F. (1978). *Les oiseaux de Chine*. Les Éditions du Pacifique, Papeete, Tahiti.

Evans, P.R. (1975). Notes on the feeding of waders on Heron Islands. *The Sunbird*, 6, 25–30.

Evans, P.R. (1976). Energy balance and optimal foraging strategies in shorebirds: some implications for their distributions and movements in the nonbreeding season. *Ardea*, 64, 117–139.

Evans, P.R. (1979). Reclamation of intertidal land: some effects on Shelduck and Wader populations in the Tees estuary. *Verhandlungen der ornithologischen Gesellschaft Bayern*, 23, 147–168.

Evans, P.R. (1984). Introduction. In *Coastal waders and wildfowl in winter*, (ed.

P.R. Evans, J.D. Goss-Custard and W.G. Hale), pp. 3–7. Cambridge University Press, Cambridge.

Evans, P.R. (1991). Introductory remarks: habitat loss—effects on shorebird populations. *Acta XX. Congressus Internationalis Ornithologici*, 2197–2198.

Evans, P.R. and Dugan, P.J. (1984). Coastal birds: numbers in relation to food resources. In *Coastal waders and wildfowl in winter*, (ed. P.R. Evans, J.D. Goss-Custard and W.G. Hale), pp. 8–28. Cambridge University Press, Cambridge.

Evans, P.R. and Moon, S.J. (1981). Heavy metals in shorebirds and their prey in North-east England. In *Heavy metals in Northern England: environmental and biological Aspects*, (ed. P.J. Say and B.A. Whitton), pp. 181–190. University of Durham, Durham.

Evans, P.R. and Smith, P.C. (1975). Studies on shorebirds at Lindisfarne, Northumberland. 2. Fat and pectoral muscle as indicators of body condition in the Bar-tailed Godwit. *Wildfowl*, 26, 64–76.

Evans, P.R., Herdson, D.M., Knights, P.J. and Pienkowski, M.W. (1979). Short-term effects of reclamation of part of Seal Sands, Teesmouth, on wintering waders and Shelduck. I. Shorebird diets, invertebrate densities, and the impact of predation on the invertebrates. *Oecologia*, 41, 183–206.

Evans, P.R., Uttley, J.D., Davidson, N.C. and Ward, P. (1987). Shorebirds (*Charadrii* and *Scolopaci*) as agents of transfer of heavy metals within and between estuarine ecosystems. In *Pollutant transport and fate in ecosystems*, (ed. P.J. Coughtrey, M.H. Martin and M.H. Unsworth), pp. 337–352. Blackwell Scientific Publications, Oxford.

Evans, R.M. (1970). Imprinting and mobility in young Ring-billed Gulls, *Larus delawarensis*. *Animal Behavior Monographs*, 3, 193–248.

Everaarts, J.M., Buck, A. de, Hillebrand, M.T.J. and Boon, J.P. (1991). Residues of chlorinated biphenyl congeners and pesticides in brain and liver of the Oystercatcher (*Haematopus ostralegus*) in relation to age, sex and biotransformation capacity. *Science of Total Environment*, 100, 483–499.

Falla, R.A., Sibson, R.B. and Turbott, E.G. (1979). *The new guide to the birds of New Zealand and outlying islands*. Dai Nippon, Hong Kong.

Feare, C.J. (1967). The effect of predation by shore-birds on a population of Dogwhelks *Thais lapillus*. *Ibis*, 109, 474.

Feare, C.J. (1971). Predation of limpets and Dogwhelks by Oystercatchers. *Bird Study*, 18, 121–129.

Franklin, A. and Pickett, G.D. (1978). Studies on the indirect effects of fishing on stocks of cockles, *Cardium edule*, in the Thames Estuary and Wash. *Fishery Research Technical Report*, MAFF, Directorate of Fisheries Research Lowestoft No 42, 1–9.

Fretwell, S.D. (1972). *Populations in a seasonal environment*. Princeton University Press, New Jersey.

Fretwell, S.D. and Lucas, H.L. (1969). On territorial behavior and other factors influencing habitat distribution in birds. I. Theoretical development. *Acta Biotheoretica*, 19, 16–36.

Gavrin, V.F., Dolgushin, I.A., Korelov, M. N. and Kuzmina, M.A. (1962). [*The birds of Kazakhstan*]. (In Russian). Academy of Sciences, Alma-Ata.

Gendron, R.P. (1986). Searching for cryptic prey: evidence for optimal search rates and the formation of search images in quality. *Animal Behaviour*, 34, 898–912.

Gendron, R.P. and Staddon, J.E.R. (1983). Searching for cryptic prey: the effect of search rate. *American Naturalist*, **121**, 172–186.

Gerritsen, A.F.C., Heezik, Y.M. van and Swennen, C. (1983). Chemoreception in two further *Calidris* species (*Calidris maritima* and *Calidris canutus*). *Netherlands Journal of Zoology*, **33**, 485–496.

Gibbons, D.W., Reid, J.B. and Chapman, R.A. (1993). *The new atlas of breeding birds in Britain and Ireland: 1988–1991*. Poyser, London.

Gibson, F. (1978). Ecological aspects of the time budget of the American Oystercatcher. *American Midlands Naturalist*, **99**, 65–82.

Glutz v. Blotzheim, U.N., Bauer, K.M. and Bezzel, E. (1975). *Handbuch der Vögel Mitteleuropas*, Vol. **6/1**. Akademische Verlagsgesellschaft, Wiesbaden.

Goater, C.P. (1988). Patterns of helminth parasitism in the Oystercatcher, *Haematopus ostralegus*, from the Exe estuary, England. Unpublished Ph.D. thesis. University of Exeter.

Goater, C.P. (1993). Population biology of *Meiogymnophallus minutus* (*Trematoda: Gymnophallidae*) in cockles *Cerastoderma edule* from the Exe estuary, England. *Journal of the Marine Biological Association*, **73**, 163–177.

Goater, C.P., Goss-Custard J.D. and Kennedy, C.R. (1995). Population biology of two species of helminth in Oystercatchers, *Haematopus ostralegus*. *Canadian Journal of Zoology*, **75**, 296–300.

Goede, A.A. (1985). Mercury, selenium, arsenic and zinc in waders from the Dutch Wadden Sea. *Environmental Pollution*, (Ser. A) **37**, 287–309.

Goede, A.A. (1993). Variation in the energy intake of captive Oystercatchers (*Haematopus ostralegus*). *Ardea*, **81**, 89–97.

Goede, A.A. and Voogt, P. de (1985). Lead and cadmium in waders from the Dutch Wadden Sea. *Environmental Pollution*, (Ser. A) **37**, 311–322.

Gompel, J. van (1987). Mortaliteit van waadvogels aan de Belgische kust tijdens de koudeperiode januari-february 1985. *Oriolus*, **53**, 175–186.

Gosler, A. (1987). Aspects of the bill morphology in relation to ecology in the Great Tit, *Parus major*. Unpublished D.Phil. thesis. University of Oxford.

Goss-Custard, J.D. (1970). Feeding dispersion in some overwintering wading birds. In *Social behaviour in birds and mammals*, (ed. J.H. Crook), pp. 3–35. Academic Press, New York.

Goss-Custard, J.D. (1977a). The energetics of prey selection by Redshank, *Tringa totanus* (L.), in relation to prey density. *Journal of Animal Ecology*, **46**, 1–19.

Goss-Custard, J.D. (1977b). The ecology of the Wash. III. Density-related behaviour and the possible effects of a loss of feeding grounds on wading birds (*Charadrii*). *Journal of Applied Ecology*, **14**, 721–739.

Goss-Custard, J.D. (1977c). Predator responses and prey mortality in redshank, *Tringa totanus*, and a preferred prey, *Corophions volutator* (Pallas). *Journal of Animal Ecology*, **46**, 21–35.

Goss-Custard, J.D. (1980). Competition for food and interference among waders. *Ardea*, **68**, 31–52.

Goss-Custard, J.D. (1985). Foraging behaviour of wading birds and the carrying capacity of estuaries. In *Behavioural ecology: ecological consequences of adaptive behaviour*, (ed. R.M. Sibly and R.H. Smith), pp. 169–188. Blackwell Scientific Publications, Oxford.

Goss-Custard, J.D. (1993). The effect of migration and scale on the study of bird populations: 1991 Witherby Lecture. *Bird Study*, **40**, 81–96.

Goss-Custard, J.D. (1995). Effect of habitat loss and habitat change on estuarine shorebird populations. *Coastal Zone Topics: Process, Ecology and Management*, **1**, 61–67.

Goss-Custard, J.D. and Charman, K. (1976). Predicting how many wintering waterfowl an area can support. *Wildfowl*, **27**, 157–158.

Goss-Custard, J.D. and Durell, S.E.A. le V. dit (1983). Individual and age differences in the feeding ecology of Oystercatchers, *Haematopus ostralegus*, wintering on the Exe. *Ibis*, **125**, 155–171.

Goss-Custard, J.D. and Durell, S.E.A. le V. dit (1984). Feeding ecology, winter mortality and the population dynamics of Oystercatchers, *Haematopus ostralegus*, on the Exe estuary. In *Coastal waders and wildfowl in winter*, (ed. P.R. Evans, J.D. Goss-Custard and W.G. Hale), pp. 190–208. Cambridge University Press.

Goss-Custard, J.D. and Durell, S.E.A. le V. dit (1987a). Age-related effects in Oystercatchers, *Haematopus ostralegus*, feeding on mussels, *Mytilus edulis*. I. Foraging efficiency and interference. *Journal of Animal Ecology*, **56**, 521–536.

Goss-Custard, J.D. and Durell, S.E.A. le V. dit (1987b). Age-related effects in Oystercatchers, *Haematopus ostralegus*, feeding on mussels, Mytilus edulis. II. Aggression. *Journal of Animal Ecology*, **56**, 537–548.

Goss-Custard, J.D. and Durell, S.E.A. le V. dit (1987c). Age-related effects in Oystercatchers, *Haematopus ostralegus*, feeding on mussels, *Mytilus edulis*. III. The effect of interference on overall intake rate. *Journal of Animal Ecology*, **56**, 549–558.

Goss-Custard, J.D. and Durell, S.E.A. le V. dit (1988). The effect of dominance and feeding method on the intake rates of Oystercatchers, *Haematopus ostralegus*, feeding on mussels. *Journal of Animal Ecology*, **57**, 827–844.

Goss-Custard, J.D. and Durell, S.E.A. le V. dit (1990). Bird behaviour and environmental planning: approaches in the study of wader populations. *Ibis*, **132**, 273–289.

Goss-Custard, J.D. and Sutherland, W.J. (1984). Feeding specialization in Oystercatchers *Haematopus ostralegus*. *Animal Behaviour*, **32**, 299–301.

Goss-Custard, J.D. and Verboven, N. (1993). Disturbance and feeding shorebirds on the Exe estuary. *Wader Study Group Bulletin*, **68**, 59–66.

Goss-Custard, J.D. and Willows, R. (1996). Modelling the responses of mussel *Mytilus edulis* and Oystercatcher *Haematopus ostralegus* populations to environmental change. In *Aquatic predators and their prey* (ed. S.P.R. Greenstreet and M.L. Tasker). Blackwells Scientific Publications, Oxford. (In press.)

Goss-Custard, J.D., Jenyon, R.A., Jones, R.E., Newbery, P.E. and Williams, R. le B. (1977a). The ecology of the Wash. II. Seasonal variation in the feeding conditions of wading birds (*Charadrii*). *Journal of Applied Ecology*, **14**, 701–719.

Goss-Custard, J.D., Jones, R.E. and Newbery, P.E. (1977b). The ecology of the Wash. I. Distribution and diet of wading birds (*Charadrii*). *Journal of Applied Ecology*, **14**, 681–700.

Goss-Custard, J.D., Kay, D.G. and Blindell, R.M. (1977c). The density of migratory and overwintering Redshank, *Tringa totanus* (L.), and Curlew, *Numenius arquata* (L.), in relation to the density of their prey in South-east England. *Estuarine and Coastal Marine Science*, **5**, 497–510.

Goss-Custard, J.D., McGrorty, S., Reading, C.J. and Durell, S.E.A. le V. dit (1980). Oystercatchers and mussels on the Exe Estuary. *Essays on the Exe Estuary*. Devon Association, Special Volume 2, 161–185.

Goss-Custard, J.D., McGrorty, S., Reading, C.J. and Durell, S.E.A. le V. dit (1981). Factors affecting the occupation of mussel *Mytilus edulis* beds by Oystercatchers *Haematopus ostralegus* on the Exe estuary, Devon. In *Feeding and survival strategies of estuarine organisms*, (ed. N.V. Jones and W.J. Wolff), pp. 217–229. Plenum Publishing Corporation, New York.

Goss-Custard, J.D., Durell, S.E.A. le V. dit and Ens, B.J. (1982a). Individual differences in aggressiveness and food stealing among wintering Oystercatchers, *Haematopus ostralegus* L. *Animal Behaviour*, 30, 917–928.

Goss-Custard, J.D., Durell, S.E.A. le V. dit, McGrorty, S. and Reading, C.J. (1982b). Use of mussel, *Mytilus edulis* beds by Oystercatchers *Haematopus ostralegus* according to age and population size. *Journal of Animal Ecology*, 51, 543–554.

Goss-Custard, J.D., Durell, S.E.A. le V. dit, Sitters, H.P. and Swinfen, R. (1982c). Age-structure and survival of a wintering population of Oystercatchers. *Bird Study*, 29, 83–98.

Goss-Custard, J.D., Clarke, R.T. and Durell, S.E.A. le V. dit (1984). Rates of food intake and aggression of Oystercatchers, *Haematopus ostralegus*, on the most and least preferred mussel, *Mytilus edulis*, beds on the Exe estuary. *Journal of Animal Ecology*, 53, 233–245.

Goss-Custard, J.D., Cayford, J.T., Boates, J.S. and Durrel, S.E.A. le V. dit (1987). Field tests of the accuracy of estimating prey size from bill-length in Oystercatchers, *Haematopus ostralegus*, feeding on mussels, *Mytilus edulis*. *Animal Behaviour*, 35, 1078–1083.

Goss-Custard, J.D., Warwick, R.M., Kirby, R., McGrorty, S., Clarke, R.T., Pearson, B., Rispin, W.E., Durell, S.E.A. Le V. dit and Rose, R.J. (1991). Towards predicting wading bird densities from predicted prey densities in a post-barrage Severn Estuary. *Journal of Applied Ecology*, 28, 1004–1026.

Goss-Custard, J.D., Caldow, R.W.G. and Clarke, R.T. (1992). Correlates of the density of foraging Oystercatchers, *Haematopus ostralegus*, at different population sizes. *Journal of Animal Ecology*, 61, 159–173.

Goss-Custard, J.D., West, A.D. and Durell, S.E.A. le V. dit (1993). The availability and quality of the mussel (*Mytilus edulis*) prey of Oystercatchers (*Haematopus ostralegus*). *Netherlands Journal of Sea Research*, 31, 419–439.

Goss-Custard, J.D., Caldow, R.W.G., Clarke, R.T., Durell, S.E.A. le V. dit and Sutherland, W.J. (1995a). Deriving population parameters from individual variations in foraging behaviour: I. Empirical game theory distribution model of Oystercatchers *Haematopus ostralegus* feeding on mussels *Mytilus edulis*. *Journal of Animal Ecology*, 64, 265–276.

Goss-Custard, J.D., Caldow, R.W.G., Clarke, R.T., Durell, S.E.A. le V. dit and West, A.D. (1995b). Consequences of habitat loss and change in wintering migratory birds: predicting the local and global effects from studies of individuals. *Ibis*, 137, Suppl. 1, 56–66.

Goss-Custard, J.D., Caldow, R.W.G., Clarke, R.T. and West, A.D. (1995c). Deriving population parameters from individual variations in foraging behaviour: II. Model tests and population parameters. *Journal of Animal Ecology*, 64, 277–289.

Goss-Custard, J.D., Clarke, R.T., Briggs, K.B., Ens, B.J., Exo, K.-M., Smit, C., *et al.* (1995d). Population consequences of winter habitat loss in a migratory shorebird: I. Estimating model parameters. *Journal of Applied Ecology*, **32**, 317–333.

Goss-Custard, J.D., Clarke, R.T., Durell, S.E.A. le V. dit, Caldow, R.W.G. and Ens, B.J. (1995e). Population consequences of habitat loss and change in wintering migratory birds: II. Model predictions. *Journal of Applied Ecology*, **32**, 334–348.

Goss-Custard, J.D., McGrorty, S. and Durell, S.E.A. le V. dit. (1996a). The effect of the European Oystercatchers on shellfish populations. *Ardea*, **84**. (In press.)

Goss-Custard, J.D., West, A., Caldow, R.W.G., Durell, S.E.A. le V. dit and McGrorty, S. (1996b). An empirical optimality model to predict the intake rates of Oystercatchers *Haematopus ostralegus* feeding on mussels *Mytilus edulis*. *Ardea*, **84**. (In press.)

Gould, S.J. and Lewontin, R.C. (1979). The spandrels of San Marco and the Panglossian paradigm: a critique of the adaptationist programme. *Proceedings of the Royal Society of London B*, **205**, 581–598.

Grafen, A. (1988). On the uses of data on lifetime reproductive success. In *Reproductive success, studies of individual variation in contrasting breeding systems*, (ed. T.H. Clutton-Brock), pp. 454–471. Chicago University Press.

Granéli, E. (1987). Nutrient limitation of phytoplankton biomass in a brackish water bay highly influenced by river discharge. *Estuarine Coastal Shelf Science*, **25**, 555–565.

Granéli, E., Wallström, K., Larsson, U., Granéli, W. and Elmgren, R. (1990). Nutrient limitation of primary production in the Baltic Sea area. *Ambio*, **19**, 142–151.

Grant, P.R. (1986). *Ecology and evolution of Darwin's finches*. Princeton University Press.

Gray, J.S. (1992). Eutrophication in the sea. In *Marine eutrophication and population dynamics*, (ed. G. Colombo, I. Ferrari, V.U. Ceccherelli and R. Rossi), pp. 3–15. Olsen and Olsen, Fredensborg.

Greenberg, R. (1980). Demographic aspects of long-distance migration. In *Migrant birds in the Neotropics: behaviour, distribution and conservation*, (ed. A. Keast and E.S. Morton), pp. 493–504. Smithsonian Institution, Washington D.C.

Greenhalgh, M.E. (1969). The breeding of the Oystercatcher in north-west England. *The Naturalist*, **94**, 43–47.

Greenhalgh, M.E. (1973). A comparison of breeding success of the Oystercatcher. *The Naturalist*, **926**, 87–88.

Grolle, H. (1987). Warmtecompensatie door lopen bij de Scholekster (*Haematopus ostralegus*) en de Steenloper (*Arenaria interpres*). Unpublished student thesis. University of Groningen.

Groves, S. (1984). Chick growth, sibling rivalry, and chick production in American Black Oystercatchers. *Auk*, **101**, 525–531.

Gustafsson, L. (1988). Foraging behaviour of individual Coal Tits, *Parus ater*, in relation to their age, sex and morphology. *Animal Behaviour*, **36**, 696–704.

Habekotté, B. (1987). Scholeksters en slijkgapers. Unpublished student thesis. University of Groningen.

Hale, W.G. (1982). *Waders*. Collins, London.

Hall, K.R.L. (1959). Observations on the nest-sites and nesting behaviour of the Black Oystercatcher *Haematopus moquini* in the Cape Peninsula. *Ostrich*, 30, 143–154.

Hancock, D.A. (1971). The role of predators and parasites in a fishery for the mollusc *Cardium edule* L. In *Dynamics of populations*, (ed. P.J. den Boer and G.R. Gradwell), pp. 419–439. Centre for Agriculture, Publicity and Documentation, Wageningen.

Harris, M.P. (1967). The biology of Oystercatchers *Haematopus ostralegus* on Skokholm Islands, S. Wales. *Ibis*, 109, 180–193.

Harris, M.P. (1969). Effect of laying date on chick production in Oystercatchers and Herring Gulls. *British Birds*, 62, 70–75.

Harris, M.P. (1970). Territory limiting the size of the breeding population of the Oystercatcher (*Haematopus ostralegus*)—a removal experiment. *Journal of Animal Ecology*, 39, 707–713.

Harris, M.P. (1974). *A field guide to the birds of Galapagos*. Collins, London.

Harris, M.P., Safriel, U.N., Brooke, M.L. de and Britton, C.K. (1987). The pair bond and divorce among Oystercatchers *Haematopus ostralegus* on Skokholm Island, Wales. *Ibis*, 129, 45–57.

Hartwick, E.B. (1974). Breeding ecology of the Black Oystercatcher (*Haematopus bachmani* Audubon). *Syesis*, 7, 83–92.

Hartwick, E.B. (1976). Foraging strategy of the Black Oystercatcher (*Haematopus bachmani* Audubon). *Canadian Journal of Zoology*, 54, 142–155.

Hartwick, E.B. and Blaylock, W. (1979). Winter ecology of a Black Oystercatcher population. *Studies in Avian Biology*, 2, 207–215.

Hassell, M.P. (1971). Mutual interference between searching insect parasites. *Journal of Animal Ecology*, 40, 473–486.

Hassell, M.P. (1978). *The dynamics of arthropod predator-prey systems*. Princeton University Press, New Jersey.

Hassel, M.P. and May, R.M. (1985). From individual behaviour to population dynamics. In *Behavioural ecology: ecological consequences of adaptive behaviour*, (ed. R.M. Sibly and R.H. Smith), pp. 3–32. Blackwell Scientific Publications, Oxford.

Haverschmidt, F. (1946). Notes on the nest-sites of Oystercatchers and Long-eared Owl as a hole breeder. *British Birds*, 39, 334–33.

Hayman, P., Marchant, J. and Prater, T. (1986). *Shorebirds: an identification guide to the waders of the world*. Christopher Helm, London.

Heezik, Y.M. van, Gerritsen, A.F.C. and Swennen, C. (1983). The influence of chemoreception on the foraging behaviour of two species of sandpiper, *Calidris alba* (Pallas) and *Calidris alpina* (L.). *Netherlands Journal of Sea Research*, 17, 47–56.

Heg, D., Ens, B.J., Burke, T., Jenkins, L. and Kruijt, J.P. (1993). Why does the typically monogamous Oystercatcher *Haematopus ostralegus* engage in extra-pair copulations? *Behaviour*, 126, 247–289.

Heiligenberg, T. van den (1984). De ecologische effecten van winnen van wad-pieren en andere bodemdieren in het intergetijdengebied. *Report Rijksinstituut voor Natuurbeheer*, (RIN)84/3, Texel.

Heinrich, B. (1979). Majoring and minoring by foraging Bumblebees, *Bombus vagans*: an experimental analysis. *Ecology*, 60, 245–255.

Helbing, G.L. (1977). Maintenance activities of the Black Oystercatcher,

Haematopus bachmani Audubon, in northwestern California. Unpublished M.Sc. thesis. Humbolt State University, California.

Heppleston, P.B. (1970). Anatomical observations on the bill of the Oyster-catcher (*Haematopus ostralegus occidentalis*) in relation to feeding behaviour. *Journal of Zoology London*, **161**, 519–524.

Heppleston, P.B. (1971a). Feeding techniques of the Oystercatcher. *Bird Study*, **18**, 15–20.

Heppleston, P.B. (1971b). The feeding ecology of Oystercatchers (*Haematopus ostralegus* L.) in winter in Northern Scotland. *Journal of Animal Ecology*, **40**, 651–672.

Heppleston, P.B. (1972). The comparative breeding ecology of Oystercatchers (*Haematopus ostralegus*) in inland and coastal habitats. *Journal of Animal Ecology*, **41**, 23–51.

Heppleston, P.B. (1973). The distribution and taxonomy of Oystercatchers. *Notornis*, **20**, 102–112.

Hockey, P.A.R. (1981a). Feeding techniques of the African Black Oystercatcher *Haematopus moquini*. In *Proceedings of the symposium on birds of the sea and shore 1979*, (ed. J. Cooper), pp. 99–115. African Seabird Group, Cape Town.

Hockey, P.A.R. (1981b). Morphometrics and sexing of the African Black Oyster-catcher. *Ostrich*, **52**, 244–247.

Hockey, P.A.R. (1982). The taxonomic status of the Canary Islands Oyster-catcher *Haematopus meadewaldoi*. *Bulletin of the British Ornithologists' Club*, **102**, 77–83.

Hockey, P.A.R. (1983a). Aspects of the breeding biology of the African Black Oystercatcher *Haematopus moquini*. *Ostrich*, **54**, 26–35.

Hockey, P.A.R. (1983b). The distribution, population size, movements and conservation of the African Black Oystercatcher *Haematopus moquini*. *Biological Conservation*, **25**, 233–262.

Hockey, P.A.R. (1984). Growth and energetics of the African Black Oyster-catcher *Haematopus moquini*. *Ardea*, **72**, 111–117.

Hockey, P.A.R. (1985). Observations on the communal roosting of African Black Oystercatchers. *Ostrich*, **56**, 52–57.

Hockey, P.A.R. (1987). The influence of coastal utilization by man on the pre-sumed extinction of the Canarian Black Oystercatcher *Haematopus meade-waldoi*. *Biological Conservation*, **39**, 49–62.

Hockey, P.A.R. and Branch, G.M. (1983). Do Oystercatchers influence Limpet shell shape? *Veliger*, **26**, 139–141.

Hockey, P.A.R. and Cooper, J. (1980). Paralytic shellfish poisoning—a control-ling factor in Black Oystercatcher populations? *Ostrich*, **51**, 188–190.

Hockey, P.A.R. and Cooper, J. (1982). Occurrence of the European Oyster-catcher *Haematopus ostralegus* in Southern Africa. *Ardea*, **70**, 55–58.

Hockey, P.A.R. and Erkom Schurink, C. van (1992). The invasive biology of the mussel *Mytilus galloprovincialis* on the southern African coast. *Transac-tions of the Royal Society of South Africa*, **48**, 123–139.

Hockey, P.A.R. and Underhill, L.G. (1984). Diet of the African Black Oyster-catcher on rocky shores: spatial, temporal and sex-related variation. *South African Journal of Zoology*, **19**, 1–11.

Hoekstra, R. (1988). Effect van loopsnelheid bij het fourageren van Scholeksters

(*Haematopus ostralegus*) op *Nereis diversicolor*. Unpublished student thesis, University of Groningen.

Högstedt, H. (1980). Evolution of clutch size in birds: adaptive variation in relation to territory quality. *Science*, **210**, 1148–1150.

Holgersen, H. (1980). Bird ringing report 1976–78, Stavanger Museum. *Sterna*, **17**, 37–82.

Holgersen, H. (1982). Bird ringing report 1979–80, Stavanger Museum. *Sterna*, **17**, 85–123.

Holling, C.S. (1959). Some characteristics of simple types of predation and parasitism. *Canadian Entomologist*, **91**, 385–398.

Holmes, R.T. and Pitelka, F.A. (1968). Food overlap among coexisting sandpipers on northern Alaskan tundra. *Systematic in Zoology*, **17**, 305–318.

Holmgren, N. and Lundberg, S. (1993). Despotic behaviour and the evolution of migration patterns in birds. *Ornis Scandinavica*, **24**, 103–109.

Hooijmeijer, J.C.E.W. (1991). Onderzoek naar de verstoring op het gedrag en het voorkomen van Scholeksters (*Haematopus ostralegus*) in de Mokbaai op Texel. Report *Rijksinstituut voor Natuurbeheer* (RIN) 91/27, Texel.

Horwood, J.W. and Goss-Custard, J.D. (1977). Predation by the Oystercatcher, *Haematopus ostralegus* (L.), in relation to the cockle, *Cerastoderma edule* (L.), fishery in the Burry Inlet, South Wales. *Journal of Applied Ecology*, **14**, 139–158.

Hosper U.G. (1978). Fourageerstrategie en voedselopname van Scholeksters (*Haematopus ostralegus*) in het binnenland. Unpublished student thesis, University of Groningen.

Houston, A.I. (1993). The importance of state. In *Diet selection: an interdisciplinary approach to foraging behaviour*, (ed. R.N. Hughes), pp. 10–31. Blackwell Scientific Publications, Oxford.

Howe, M.A. (1982). Social organization in a nesting population of Eastern Willets (*Catoptrophorus semipalmatus*). *Auk*, **99**, 88–102.

Hoyt, D.F. (1979). Practical methods of estimating volume and fresh weight of bird eggs. *Auk*, **96**, 73–77.

Hudson, P.J. (1994). Parasitic infections of birds. *Trends in Ecology and Evolution*, **10**, 3–4.

Hughes, R.N. (1969). A study of feeding in *Scrobicularia plana*. *Journal of the Marine Biological Association of the United Kingdom*, **49**, 805–823.

Hughes, R.N. (1970). Population dynamics of the bivalve *Scrobicularia plana* (da Costa) on an intertidal mud-flat in North Wales. *Journal of Animal Ecology*, **39**, 333–356.

Hull, S.C. (1987). Macroalgal mats and species abundance: a field experiment. *Estuarine and Coastal Shelf Science*, **25**, 519–532.

Hulscher, J.B. (1964a). Hoe een Scholekster Strandkrabben ving. *De Levende Natuur*, **67**, 49–52.

Hulscher, J.B. (1964b). Scholeksters en Lamellibranchiaten in de Waddenzee. *De Levende Natuur*, **67**, 80–85.

Hulscher, J.B. (1964c). Scholeksters en wormen. *De Levende Natuur*, **67**, 97–102.

Hulscher, J.B. (1973). Enkele waarnemingen over de wegtrek van de Scholekster van de broedgebieden na de broedtijd in Friesland. *Vanellus*, **26**, 35–39.

Hulscher, J.B. (1974). An experimental study of the food intake of the Oystercatcher *Haematopus ostralegus* L. in captivity during the summer. *Ardea*, **62**, 155–171.

Hulscher, J.B. (1975a). Het voorkomen in de winter en het begin van de voorjaarstrek van de Scholekster in het binnenland van Friesland. *Vanellus*, **28**, 141–147.

Hulscher, J.B. (1975b). De trek van de Scholekster in Friesland aan de hand van ringgegevens. *De Levende Natuur*, **78**, 218–230.

Hulscher, J.B. (1976a). Scholekster. In *Vogels in Friesland*, Vol. I. (ed. D.T.E. van der Ploeg, W. de Jong, M.J. Swart, B. van der Veen, J.A. de Vries, J.H.P. Westhof and A.G. Witteveen), pp. 416–443. De Tille, Leeurwarden.

Hulscher, J.B. (1976b). Scholeksters verlieten het binnenland tijdens vorstperiode 26 jan.–7 febr. '76. *Vanellus*, **29**, 47–49.

Hulscher, J.B. (1976c). Localisation of cockles (*Cardium edule* L.) by the Oystercatcher (*Haematopus ostralegus* L.) in darkness and daylight. *Ardea*, **64**, 292–310.

Hulscher, J.B. (1977a). The progress of wing-moult of Oystercatchers *Haematopus ostralegus* at Drachten, Netherlands. *Ibis*, **119**, 507–512.

Hulscher, J.B. (1977b). Wat verkeersslachtoffers van Scholeksters ons nog kunnen leren. *Vanellus*, **30**, 170–179.

Hulscher, J.B. (1980). Oystercatchers (*Haematopus ostralegus* L.). In *Birds of the Wadden Sea*, (ed. C.J. Smit and W.J. Wolff), pp. 92–104. Balkema, Rotterdam.

Hulscher, J.B. (1982). The Oystercatcher *Haematopus ostralegus* as a predator of the bivalve *Macoma balthica* in the Dutch Wadden Sea. *Ardea*, **70**, 89–152.

Hulscher, J.B. (1983). Oystercatcher (*Haematopus ostralegus* L.). In *Ecology of the Wadden Sea*, (ed. W.J. Wolff), pp. 6/92–6/104. Balkema, Rotterdam.

Hulscher, J.B. (1985). Growth and abrasion of the Oystercatcher bill in relation to dietary switches. *Netherlands Journal of Zoology*, **35**, 124–154.

Hulscher, J.B. (1988). Mossel doodt Scholekster *Haematopus ostralegus*. *Limosa*, **61**, 42–45.

Hulscher, J.B. (1989). Sterfte en overleving van Scholeksters *Haematopus ostralegus* bij strenge vorst. *Limosa*, **62**, 177–181.

Hulscher, J.B. (1990). Survival of Oystercatchers during hard winter weather. *Ring*, **13**, 167–172.

Hulscher, J.B. (1993). Uitzonderlijk grote aantallen Scholeksters in het binnenland gedurende de winter van 1992/93. *Limosa*, **66**, 117–123.

Hulscher, J.B. and Ens, B.J. (1991). Somatic modifications of feeding system structures due to feeding on different foods with emphasis on changes in bill shape in Oystercatchers. *Acta XX Congressus Internationalis Ornithologici*, 889–896.

Hulscher, J.B. and Ens, B.J. (1992). Is the bill of the male Oystercatcher a better tool for attacking mussels than the bill of the female? *Netherlands Journal of Zoology*, **42**, 85–100.

Hulscher, J.B., Bunskoeke, E.J., Alting, E. and Ens, B.J. (1996). Subtle differences between male and female Oystercatchers in feeding on the bivalve mollusc *Macoma balthica*. *Ardea*, **84**. (In press.)

Hummel, H. (1985). Food intake of *Macoma balthica* (*Mollusca*) in relation to seasonal changes in its potential food on a tidal flat in the Dutch Wadden Sea. *Netherlands Journal of Sea Research*, **19**, 52–76.

Hummel, H., Bogaards, R.H., Nieuwenhuize, J., Wolf, L. de and Liere, J.M. van

(1990). Spatial and seasonal differences in the PCB content of the mussel *Mytilus edulis*. *Science of Total Environment*, **92**, 155–63.

Humphrey, P.S., Bridge, D., Reynolds, P.W. and Peterson, R.T. (1970). *Birds of Isla Grande (Tierra del Fuego)*. Smithsonian Institution, Washington D.C.

Huston, M., DeAngelis, D. and Post, W. (1988). New computer models unify ecological theory. *Bioscience*, **38**, 682–691.

Hutton, M. (1981). Accumulation of heavy metals and selenium in three seabird species from the United Kingdom. *Environmental Pollution* (Ser. A) **26**, 129–145.

Impe, J. van (1985). Estuarine pollution as a probable cause of an increase of estuarine birds. *Marine Pollution Bulletin*, **16**, 271–276.

Jackson, M.J. and James, R. (1979). The influence of bait digging on cockle, *Cerastoderma edule*, populations in North Norfolk. *Journal of Applied Ecology*, **16**, 671–679.

Jakobsen, A.B. and Rasmussen, L.M. (1989). Langli/Skallingen Vadehavet. *Arsrapport over observationer* 1987. Miljominiseriet Skov-og Naturstyrelsen.

Jeffrey, R.G. (1987). Influence of human disturbance on the nesting success of African Black Oystercatchers. *South African Journal of Wildlife Research*, **17**, 71–72.

Jehl, J.R. (1968). Relationships in the *Charadrii* (shorebirds): a taxonomic study based on colour patterns of the downy young. *Memories of the San Diego Society of Natural History*, **3**, 1–54.

Jehl, J.R. (1973). Breeding biology and systematic relationships of the Stilt Sandpiper. *Wilson Bulletin*, **85**, 115–147.

Jehl, J.R. (1978). A new hybrid Oystercatcher from South America, *Haematopus leucopodus x H. ater*. *Condor*, **80**, 344–346.

Jehl, J.R., Rumboll, M.A.E. and Winter, J.P. (1973). Winter bird populations of Golfo San Jose, Argentina. *Bulletin of the British Ornithologists' Club*, **93**, 56–63.

Jenkins, D., Watson, A. and Miller, G.R. (1963). Population studies on Red Grouse *Lagopus lagopus scoticus* in north-east Scotland. *Journal of Animal Ecology*, **32**, 317–376.

Johnsgard, P.A. (1981). *The plovers, sandpipers, and snipes of the world*. University of Nebraska Press, Lincoln.

Johnson, A.W. and Goodall, J.D. (1965). *The birds of Chile*, Vol. I. Platt Establecimientos Gráficos S.A., Buenos Aires.

Johnson, C. (1985). Patterns of seasonal weight variation in waders on the Wash. *Ringing and Migration*, **6**, 19–32.

Jones, A. (1979a). Notes on the behaviour of Variable Oystercatchers. *Notornis*, **26**, 47–52.

Jones, A.M. (1979b). Structure and growth of a high-level population of *Cerastoderma edule (Lamellibranchiata)*. *Journal of the Marine Biological Association of the United Kingdom*, **59**, 277–287.

Kaufmann, J.H. (1983). On the definitions and functions of dominance and territoriality. *Biological Review*, **58**, 1–20.

Keighley, J. (1949). Oystercatchers. *Skokholm Bird Observatory Report* 1948, 6–9.

Keijl, G.O. and Mostert, C. (1988). Vorsttrek van Scholeksters *Haematopus ostralegus* langs de kust in 1987. *Sula*, **2**, 113–118.

Kenyon, K.W. (1949). Observations on behaviour and populations of Oyster-catchers in Lower California. *Condor*, 51, 193–199.

Kersten, M. and Brenninkmeijer, A. (1995). Growth rate, fledging success and post-fledging survival of juvenile Oystercatchers. *Ibis*, 137, 396–404.

Kersten, M. and Piersma, T. (1987). High levels of energy expenditure in shore-birds; metabolic adaptations to an energetically expensive way of life. *Ardea*, 75, 175–187.

Kersten, M. and Visser, W. (1996a). The rate of food processing in the Oyster-catcher: food intake and energy expenditure constrained by a digestive bottle-neck. *Functional Ecology*. (In press.)

Kersten, M. and Visser, W. (1996b). Food intake of Oystercatchers during day and night measured with an electronic nest balance. *Ardea*, 84: (In press.)

King, W.B. (1981). *Endangered birds of the world*. Smithsonian Institution, Washington, D.C.

Kirby, J.S., Clee, C. and Seager, V. (1993). Impact and extent of recreational disturbance to wader roosts on the Dee estuary, some preliminary results. *Wader Study Group Bulletin*, 68, 53–58.

Klomp, H. (1980). Fluctuations and stability in Great Tit populations. *Ardea*, 68, 205–224.

Koehl, M.A.R. (1989). Discussion: from individuals to populations. In *Perspect-ives in ecological theory*, (ed. J. Roughgarden, R.M. May and S.A. Levin), pp. 39–53. Princeton University Press, New Jersey.

Koeman, J.H. (1971). Het voorkomen en de toxicologische betekenis van enkele chloorkoolwaterstoffen aan de Nederlandse kust in de periode van 1965 tot 1970. Ph.D thesis. University of Utrecht.

Koeman, J.H., Bothof, T., Vries, R. de, Velzen-Blad, H. van and Vos, J.G. (1972). The impact of persistent pollutants on piscivorous and molluscivorous birds. *TNO-nieuws*, 27, 561–569.

Koene, P. (1978). De Scholekster: aantalseffecten op de voedselopname. Unpub-lished student thesis. University of Groningen.

Koepff, C. and Dietrich, K. (1986). Störungen von Küstenvögeln durch Wasser-fahrzeuge. *Vogelwarte*, 33, 232–248.

Koopman, K. (1987a). Verschillen in overwinteringsgebied van Friese Scholeksters *Haematopus ostralegus*. *Limosa*, 60, 179–183.

Koopman, K. (1987b). Scholekster-sterfte aan de Friese Waddenkust in februari 1986. *Twirre*, 10, 8–15.

Koopman, K. (1992). Biometrie, gewichtsverloop en handpenrui van een binnen-landse populatie Scholeksters *Haematopus ostralegus*. *Limosa*, 65, 103–108.

Krebs, C.J. (1991). The experimental paradigm and long-term population studies. *Ibis*, 133, 53–58.

Krebs, J.R. and Davies, N.B. (1991). *Behavioural ecology: an evolutionary approach*. (3rd edn). Blackwell Scientific Publications, Oxford.

Krebs, J.R. and Kacelnik, A. (1991). Decision-making. In *Behavioural ecology: an evolutionary approach*, (ed. J.R. Krebs and N.B. Davies), pp. 105–136. Blackwell Scientific Publications, Oxford.

Kristensen, J. (1957). Differences in density and growth in a cockle population in the Dutch Wadden Sea. *Archives Neerlandaises de Zoologie*, 12, 351–453.

Kubiak, T.J., Harris, H.J., Smith, L.M., Schwartz, T.R., Stalling, D.L., Trick, J.A., *et al.* (1989). Microcontaminants and reproductive impairment in the

Forster's Tern on Green Bay, Lake Michigan—1983. *Archives of Environmental Contamination and Toxicology*, 18, 706–727.

Kus, B.E., Ashman, P., Page, G.W. and Stenzel, L.E. (1984). Age-related mortality in a wintering population of Dunlin. *Auk*, 101, 69–73.

Kushlan, J.A. (1978). Nonrigorous foraging by robbing egrets. *Ecology*, 59, 649–653.

Kushlan, J.A. (1979). Short-term energy maximization of egret foraging. *Ecology*, 60, 645–646.

Lack, D. (1954). *Natural regulation of animal numbers.* Oxford University Press, Oxford.

Lack, D. (1968). *Ecological adaptations for breeding in birds.* Methuen and Co., London.

Lambeck, R.H.D. (1991). Changes in abundance, distribution and mortality of wintering Oystercatchers after habitat loss in the Delta area, SW Netherlands. *Acta XX Congressus Internationalis Ornithologici*, 2208–2218.

Lambeck, R.H.D. and Wessel, E.G.J. (1991). Seasonal variation in mortality of the Oystercatcher (*Haematopus ostralegus*). In *Progress report: Delta institute for hydrobiological research*, (ed. C.H.R. Heip and E.S. de Bruijn), pp. 22–24. Koninklijke Nederlandse Akademie van Wetenschappen, Amsterdam.

Lambeck, R.H.D. and Wessel, E.G.J. (1993). A note on Oystercatchers from the Varangerfjord, NE Norway. *Wader Study Group Bulletin*, 66, 74–79.

Lambeck, R.H.D., Hannewijk, A. and Brummelhuis E.B.M. (1988). Een bestandsopname in november 1987 van de Kokkel (*Cerastoderma edule*) op twee platen in de Oosterschelde: mogelijke effekten van visserij. *Mimeographed Internal Report Delta Instituut*, No. **1988–6**, Yerseke.

Lambeck, R.H.D., Sandee, A.J.J. and Wolf, L. de (1989). Long-term patterns in the wader usage of an intertidal flat in the Oosterschelde (SW Netherlands) and the impact of the closure of an adjacent estuary. *Journal of Applied Ecology*, 26, 419–431.

Lambeck, R.H.D., Nieuwenhuize, J. and Liere, J.M. van (1991). Polychlorinated biphenyls in Oystercatchers (*Haematopus ostralegus*) from the Oosterschelde (Dutch Delta area) and the Western Wadden Sea, that died from starvation during severe winter weather. *Environmental Pollution*, 71, 1–16.

Lambeck, R.H.D., Bianki, V.V., Schekkermann, H., Wessel, E.G.J., Herman, P.M.J. and Koryakin, A.S. (1995). Biometrics and migration of Oystercatchers (*Haematopus ostralegus*) from the White Sea Region (NW Russia). *Ringing and Migration*, 16, 140–158.

Lane, B. and Davies, J. (1987). *Shorebirds in Australia.* Nelson, Melbourne.

Langston, W.J. (1978). Accumulation of polychlorinated biphenyls in the cockle *Cerastoderma edule* and the tellin *Macoma balthica*. *Marine Biology*, 45, 265–272.

Larson, S. (1957). The suborder *Charadrii* in arctic and boreal areas during the Tertiary and Pleistocene. *Acta Vertebratica*, 1, 1–84.

Latesteijn, H.C. van and Lambeck, R.H.D. (1986). The analysis of monitoring data with the aid of time-series analysis. *Environmental Monitoring Assessment*, 7, 287–297.

Lauro, B. and Burger, J. (1989). Nest-site selection of American Oystercatchers (*Haematopus palliatus*) in salt marshes. *Auk*, 106, 185–192.

Laursen, K., Gram, I. and Frikke, J. (1984). Trëkkende vandfugle ved det frem-skudte dige ved Højer, 1982. *Danske Vildtundersøgelser*, **37**, 1–36.

Leewis, R.J., Baptist, H.J.M. and Meininger, P.L. (1984). The Dutch Delta area. In *Coastal waders and wildfowl in winter*, (ed. P.R. Evans, J.D. Goss-Custard and H.W. Hale), pp. 253–260. Cambridge University Press.

Legg, K. (1954). Nesting and feeding of the Black Oystercatcher near Monterey, California. *Condor*, **56**, 359–360.

Leopold, M.F., Swennen, C. and Bruijn, L.L.M. de (1989). Experiments on selection of feeding site and food size in Oystercatchers *Haematopus ostralegus* of different social status. *Netherlands Journal of Sea Research*, **23**, 333–346.

Lessels, C.M. (1991). The evolution of life histories. In *Behavioural ecology: an evolutionary approach*, (ed. J.R. Krebs and N.B. Davies), pp. 32–68. Blackwell Scientific Publications, Oxford.

Lessells, C.M. and Stephens, D.W. (1983). Central place foraging: single-prey loaders again. *Animal Behaviour*, **31**, 238–243.

Lévêque, R. (1964). Notes sur la reproduction des oiseaux aux îles Galapagos. *Alauda*, **32**, 5–44.

Levings, S.C., Garrity, S.D. and Ashkenas, L.R. (1986). Feeding rates and prey selection of Oystercatchers in the Pearl Islands of Panama. *Biotropica*, **18**, 62–71.

Lewis, S.A., Becker, P.H. and Furness, R.W. (1993). Mercury levels in eggs, tissues, and feathers of Herring Gulls *Larus argentatus* from the German Wadden Sea Coast. *Environmental Pollution*, **80**, 293–299.

L'Hyver, M.-A. (1985). Intraspecific variation in nesting phenology, clutch size and egg size in the Black Oystercatcher *Haematopus bachmani*. Unpublished M.Sc. thesis. University of Victoria.

Lind, H. (1965). Parental feeding in the Oystercatcher. *Dansk Ornithologisk Forenings Tidsskrift*, **59**, 1–31.

Linders, A.M. (1985). Een inleidend onderzoek naar de relatie tussen de fouragerende Scholekster (*Haematopus ostralegus*) en de getijdecyclus en hat aanbod van mosselen. Unpublished Student thesis. Rijkswaterstaat Deltadienst, Middelburg.

Łomnicki, A. (1978). Individual differences between animals and the natural regulation of their numbers. *Journal of Animal Ecology*, **47**, 461–475.

Łomnicki, A. (1980). Regulation of population density due to individual differences and patchy environment. *Oikos*, **35**, 185–193.

Łomnicki, A. (1982). Individual heterogeneity and population regulation. In *Current problems in sociobiology*, (ed. King's College Sociobiology Group), pp. 153–167. Cambridge University Press.

Loye, J.E. and Zuk, M. (1991). *Bird-parasite interactions: ecology, evolution and behaviour*. Oxford University Press, Oxford.

Lunais, B. (1975). Caractéristique et signification du comportement prédateur de l'Huîtrier-pie *Haematopus ostralegus* sur l'Huître de culture *Crassostrea gigas*. Memoires DEA Ecologie de Université de Tours.

Maagaard, L. and Jensen, K.T. (1994). Prey size selection, intake rate and distribution of staging Oystercatchers (*Haematopus ostralegus*) feeding on an intertidal mussel bed (*Mytilus edulis*). *Ophelia*, Supplement, **6**, 201–215.

MacArthur, R.H. and Pianka, E.R. (1966). An optimal use of a patchy environment. *American Naturalist*, **100**, 603–609.

Makkink, G.F. (1942). Contribution to the knowledge of the behaviour of the Oystercatcher (*Haematopus ostralegus* L.). *Ardea*, 31, 23–74.

Marchant, S. and Higgins, P.J. (ed.) (1993). *Handbook of Australian, New Zealand and Antarctic Birds*. Oxford University Press, Melbourne.

Marcström, V. and Mascher, J.W. (1979). Weights and fat in Lapwings *Vanellus vanellus* and Oystercatchers *Haematopus ostralegus* starved to death during a cold spell in spring. *Ornis Scandinavica*, 10, 235–240.

Martin, A.P. (1991). Feeding ecology of birds on the Swartkops Estuary, South Africa. Unpublished Ph.D. thesis. University of Port Elizabeth.

Martinez, A., Motis, A., Matheu, E. and Llimona, F. (1983). Data on the breeding biology of the Oystercatcher *Haematopus ostralegus* L. in the Ebro Delta. *Ardea*, 71, 229–234.

Masman, D. and Klaassen, M. (1987). Energy expenditure during free flight in trained and free-living Eurasian Kestrels (*Falco tinnunculus*). *Auk*, 104, 603–616.

Maxson, S.J. and Bernstein, N.P. (1984). Breeding season time budget of the Southern Black-backed Gull in Antarctica. *Condor*, 86, 401–409.

May, R.M. (1981). *Theoretical ecology: principles and applications*. Blackwell Scientific Publications, Oxford.

Maynard Smith, J. (1977). Parental investment: a prospective analysis. *Animal Behaviour*, 25, 1–9.

Maynard Smith, J. (1982). *Evolution and the theory of games*. Cambridge University Press, Cambridge.

McGrorty, S., Clarke, R.T., Reading, C.J. and Goss-Custard, J.D. (1990). Population dynamics of the mussel *Mytilus edulis*: density changes and regulation of the population in the Exe estuary, Devon. *Marine Ecology Progress Series*, 67, 157–169.

McKean, J.L. (1978). Some remarks on the taxonomy of Australasian Oystercatchers, *Haematopus* spp. *Sunbird*, 9, 3–6.

McLusky, D.S. and Elliott, M. (1981). The feeding and survival strategies of estuarine molluscs. In *Feeding and survival strategies of estuarine organisms*, (ed. N.V. Jones and W.J. Wolff), pp. 193–216. Plenum Press, New York.

McLusky, D.S., Anderson, F.E. and Wolfe-Murphy, S. (1983). Distribution and recovery of *Arenicola marina* and other benthic fauna after bait digging. *Marine Ecology Progress Series*, 11, 173–179.

McNeil, R., Drapeau, P. and Goss-Custard, J.D. (1992). The occurrence and adaptive significance of nocturnal habits in waterfowl. *Biological Reviews*, 67, 381–419.

Mead, C.J. and Clark, J.A. (1993). Report on bird ringing in Britain and Ireland for 1991. *Ringing and Migration*, 14, 1–72.

Meer, J. van der (1985). De verstoring van vogels op de slikken van de Oosterschelde. *Report Rijkswaterstaat-Deltadienst*, No. 85.09, Middelburg.

Meesters, E. and Münninghoff, M. (1986). Effecten van het gedeeltelijk wegvissen van een mosselbank op de daarop fouragerende Scholeksters (*Haematopus ostralegus*). *Report Rijksinstituut voor Natuurbeheer*, Texel.

Meinertzhagen, R. (1954). *Birds of Arabia*. Oliver and Boyd, London.

Meininger, P.L. and Haperen, van (1988). Vogeltellingen in het zuidelijk Deltagebied in 1984/85–1986/87. Rijkswaterstaat–Dienst Getijdewateren, LNV–NMF, Nota GWAO-88.1010. Middelburg.

Meininger, P.L., Baptist, H.J.M. and Slob, G.J. (1984). Vogeltellingen in het

Deltagebied in 1975/76–1979/80. Rijkswaterstaat Deltadienst en Staatsbosbeheer Zeeland, Nota DDM1-84.23, Middelburg.

Meininger, P.L., Blomert, A.-M. and Marteijn, E.C.L. (1991). Watervogelsterfte in het Deltagebied, ZW-Nederland, gedurende de drie koude winters van 1985, 1986 en 1987. *Limosa*, **64**: 89–102.

Meininger, P.L., Berrevoets, C.M. and Strucker, R.C.W. (1994). Watervogeltellingen in het zuidelijk Deltagebied, 1987–91. *Rijksinstituut voor Kust en Zee*, NIOO-CEMO, Rapport, RIKZ-94.005. Middelburg.

Meire, P.M. (1991). Effects of a substantial reduction in intertidal area on numbers and densities of waders. *Acta XX Congressus Internationalis Ornithologici*, 2219–2235.

Meire, P.M. (1993). Wader populations and macrozoobenthos in a changing estuary: the Oosterschelde (The Netherlands). Report Institute for Nature Conservation No. 93.05, Hasset. Ph.D. thesis. University of Gent.

Meire, P.M. (1996a). Interactions between Oystercatchers (*Haematopus ostralegus*) and mussels (*Mytilus edulis*): implications from optimal foraging theory. *Ardea*, **84**. (In press.)

Meire, P.M. (1996b). Feeding behaviour of Oystercatchers during a period of tidal manipulation. *Ardea*, **84**. (In press.)

Meire, P.M. (1996c). Distribution of Oystercatchers (*Haematopus ostralegus*) over a tidal flat in relation to their main prey species, cockles (*Cerastoderma edule*) and mussels (*Mytilus edulis*): did it change after a substantial habitat loss? *Ardea*, **84**. (In press.)

Meire, P.M. and Ervynck, A. (1986). Are Oystercatchers (*Haematopus ostralegus*) selecting the most profitable mussels (*Mytilus edulis*)? Animal Behaviour, 34, 1427–1435.

Meire, P.M. and Kuyken, E. (1984). Relationships between the distributions of waders and the intertidal benthic fauna of the Oosterschelde. In *Coastal waders and wildfowl in winter*, (ed. P.R. Evans, J.D. Goss-Custard and W.G. Hale), pp. 57–68. Cambridge University Press.

Meire, P.M., Schekkerman, H. and Meininger, P.L. (1994a). Consumption of benthic invertebrates by waterbirds in the Oosterschelde estuary, SW Netherlands. *Hydrobiologia*, **282/283**, 525–546.

Meire, P.M., Seys, J., Buijs, J. and Coosen, J. (1994b). Spatial and temporal patterns of intertidal macrobenthic populations in the Oosterschelde: are they influenced by the construction of the storm-surge barrier? *Hydrobiologia*, **282/283**, 157–182.

Meltofte, H. (1982). Jagtlige forstyrrelser af svømme- og vadefugle. *Dansk Ornithologisk Forenings Tidsskrift*, 76, 21–35.

Meltofte, H. (1993). Vadefugletrækket gennem Danmark. *Dansk Ornithologisk Forenings Tidsskrift*, 87, 1–180.

Meltofte, H., Pihl, S. and Sørensen, B.M. (1972). Efterårstrækket af vadefugle (*Charadrii*) ved Blåvandshuk 1963–1971. *Dansk Ornithologisk Forenings Tidsskrift*, 66, 63–69.

Metcalfe, N.B. (1989). Flocking preferences in relation to vigilance, benefits and aggression costs in mixed-species shorebird flocks. *Oikos*, 56, 91–98.

Mildenberger, H. (1982). *Die Vögel des Rheinlandes*. Vol.1, Kilda Verlag, Greven.

Miller, E.H. (1985). Parental behaviour in the Least Sandpiper (*Calidris minutilla*). *Canadian Journal of Zoology*, 63, 1593–1601.

Miller, E.H. and Baker, A.J. (1980). Displays of the Magellanic Oystercatcher *Haematopus leucopodus*. *Wilson Bulletin*, 92, 149–168.

Minchella, D.J. and Scott, M.E. (1991). Parasitism: A cryptic determinant of animal community structure. *Trends in Ecology and Evolution*, 6, 250–254.

Mitchell, J.R., Moser, M.E. and Kirby, J.S. (1988). Declines in midwinter counts of waders roosting on the Dee estuary. *Bird Study*, 35, 191–198.

Mitchell, W.A. and Valone, T.J. (1990). The optimization research program: studying adaptations by their function. *Quarterly Review of Biology*, 65, 43–52.

Møller, H.S. (1978). Fuglejagten i Danmark 1961–1975. *Proceedings of the First Nordic Congress of Ornithology*. *Anser* (Lund), Suppl. 3, 177–183.

Morrell, S.H., Huber, H.R., Lewis, T.J. and Ainley, D.G. (1979). Feeding ecology of Black Oystercatchers on South Farallon Island, California. *Studies in Avian Biology*, 2, 185–186.

Moser, M.E. (1988). Limits to the number of Grey Plovers *Pluvialis squatarola* wintering on British estuaries: an analysis of long term population trends. *Journal of Applied Ecology*, 25, 473–485.

Myers, J.P. (1984). Spacing behaviour of non-breeding shorebirds. In *Behaviour of marine animals*, vol. VI, (ed. J. Burger), pp. 271–321. Plenum Press, New York.

Myers, J.P., Connors, P.G. and Pitelka, F.A. (1979). Territory size in wintering Sanderlings the effects of prey abundance and intruder density. *Auk*, 96, 551–555.

Nedderman, R.M. (1953). Notes on the nesting of Oystercatchers on Skokholm in 1953. *Skokholm Bird Observatory Report*, 1953, 27–29.

Newman, M. (1989). Oystercatcher affairs. *Stilt*, 15, 9.

Newton, I. (1980). Food limitation in birds. *Ardea*, 68, 3–30.

Newton, I. (ed.) (1989). *Lifetime reproduction in birds*. Academic Press, London.

Newton, I. (1992). Experiments on the limitation of bird numbers by territorial behaviour. *Biological Reviews*, 67, 129–173.

Nice, M.M. (1962) Development of behaviour in precocial birds. *Transactions of the Linnean Society New York*, 8, 1–211.

Nichols, J.D. (1991). Responses of North American duck populations to exploitation. In *Bird population studies*, (ed. C.M. Perrins, J.-D. Lebreton and G.J.M. Hirons), pp. 360–372. Oxford University Press, Oxford.

Nicholls, D.J., Tubbs, C.R.. and Haynes, F.N. (1981). The effect of green algal mats on intertidal macrobenthic communities and their predators. *Kieler Meeresforschungen*, Sonderheft 5, 511–520.

Nienhuis, P.H. and Smaal, A.C. (1994). The Oosterschelde estuary, a case-study of a changing ecosystem: an introduction. *Hydrobiologia*, 282/283, 1–14.

Nol, E. (1984). Reproductive strategies in the Oystercatchers (*Aves: Haematopodidae*). Unpublished Ph.D. thesis. University of Toronto.

Nol, E. (1985). Sex rôles in the American Oystercatcher. *Behaviour*, 95, 232–260.

Nol, E. (1986). Incubation period and foraging technique in shorebirds. *American Naturalist*, 128, 115–119.

Nol, E. (1989). Food supply and reproductive performance of the American Oystercatcher in Virginia. *Condor*, 91, 429–435.

Nol, E., Baker, A.J. and Cadman, M.D. (1984). Clutch initiation dates, clutch size, and egg size of the American Oystercatcher in Virginia. *Auk*, 101, 855–867.

Nordberg, S. (1950). Researches on the bird fauna of the marine zone in the Åland Archipelago. *Acta Zoologica Fennica*, **63**, 1–62.

North, P.M. (ed.) (1987). Ringing recovery analytical methods. *Acta Ornithologica*, **23**, 1–175.

Norton-Griffiths, M.N. (1967). Some ecological aspects of the feeding behaviour of the Oystercatcher *Haematopus ostralegus* on the Edible Mussel *Mytilus edulis*. *Ibis*, **109**, 412–424.

Norton-Griffiths, M.N. (1968). The feeding behaviour of the Oystercatcher (*Haematopus ostralegus*). Unpublished D.Phil. thesis. University of Oxford.

Norton-Griffiths, M. (1969). The organization, control and development of parental feeding in the Oystercatcher (*Haematopus ostralegus*). *Behaviour*, **34**, 55–114.

O'Connor, R.J. (1980). Pattern and process in Great Tit (*Parus major*) populations in Britain. *Ardea*, **68**, 165–184.

O'Connor, R.J. (1984). *The growth and development of birds*. Wiley, Chichester.

O'Connor, R.J. and Brown, R.A. (1977). Prey depletion and foraging strategy in the Oystercatcher *Haematopus ostralegus*. *Oecologia*, **27**, 75–92.

O'Connor, R.J. and Cawthorne, A. (1982). How Britain's birds survived the winter. *New Scientist*, **25**, 786–788.

Ólaffson, E.B. (1988). Inhibition of larval settlement to a soft bottom benthic community by drifting algal mats: an experimental test. *Marine Biology*, **97**, 571–574.

Olson, S.L. and Steadman, D.W. (1979). The fossil record of the *Glareolidae* and *Haematopodidae* (*Aves: Charadriiformes*). *Proceedings of the Biological Society of Washington*, **91**, 972–981.

Orians, G.H. (1969). Age and hunting success in the Brown Pelican (*Pelicanus occidentalis*). *Animal Behaviour*, **17**, 316–319.

Orians, G.H. and N.E. Pearson (1979). On the theory of central place foraging. In *Analysis of ecological systems*, (ed. D.J. Horn, R.D. Mitchell and G.R. Stairs), pp. 154–177. Ohio State University Press, Columbus.

Osborn, D., Every, W.J. and Bull, K.R. (1983). The toxicity of trialkyl lead compounds to birds. *Environmental Pollution* (Ser. A), **31**, 261–275.

Page, G. and Whitacre, D.F. (1975). Raptor predation on wintering shorebirds. *Condor*, **77**, 73–83.

Paine, R.T., Wootton, J.T. and Boersma, P.D. (1990). Direct and indirect effects of Peregrine Falcon predation on seabird abundance. *Auk*, **107**, 1–9.

Parker, G.A. (1974). Assessment strategy and the evolution of animal conflicts. *Journal of Theoretical Biology*, **47**, 223–243.

Parker, G.A. (1982). Phenotype-limited evolutionarily stable strategies. In *Current problems in sociobiology*, (ed. King's College Sociobiology Group), pp. 173–201. Cambridge University Press, Cambridge.

Parker, G.A. and Sutherland, W.J. (1986). Ideal free distributions when individuals differ in competitive ability: phenotype-limited ideal free models. *Animal Behaviour*, **34**, 1222–1242.

Parslow, J.L.F. (1973). Mercury in waders from the Wash. *Environmental Pollution*, **5**, 295–304.

Partridge, L. and Green, P. (1985). Instraspecific feeding specializations and population dynamics. In *Behavioural ecology: ecological consequences of adaptive behaviour*, (ed. R.M. Sibly and R.H. Smith), pp. 207–226. Blackwell Scientific Publications, Oxford.

Pearson,T.H. and Rosenberg, R. (1978). Macrobenthic succession in relation to organic enrichment and pollution of the marine environment. *Oceanography and Marine Biology Annual Review*, **16**, 229–311.

Perdeck, A.C. (1977). The analysis of ringing data: pitfalls and prospects. *Vogelwarte*, **29**, 33–44.

Perrins, C.M. and Moss, D. (1975). Reproductive rates in the Great Tit. *Journal of Animal Ecology*, **44**, 659–706.

Peters, J.L. (1934). *Checklist of the birds of the world*, Vol. II, Harvard University Press.

Peters, W.D. and Grubb, P. (1983). An experimental analysis of sex-specific foraging in the Downy Woodpecker *Dicoides pubescens*. *Ecology*, **64**, 1437–1443.

Phillips, D.J.H. (1977). The use of biological indicator organisms to monitor trace metal pollution in marine and estuarine environments. *Environmental Pollution*, **13**, 281–317.

Picozzi, N. and Catt, D.C. (1989). Early return of Oystercatchers to north-east Scotland in spring 1989. *Scottish Birds*, **15**, 182–183.

Pienkowski, M.W. and Evans, P.R. (1985). The role of migration in the population dynamics of birds. In *Behavioural ecology: ecological consequences of adaptive behaviour*, (ed. R.M. Sibly and R.H. Smith), pp. 331–352. Blackwell Scientific Publications, Oxford.

Piersma, P. (1987). Production by intertidal benthic animals and limits to their predation by shorebirds: a heuristic model. *Marine Ecology Progress Series*, **38**, 187–196.

Piersma, T., Goeij, P. de and Tulp, I. (1993a). An evaluation of intertidal feeding habitats from a shorebird perspective: towards relevant comparisons between temperate and tropical mudflats. *Netherlands Journal of Sea Research*, **31**, 503–512.

Piersma, T., Koolhaas, A. and Dekinga, A. (1993b). Interactions between stomach structure and diet choice in shorebirds. *Auk*, **110**, 552–564.

Piersma, R., Tulp, I. and Schekkerman, H. (1996). Final countdown of waders during starvation: terminal use of nutrients in relation to structural size and concurrent energy expenditure. *Ardea*, **83**. (In press.)

Pieters, T., Storm, C., Walhout, V.T. and Ysebaert, T. (1991). Het Schelde estuarium, méér dan een vaarweg. Rijkswaterstaat, Nota GWWS-91.081, Middelburg.

Pilcher, R.E.M. (1964). Effects of the cold winter of 1962–63 on birds of the north coast of the Wash. *Wildfowl Trust Annual Report*, **15**, 23–26.

Piper, W.H. and Wiley, R.H. (1989). Correlates of dominance in wintering White-throated Sparrows: age, sex and location. *Animal Behaviour*, **37**, 298–310.

Pleines, S. (1990). Siedlungsdichte, Brutbiologie und Populationsentwicklung des Austernfischers (*Haematopus ostralegus*) auf der Nordseeinsel Mellum. Unpublished Diploma thesis. University of Köln.

Polman, G.J. (1988). Het effect van een experimenteel aangelegde mosselbank op dichtheid, prooikeuze en agressief gedrag van Scholeksters. Unpublished student thesis. University of Groningen.

Prater, A.J. (1981). *Estuary birds of Britain and Ireland*. Poyser, Carlton.

Price, T. (1987). Diet variation in a population of Darwin's Finches. *Ecology*, 68, 1015–1028.

Prŷs-Jones, R.P., Underhill, L.G. and Waters, R.J. (1994). Index numbers for waterbird populations. II. Coastal wintering waders in the United Kingdom, 1970/71–1990/91. *Journal of Applied Ecology*, 31, 481–492.

Purdy, M.A. (1985). Parental behaviour and rôle differentiation in the Black Oystercatcher *Haematopus bachmani*. Unpublished M.Sc. thesis, University of Victoria.

Purdy, M.A. and Miller, E.H. (1988). Time budget and parental behaviour of breeding American Black Oystercatchers (*Haematopus bachmani*) in British Columbia. *Canadian Journal of Zoology*, 66, 1742–1751.

Quinn, J.L. and Kirby, J.S. (1993). Oystercatchers feeding on grasslands and sandflats in Dublin Bay. *Irish Birds*, 5, 35–44.

Radach, G. and Moll, A. (1990). State-of-the-art in algal bloom modelling. *Water Pollution Research Report*, 12, 115–149.

Raffaelli, D., Hull, S. and Milne, H. (1989). Long-term changes in nutrients, weed mats and shorebirds in an estuarine system. *Cahier Biologique Marine*, 30, 259–270.

Rands, M.R.W. and Barkham, J.P. (1981). Factors controlling within-flocks feeding densities in three species of wading bird. *Ornis Scandinavica*, 12, 28–36.

Rankin, G.D. (1979). Aspects of the breeding biology of wading birds (Charadrii) on a saltmarsh. Unpublished Ph.D. thesis. University of Durham.

Reading, C.J. and McGrorty, S. (1978). Seasonal variations in the burying depth of *Macoma balthica* (L.) and its accessibility to wading birds. *Estuarine and Coastal Marine Science*, 6, 135–144.

Rehfisch, M.M., Langston, R.H.W., Clark, N.A. and Forrest, C. (1993). A guide to the provision of refuges for waders. An analysis of Wash Wader Ringing Group data. *BTO Research Report*, 120, Thetford.

Reijnders, P.J.H. (1986). Reproductive failure in Common Seals feeding on fish from polluted waters. *Nature*, 324, 456–457.

Revier, H. (1993). Mosselen en kokkels duur betaald. *Waddenbulletin*, 93, 150–153.

Ricklefs, R.E. (1967). A graphical method of fitting equations to growth curves. *Ecology*, 48, 978–983.

Ricklefs, R.E. (1968). Patterns of growth in birds. *Ibis*, 110, 419–451.

Ricklefs, R.E. (1969). Preliminary models for growth rates of altricial birds. *Ecology*, 50, 1031–1039.

Ricklefs, R.E. (1983). Avian postnatal development. In *Avian Biology*, Vol. VII, (ed. J.R. King and K.C. Parkes), pp. 1–83. Academic Press, New York.

Ridgway, R. (1919). The birds of North and Middle America. *Bulletin of the United States National Museum*, 50, 8.

Ritchie, M.R. (1988). Individual variation in the ability of Columbian Ground Squirrels to select an optimal diet. *Evolutionary Ecology*, 2, 232–252.

Rittinghaus, H. (1950). Über das Verhalten eines vom Sandregenpfeifer (*Ch. hiaticula*) ausgebrüteten und geführten Seeregenpfeifers (*Ch. alexandrinus*). *Vogelwarte*, 15, 187–192.

Rohwer, S.A. and Ewald, P.H. (1981). The cost of dominance and the advantage of subordination in a badge signalling system. *Evolution*, 35, 441–454.

Rohwer, S. and Rohwer, F.C. (1978). Status signalling in Harris Sparrows: experimental deceptions achieved. *Animal Behaviour*, 26, 1012–1022.

Rosenberg, R. (1985). Eutrophication—the future marine coastal nuisance. *Marine Pollution Bulletin*, 16, 227–231.

Roughgarden, J. (1972). Evolution of niche width. *American Naturalist*, 106, 683–718.

Roughgarden, J. (1974). Niche width: biogeographic patterns among *Analis* lizard populations. *American Naturalist*, 108, 429–442.

Runde, O.J. (1984). Bird ringing report 1981, Stavanger Museum. *Sterna*, 17, 129–154.

Runde, O.J. (1985). Bird ringing report 1982, Stavanger Museum. *Sterna*, 17, 157–186.

Runde, O.J. (1987). Bird ringing report 1983, Stavanger Museum. *Sterna*, 17, 197–224.

Runde, O.J. and Barrett, R.T. (1981). Variation in egg size and incubation period in the Kittiwake *Rissa tridactyla* in Norway. *Ornis Scandinavica*, 12, 80–86.

Rydzewski, W. (1978). The longevity of ringed birds. *Ring*, 96, 218–270.

Saeijs, H.L.F. and Baptist, H.J.M. (1977). Vogels in de Deltawateren van Zuid-West Nederland. Overzicht simultaantellingen. Deltadienst Rijkswaterstaat - Hoofdafdeling Milieu en Inrichting, Nota 77–34, Middelburg.

Saeijs, H.L.F. and Baptist, H.J.M (1980). Coastal engineering and European wintering wetland birds. *Biological Conservation*, 17, 63–83.

Safriel, U.N. (1967). Population and food study of the Oystercatcher. Unpublished D. Phil. thesis. University of Oxford.

Safriel, U.N. (1975). On the significance of clutch size in nidifugous birds. *Ecology*, 56, 703–708.

Safriel, U.N. (1981). Social hierarchy among siblings in broods of the Oystercatcher *Haematopus ostralegus*. *Behavioural Ecology and Sociobiology*, 9, 59–63.

Safriel, U.N. (1982). Effects of disease on social hierachy of young Oystercatchers. *British Birds*, 75, 365–369.

Safriel, U.N. (1985). 'Diet dimorphism' within an Oystercatcher *Haematopus ostralegus* population—adaptive significance and effects on recent distribution dynamics. *Ibis*, 127, 287–305.

Safriel, U.N. and Harris, M.P. (1985). Spatial and seasonal distribution of 'dermatitis', a disease of Oystercatchers on Skokholm and its ecological effects. *Acta XVIII Congressus Internationalis Ornithologici*, 1168.

Safriel, U.N., Harris, M.P., Brooke, M. de L. and Britton, C.K. (1984). Survival of breeding Oystercatchers *Haematopus ostralegus*. *Journal of Animal Ecology*, 53, 867–877.

Salomonsen, F. (1955). The evolutionary significance of bird-migration. *Dansk Biologisk Meddelelser*, 22, 1–62.

Scheiffarth, G. (1989). Aktivitäts- und Verhaltensmuster des Austernfischers *Haematopus ostralegus* L. zur Brutzeit, unter besonderer Berücksichtigung des Territorial- und Nahrungssucheverhaltens. Unpublished diploma thesis. University of Köln.

Schekkerman, H., Meininger, P.L. and Meire, P.M. (1994). Changes in the waterbird populations of the Oosterschelde, SW Netherlands, as a result of large scale coastal engineering works. *Hydrobiologia*, 282/283, 509–524.

Schnakenwinkel, G. (1970). Studien an der Population des Austernfischers auf Mellum. *Vogelwarte*, **25**, 336–355.

Schönwetter, M. (1962). *Handbuch der Oologie*. Akademie Verlag, Berlin.

Sears, J. (1988). Assessment of body condition in live birds: measurements of protein and fat reserves in the Mute Swan, *Cygnus olor*. *Journal of Zoology, London*, **216**, 295–308.

Seed, R. (1968). Factors influencing shell shape in the mussel *Mytilus edulis*. *Journal of the Marine Biological Association of the United Kingdom*, **48**, 561–584.

Seys, J., Meire, P.M., Coosen, J.C. and Craeymeersch, J.A.M. (1994). Long-term changes (1979–89) in the intertidal macrozoobenthos of the Ooster-schelde estuary: are patterns in total density, biomass and diversity induced by the construction of the storm-surge barrier? *Hydrobiologia*, **282/283**, 251–264.

Sharrock, J.T.R. (1976). *The atlas of breeding birds in Britain and Ireland*. Poyser, Berkhamsted.

Shields, W.M. (1977). The social significance of avian winter plumage variability: a comment. *Evolution*, **31**, 905–907.

Shkedy, Y. and Safriel, U.N. (1992). Nest predation and nestling growth rate of two lark species in the Negev Desert, Israel. *Ibis*, **134**, 268–272.

Sibley, C.G. and Ahlquist, J.E. (1990). *Phylogeny and classification of birds: a study in molecular evolution*. Yale University Press, New Haven.

Sibley, C.G. and Monroe, B.L. jr. (1990). *Distribution and taxonomy of birds of the world*. Yale University Press, New Haven.

Sibly, R.M. and Smith, R.H. (1985). *Behavioural ecology: ecological consequences of adaptive behaviour*. Blackwell Scientific Publications, Oxford.

Sibson, R.B. (1966). Increasing numbers of South Island Pied Oystercatchers visiting northern New Zealand. *Notornis*, **13**, 94–97.

Siegfried, W.R. (1994). *Rock shores: exploitation in Chile and South Afrika*. Springer, Berlin.

Sinclair, A.R.E. (1989). Population regulation in animals. In *Ecological concepts: the contribution of ecology to an understanding of the natural world*, (ed. J.M. Cherrett), pp. 197–241. Blackwell Scientific Publications, Oxford.

Skeel, M.A. (1978). Vocalizations of the Whimbrel on its breeding grounds. *Condor*, **80**, 194–202.

Smit, C.J. and Piersma, T. (1989). Numbers, midwinter distribution and migration of wader populations using the East Atlantic Flyway. In *Flyways and reserve networks for waterbirds*, (ed. H. Boyd and J.-Y. Pirot), *IWRB Special Publication*, 9, pp. 24–63.

Smit, C.J. and Visser, G.J.M., (1993). Effects of disturbance on shorebirds: a summary of existing knowledge from the Dutch Wadden Sea and Delta area. *Wader Study Group Bulletin*, **68**, 6–19.

Smit, T., Bakhuizen, T. and Moraal, L.G. (1988). Metallisch lood als bron van loodvergiftiging in Nederland. *Limosa*, **61**, 175–178.

Soulsby, P.G., Lowthion, D. Houston, M. and Montgomery, H.A.C. (1985). The role of sewage effluent in the accumulation of macroalgal mats on intertidal mudflats in two basins in southern England. *Netherlands Journal of Sea Research*, **19**, 257–263.

Speakman, J.R. (1984). The energetics of foraging in wading birds (Charadrii). Unpublished Ph.D. thesis. University of Stirling.

Speakman, J.R. (1987). Apparent absorption efficiencies of Redshank and Oystercatcher: implications for the predictions of optimal foraging models. *American Naturalist*, **130**, 677–691.

Stearns, S. (1992). *The evolution of life histories*. Oxford University Press.

Stephens, D.W. and Krebs, J.R. (1986). Foraging theory. Princeton University Press, New Jersey

Stock, M., Leopold, M.F. and Swennen, C. (1987a). Rastverhalten, Revierbesetzung und Siedlungsdichte des Austernfischers—*Haematopus ostralegus*—auf der Hallig Langeness (Schleswig-Holstein, BRD). *Ökologie der Vögel*, **9**, 31–45.

Stock, M., Strotmann, J., Witte, H. and Nehls, G. (1987b). Jungvögel sterben im harten Winter zuerst: Winterverluste beim Austernfischer, *Haematopus ostralegus*. *Journal für Ornithologie*, **128**, 325–331.

Stock, M., Herber, R.F.M. and Gerón, H.M.A. (1989). Cadmium levels in Oystercatchers *Haematopus ostralegus* from the German Wadden Sea. *Marine Ecology Progress Series*, **53**, 227–234.

Stralen, M.R. van, Kesteloo-Hendrikse, J.J. and Brand, C.M. (1991). Bestandsgrootte en visserijmortaliteit van kokkels in de Oosterschelde in 1989. *Report Rijksinstituut voor Visserijonderzoek*, AQ 91–02, IJmuiden.

Strauch, J.G. (1978). The phylogeny of the *Charadriiformes* (*Aves*): a new estimate using the method of character compatibility analysis. *Transactions of the Zoological Society of London*, **34**, 263–345.

Summers, R.W. and Cooper, J. (1977). The population, ecology and conservation of the Black Oystercatcher *Haematopus moquini*. *Ostrich*, **48**, 28–40.

Sutherland, W.J. (1982a). Do Oystercatchers select the most profitable cockles? *Animal Behaviour*, **30**, 857–861.

Sutherland, W.J. (1982b). Spatial variation in the predation of cockles by Oystercatchers at Traeth Melynog, Anglesy. I. The cockle population. *Journal of Animal Ecology*, **51**, 481–489.

Sutherland, W.J. (1982c). Spatial variation in the predation of cockles by Oystercatchers at Traeth Melynog, Anglesey. II. The pattern of mortality. *Journal of Animal Ecology*, **51**, 491–500.

Sutherland, W.J. (1982d). Food supply and dispersal in the determination of wintering population levels of Oystercatchers, *Haematopus ostralegus*. *Estuarine and Coastal Shelf Science*, **14**, 223–229.

Sutherland, W.J. (1983). Aggregation and the ideal free distribution. *Journal of Animal Ecology*, **52**, 821–828

Sutherland, W.J. (1987). Why do animals specialize? *Nature*, **325**, 483–484.

Sutherland, W.J. (1992). Game theory models of functional and aggregative responses. *Oecologia*, **90**, 150–152

Sutherland, W.J. (1996). *From individual behaviour to population ecology*. Oxford University Press, Oxford.

Sutherland, W.J. and Anderson, C.W. (1987). Six ways in which a foraging predator may encounter options with different variances. *Biological Journal of the Linnean Society*, **30**, 99–114.

Sutherland, W.J. and Ens, B.J. (1987). The criteria determining the selection of mussels *Mytilus edulis* by Oystercatchers *Haematopus ostralegus*. *Behaviour*, **103**, 187–202.

Sutherland, W.J. and Koene, P. (1982). Field estimates of the strength of interference between Oystercatchers *Haematopus ostralegus*. *Oecologia*, 55, 108–109.

Sutherland, W.J. and Parker, G.A. (1985). Distribution of unequal competitors. In *Behavioural ecology: ecological consequences of adaptive behaviour*, (ed. R.M. Sibly and R.H. Smith), pp. 255–274. Blackwell Scientific Publications, Oxford.

Swennen, C. (1972). Chlorinated hydrocarbons attacked the Eider population in the Netherlands. *TNO Nieuws*, 27, 556–560.

Swennen, C. (1984). Differences in quality of roosting flocks of Oystercatchers. In *Coastal waders and wildfowl in winter*, (ed. P.R. Evans, J.D. Goss-Custard and W.G. Hale), pp. 177–189. Cambridge University Press, Cambridge.

Swennen, C. (1990). Oystercatchers feeding on Giant Bloody Cockles on the Banc d'Arguin, Mauritania. *Ardea*, 78, 53–62.

Swennen, C. (1991). Ecology and population dynamics of the Common Eider in the Dutch Wadden Sea. Ph.D. thesis. University of Groningen.

Swennen, C. and Baan, G. van der (1959). Tracking birds on tidal flats and beaches. *British Birds*, 52, 15–18.

Swennen, C. and Bruijn, L.L.M. de (1980). De dichtheid van broedterritoria van de Scholekster *Haematopus ostralegus* op Vlieland. *Limosa*, 53, 85–90.

Swennen, C. and Ching, H.L. (1974). Observations on the trematode *Parvatrema affinis*, causative agent of crawling tracks of *Macoma balthica*. *Netherlands Journal of Sea Research*, 8, 108–115.

Swennen, C. and Duiven, P. (1983). Characteristics of Oystercatchers killed by cold-stress in the Dutch Wadden Sea Area. *Ardea*, 71, 155–159.

Swennen, C., Bruijn, L.L.M. de, Duiven, P., Leopold, M.F. and Marteijn, E.C.L. (1983). Differences in bill form of the Oystercatcher *Haematopus ostralegus*; a dynamic adaptation to specific foraging techniques. *Netherlands Journal of Sea Research*, 17, 57–83.

Swennen, C., Leopold, M.F. and Bruijn, L.L.M. de (1989). Time-stressed Oystercatchers, *Haematopus ostralegus*, can increase their intake rate. *Animal Behaviour*, 38, 8–22.

Temme, M. and Ger·, W. (1988). Masse, Gewichte und mögliche Todesursachen der im Januar 1987 auf Norderney verendeten Austernfischer (*Haematopus ostralegus*). *Seevögel*, 9, 63–69.

Tensen, D. and Zoest, J. van (1983). Keuze van hoogwatervluchtplaatsen op Terschelling. Unpublished student thesis. Agricultural University Wageningen/Rijksinstituut voor Natuurbeheer, Texel.

Thelle, T. (1970). Traekket af Strandskade *Haematopus ostralegus* frå Vestnorge til Vadehavet. *Dansk Ornithologisk Forenings Tidsskrift*, 64, 229–247.

Thingstad, P. (1978). Observasjoner av tjeld fra Rinnleiret. *Vår Fuglefauna*, 1, 66–68.

Thompson, D.B.A. (1986). The economics of kleptoparasitism: optimal foraging, host and prey selection by gulls. *Animal Behaviour*, 34, 1189–1205.

Thouless, C.R. and Guinness, F.E. (1986). Conflict between red deer hinds: the winner always wins. *Animal Behaviour*, 34, 1166–1171.

Tinbergen, J.M. and Daan, S. (1990). Family planning in the Great Tit (*Parus major*): optimal clutch size as integration of parent and offspring fitness. *Behaviour*, 114, 161–190

Tinbergen, N. and Norton-Griffiths, M. (1964). Oystercatchers and mussels. *British Birds*, 57, 64–70.

Toft, G.O. (1982). Sandeel, *Ammodytidae*, a rare prey taken by the Oyster-catcher *Haematopus ostralegus*. *Fauna Norvegica*, Series C (Cinclus), 5, 95–96.

Townshend, D.J. and O'Connor, D.A. (1993). Some effects of disturbance to waterfowl from bait-digging and wildfowling at Lindisfarne National Nature Reserve, north-east England. *Wader Study Group Bulletin*, 68, 47–52.

Triplet, P. (1989a). Comparaison entre deux stratégies de recherche alimentaire de L'Huîtrier-pie *Haematopus ostralegus* en Baie de Somme. Influence des facteurs de l'environment. Unpublished Ph.D. thesis. University of Paris.

Triplet, P. (1989b). Sélectivité alimentaire liée à l'age chez l'Huîtrier-pie *Haematopus ostralegus* consummateur de *Nereis diversicolor* en baie de Somme. *Gibier Fauna Sauvage*, 6, 427–436.

Triplet, P. (1990). Selection de la taille des proies et de la zone alimentaire chez l'Huîtrier-pie *Haematopus ostralegus* consommateur de cocques *Cerasto-derma edule*. Unpublished report. Bureau de Chasse, Paris.

Triplet, P. and Etienne, P. (1991). L'Huitrier-pie (*Haematopus ostralegus*) face à une diminution de sa principale ressource alimentaire la Cocque (*Cerasto-derma edule*) en baie de Somme. *Bulletin Mensuel Office National de la Chasse*, 153, 21–28.

Triplet, P., Debacker, F. and Noyon, C. (1987). Origine et distribution des Huitriers-pies *Haematopus ostralegus* repris en France. *Bulletin Mensuel Office National de la Chasse*, 116, 38–43.

Trivers, R.L. (1972). Parental investment and sexual selection. In *Sexual selection and the descent of man*, (ed. B. Campbell), pp. 136–179. Aldine Publishing Company, Chicago.

Turpie, J.K. (1994). Comparative foraging ecology of two broad-ranging migrants, Grey Plover *Pluvialis squatarola* and Whimbrel *Numenius phaeopus* (*Aves: Charadrii*), in tropical and temperate latitudes of the western Indian Ocean. Unpublished Ph.D. thesis. University of Cape Town.

Urban, E.K., Fry, C.H. and Keith, S. (ed.) (1986). *The birds of Africa*, Vol. II, Academic Press, London.

Vader, W.J.M. (1964). A preliminary investigation into the reactions of the infauna of the tidal flats to tidal fluctuations in water level. *Netherlands Journal of Sea Research*, 2, 189–222.

Väisänen, R.A. (1977). Geographic variation in the timing of breeding and egg size in eight European species of waders. *Annales Zoologici Fennici*, 14, 1–25.

Varley, G.C. and Gradwell, G.R. (1960). Key factors in population studies. *Journal of Animal Ecology*, 29, 339–401.

Varley, G.C., Gradwell, G.R. and Hassell, M.P. (1973). *Insect population ecology*. Blackwell Scientific Publications, Oxford.

Veenstra, J. (1977). Het fourageren van de Scholekster (*Haematopus ostralegus*) in het binnenland. Unpublished student thesis. University of Groningen.

Vehrencamp, S.L., Bradburry, J.W. and Gibson, R.M. (1989). The energetic cost of display in male sage grouse. *Animal Behaviour*, 38, 885–896.

Vickery, W.L., Giraldeau, L.-A., Templeton, J.J., Kramer, D.L. and Chapman, C.A. (1991). Producers, scroungers and group foraging. *American Naturalist*, 137, 842–863.

Vines, G. (1979). Spatial distributions of territorial aggressiveness in Oyster-catchers, *Haematopus ostralegus* L. *Animal Behaviour*, 27, 300–308.

Vines, G. (1980). Spatial consequences of aggressive behaviour in flocks of Oystercatchers *Haematopus ostralegus* L. *Animal Behaviour*, **28**, 1175–1183.

Vlas, S.J. de, Bunskoeke, E.J., Ens, B.J. and Hulscher, J.B. (1996). Tidal changes between *Nereis diversicolor* and *Macoma balthica* as main prey species of the Oystercatcher *Haematopus ostralegus*. *Ardea*, **84**. (In press.)

Voous, K.H. (1965). *Atlas van de Europese vogels*. Elsevier, Amsterdam.

Wakefield, W.C. and Robertson, B.I. (1988). Breeding resource partitioning of a mixed population of Pied and Sooty Oystercatchers. *Stilt*, **13**, 39–40.

Wanink, J.H. and Zwarts, L. (1985). Does an optimally foraging Oystercatcher obey the functional response? *Oecologia*, **67**, 98–106.

Wanink, J.H. and Zwarts, L. (1993). Environmental effects on the growth rate of intertidal invertebrates and some implications for foraging waders. *Netherlands Journal of Sea Research*, **31**, 407–418.

Wanink, J.H. and Zwarts, L. (1996). Can food specialization by individual Oystercatchers be explained by differences in prey specific handling efficiencies? *Ardea*, **84**. (In press.)

Ward, D. (1993). African Black Oystercatchers (*Haematopus moquini*) feeding on Wedge Clams (*Donax serra*): the effects of non-random prey availability in the intertidal on the predictions of an optimal diet model. *Ethological and Ecological Evolution*, **5**, 457–466.

Warheit, K.I., Lindberg, D.R. and Boekelheide, R.J. (1984). Pinniped disturbance lowers reproductive success of Black Oystercatcher *Haematopus bachmani*. *Marine Ecology Progress Series*, **17**, 101–104.

Watson, A. (1980). Starving Oystercatchers in Deeside after severe snowstorm. *Scottish Birds*, **11**, 55–56.

Watt, J. (1955). Territory threat display of the Black Oystercatcher (*Haematopus unicolor unicolor*). *Notornis*, **6**, 175.

Webster, J.D. (1941a). Feeding habits of the Black Oystercatcher. *Condor*, **43**, 175–180.

Webster, J.D. (1941b). The breeding of the Black Oystercatcher. *Wilson Bulletin*, **53**, 141–156.

Wells, G.P. (1966). The Lugworm (*Arenicola*)—a study in adaptation. *Netherlands Journal of Sea Research*, **3**, 294–313.

Weston, D.P. (1990). Quantitative examination of macrobenthic community changes along an organic enrichment gradient. *Marine Ecology Progress Series*, **61**, 233–244.

Wetmore, A. (1965). *The birds of the Republic of Panama*, Part I. Smithsonian Institution, Washington DC.

Whitfield, D.P. (1985a). Social organisation and feeding behaviour of wintering Turnstone (*Arenaria interpres*). Unpublished Ph.D. thesis. University of Edinburgh.

Whitfield, D.P. (1985b). Raptor predation on wintering waders in southeast Scotland. *Ibis*, **127**, 544–558.

Whitfield, D.P. (1988). The social significance of plumage variability in wintering Turnstones *Arenaria interpres*. *Animal Behaviour*, **36**, 408–415.

Whitfield, D.P. (1990). Individual feeding specialization of wintering Turnstone *Arenaria interpres*. *Journal of Animal Ecology*, **59**, 193–212.

Whitfield, D.P., Evans, A.D. and Whitfield, P.A. (1988). The impact of raptor predation on wintering waders. *Acta XIX Congressus Internationalis Ornithologici*, 674–687.

Whitlock, R.J. (1979). The ecological significance of energy conservation during roosting for wading birds. Unpublished student thesis. University of Sterling.

Whittow, G.C. (1986). Energy metabolizm. In *Avian Physiology*, (ed. P.D. Sturkie), pp. 253–268. Springer Verlag, New York.

Widdows, J., Fieth, P. and Worrall, C.M. (1979). Relationships between seston, available food and feeding activity in the common mussel *Mytilus edulis*. *Marine Biology*, 50, 195–207.

Wiersma, P. and Piersma, T. (1994). Effects of microhabitat, flocking, climate and migratory goal on energy expenditure in the annual cycle of Red Knots. *Condor*, 96, 257–279.

Wiersma, P., Bruinzeel, L. and Piersma, T. (1993). Energiebesparing bij wadvogels: over de kieren van de kanoet. *Limosa*, 66, 41–52.

Wiley, R.H. (1981). Social structure and individual ontogenies: problems of description, mechanism, and evolution. In *Perspectives in ethology*, Vol. IV, (ed. P.P.G. Bateson and P.H. Klopfer), pp. 261–293. Plenum Press, New York.

Wilson, J.R. (1978). Agricultural influences on waders nesting on the South Uist machair. *Bird Study*, 25, 198–206.

Wilson, J.R. (1982). The wintering of shorebirds in Iceland. *Wader Study Group Bulletin*, 36, 16–19.

Wilson, J.R. and Morrison, R.I.G. (1981). Primary moult in Oystercatchers in Iceland. *Ornis Scandinavica*, 12, 211–215.

Witter, M.S. and Cuthill, I. C. (1993). The ecological costs of avian fat storage. *Philosophical Transactions of the Royal Society of London B*, 340, 73–92.

Woldhek, S. (1979). *Bird killing in the Mediterranean*. Vogelbescherming, Zeist.

Wolff, W.J. (1967). Watervogeltellingen in het gehele Nederlandse Deltagebied. *Limosa*, 40, 216–225.

Wolff, W.J. (1969). Distribution of nonbreeding waders in an estuarine area in relation to the distribution of their food organisms. *Ardea*, 57, 1–28.

Wolff, W.J., van Haperen, A.M.M., Sandee, A.J.J., Baptist, H.J.M. and Saeijs, H.L.F. (1976). The trophic role of birds in the Grevelingen estuary, as compared to their role in the saline Lake Grevelingen. In *Proceedings 10th European Symposium on Marine Biology*, (ed. G. Persoone and E. Jaspers), pp. 673–689. Universa Press, Wetteren.

Woods, R.W. (1975). *The birds of the Falkland Islands*. Nelson, Oswestry.

Ynsen, F. (1991). Karaktergetallen van de winters vanaf 1707. *Zenit*, 18, 69–73.

Zack, S. and Stutchbury, B.J. (1992). Delayed breeding in avian social systems: the role of territory quality and 'floater' tactics. *Behaviour*, 123, 194–219.

Zapata, A.R.P. (1967). Observaciones sobre aves de Puerto Deseado, Provincia de Santa Cruz. *Hornero*, 10, 351–378.

Zegers, P.M. (1973). Invloed van verstoringen op het gedrag van wadvogels. *Waddenbulletin*, 8, 3–7.

Zwarts, L. (1978). Intra- and inter-specific competition for space in estuarine bird species in a one-prey situation. *Acta XVI Congressus Internationalis Ornithologici*, 1045–1050.

Zwarts, L. (1985). The winter exploitation of Fiddler Crabs *Uca tangeri* by waders in Guinea-Bissau. *Ardea*, 73, 3–12.

Zwarts, L. (1988). De bodemfauna van de fries-groningse waddenkust. *Flevobericht*, nr. **294**, Rijksdienst IJsselmeerpolders.

Zwarts, L. (1991). Seasonal variation in body weight of the bivalves *Macoma balthica, Scrobicularia plana, Mya arenaria*, and *Cerastoderma edule* in the Dutch Wadden Sea. *Netherlands Journal Sea Research*, **28**, 231–245.

Zwarts, L. and Blomert, A.-M. (1990). Selectivity of Whimbrels feeding on Fiddler Crabs explained by component specific digestibilities. *Ardea*, **78**, 193–208.

Zwarts, L. and Blomert, A.-M. (1992). Why Knot *Calidris canutus* take medium-sized *Macoma balthica* when six prey species are available. *Marine Ecology Progress Series*, **83**, 113–128.

Zwarts, L. and Blomert, A.-M. (1996). Daily metabolized energy consumption of Oystercatchers (*Haematopus ostralegus*) feeding on larvae of the cranefly *Tipula paludosa*. *Ardea*, **84**, (In press.)

Zwarts, L. and Dirksen, S. (1990). Digestive bottleneck limits the increase in food intake of Whimbrels preparing to migrate from the Banc d'Arguin, Mauritania. *Ardea*, **78**, 257–278.

Zwarts, L. and Drent, R.H. (1981). Prey depletion and the regulation of predator density: Oystercatchers (*Haematopus ostralegus*) feeding on mussels (*Mytilus edulis*). In *Feeding and survival strategies of estuarine organisms*, (ed. N.V. Jones and W.J. Wolff), pp. 193–216. Plenum Press, New York.

Zwarts, L. and Esselink, P. (1989). Versatility of male Curlews *Numenius arquata* preying upon *Nereis diversicolor*: deploying contrasting capture modes dependent on prey availability. *Marine Ecology Progress Series*, **56**, 255–269.

Zwarts, L. and Wanink, J. (1984). How Oystercatchers and Curlews successfully deplete clams. In *Coastal waders and wildfowl in winter*, (ed. P.R. Evans, J.D. Goss-Custard and W.G. Hale), pp. 69–83. Cambridge University Press.

Zwarts, L. and Wanink, J. (1989). Siphon size and burying depth in deposit- and suspension-feeding benthic bivalves. *Marine Biology*, **100**, 227–240.

Zwarts, L. and Wanink, J. (1991). The macrobenthos fractions accessible to waders may represent marginal prey. *Oecologia*, **87**, 581–587.

Zwarts, L. and Wanink, J. (1993). How the food supply harvestable by waders in the Wadden Sea depends on the variation in energy content, body weight, biomass, burying depth and behaviour of tidal-flat invertebrates. *Netherlands Journal of Sea Research*, **31**, 441–476.

Zwarts, L. and Wanink, J.H. (1996). Oystercatchers *Haematopus ostralegus* harvest their bivalve prey when the yield is maximal. *Ardea*, **84**. (In press.)

Zwarts, L., Blomert, A.-M., Ens, B.J., Hupkes, R. and Spanje, T.M. van (1990a). Why do waders reach high feeding densities on the intertidal flats of the Banc d'Arguin, Mauritania? *Ardea*, **78**, 39–52.

Zwarts, L., Ens, B.J., Kersten, M. and Piersma, T. (1990b). Moult, mass and flight range of waders ready to take off for long-distance migration. *Ardea*, **78**, 339–364.

Zwarts, L., Ens, B.J., Goss-Custard, J.D., Hulscher, J.B. and Durell, S.E.A. le V. dit (1996a). Causes of variation in prey profitability and its consequences for the intake rate of the Oystercatcher *Haematopus ostralegus*. *Ardea*, **84**. (In press.)

Zwarts, L., Hulscher, J.B., Koopman, K., Piersma, T. and Zegers, P.M. (1996b).

Seasonal and annual variation in body weight, nutrient stores and mortality of Oystercatchers *Haematopus ostralegus*. *Ardea*, **84**. (In press.)

Zwarts. L., Hulscher, J.B., Koopman, K. and Zegers, P.M. (1996c). Discriminating the sex of Oystercatchers *Haematopus ostralegus*. *Ardea*, **84**. (In press.)

Zwarts, L., Hulscher, J.B. and Zegers, P.M. (1996d). Weight loss in Oystercatchers on the high water roost. *Ardea*, **84**. (In press.)

Index

abnormalities, anatomical 138, 139, 140
'active' prey size selection 39–50
Afrotropical Oystercatchers 252, 253
  see also Haematopus moquini
age
  body weight and 148
  dominance and 98–9, 101, 102
  of first breeding 189, 193, 196–7, 258–9,
    277
  length of summer break 192
  specialization and 66, 69, 72–3
  survival in winter 138–39
aggregation 78–80
  response 111–12, 116–29 *passim*
aggression 85–7, 195; *see also* fighting
agriculture, intensification of 359–60
airplanes 306
algal bloom 291
algal mats 291–2
Allolobophora chlorotica 236; *see also*
  earthworms
alternative food supply 143–5, 153, 333,
  344–5, 347
altricial condition 219–20
  see also parent-feeding
Amager 171, 172
anti-predator behaviour 271
Arenicola (lugworms) 8, 9, 14
  bait-digging 297, 304
  specialization in diet 57, 58
Atlantic breeding population 156, 355
  midwinter distribution 158–61, 162
  population dynamics 362, 364–77 *passim*,
    380
  spring migration 174
attack distance, threshold 92–3
Aulacomya ater 266
Australasian Oystercatchers 252–3; *see also*
  under individual names
autumn
  feeding location 114–15, 123–7
  timing of autumn migration 171–3
avoidance behaviour 90, 93–5

bait-digging 297, 304–5
beaches 7
behavioural complementarity 278
better option hypothesis 205–7
bill
  adaptations for feeding 8–11
  change in form 10–11

morphology and specialization 62–7, 71,
  74
motor patterns in detecting prey 12
risk of damage 25, 41, 54, 153
variation between species 262
bill-length
  change and feeding technique 10–11, 63, 64,
    65–7
  parent–hatchling ratio 225–6, 226
  and probing depth 31
  severe weather 179
bill-tip 10–11, 62–7
bird density
  aggregation and depletion patterns 116–31
  Exe estuary model 115, 116
  habitat loss 312, 349–50
    Delta area 312–24
    Exe estuary 340–1, 341–5
    wintering grounds 347–8
  interference and feeding 80–2, 84, 87–90,
    94–5, 110–12
  see also density dependence
Black-backed Gulls (*Larus fuscus*) 82, 84
Black-capped Chickadees (*Parus atricapillus*) 75
Black-headed Gulls (*Larus ridibundus*) 84
Black-tailed Godwit (*Limosa limosa*) 228, 229,
  359, 360
Blåvand 171, 172
blowing rings 297
body weight
  seasonal and individual variations 145–8
  survival of winter 134, 139–40, 179–80
border disputes 188, 190
boring 12
breeding
  age of first breeding 189, 193, 196–7,
    258–9, 277
  comparisons of breeding behaviour 258–9,
    277–82
  food intake rates 54
  prey activity changes in breeding season 23
  reproductive rate
    fledgling production 361–3, 364–5
    number of pairs breeding 360–1, 364
  success 282–3; *see also* chick-rearing;
    copulation; fledging success; hatching
    success;
breeding areas/habitats 258–9
  early return 174–5, 324
  impact of habitat change/loss 375–7,
    377–8
  latitude and 161, 174, 256–7, 272

breeding areas/habitats (*cont.*)
  limited space 188–95
    quality of breeding space 192–5, 197–9
    see also competition
  site fidelity 164, 264
breeding distributions 254, 256–7
breeding skills hypothesis 198
British estuaries 170, 174, 356
brooding, *see* incubation
Bumblebees (*Bombus vagans*) 75
butterfly flight 187

cadmium 293
Calidris sandpipers 228–30, 231, 233
calling 188
Carcinus maenas (shore crab) 8, 9, 19, 24, 265
career decisions 187, 187–211, 388
carrying capacity 327–51
  effect of habitat change/loss on local numbers 349–50
  Exe estuary mussel beds 333–45
    over series of winters 341–5
    within one winter 333–41
  and food limitation 345–6
  likelihood of reaching 346–9
  model to predict 329–33
caterpillars, moth 229–30, 231, 236, 237, 239
Cerastoderma edule, see cockles
Charadrii 223, 228
chasing intruders 188, 190
chick mobility 222–3
chick mortality 222
  comparisons 281, 282
  population dynamics 365, 371, 372
chick-rearing
  comparisons 277–82
  parent- and self-feeding 219–50
chlorinated hydrocarbons 294–6
clams *see Scrobicularia plana*
classification 251–3
'clubs' 188
clutch size 212, 213
  comparisons 272–3, 274–5, 276
  population dynamics 362, 365
Coal Tits (*Parus ater*) 75
coastal breeders
  population trends 355–6, 357
    habitat change/loss 372, 373, 375–7, 379
  wintering 161–3
coastal habitats: carrying capacity, *see* carrying capacity
cockle fisheries 297–302, 323
cockles (*Cerastoderma edule*) 8, 9, 19, 24, 291
  accessibility 31
  detection of 14, 14–15, 16, 17
  functional response 107–8
  impact of bait-digging 304
  night- and day-time feeding 26–8

prey depletion 125
prey size selection 37–8, 49, 50
  rejection threshold 40, 47
  touch area of shell 33
  vulnerability to cold 71, 137
collar-bearing birds 165–6
collar size 100
Common Gulls (*Larus canus*) 82, 84
competition
  for breeding territory 183, 277, 388
    early return 174–5
    reproductive rate 360–1
    social system 188–95
    territory acquisition 196–204
    see also dominance
  for food 70, 73, 107–12
    carrying capacity 329–32, 337–41
    equal effects 107–10
    seasonal changes 141
    social careers and competitive ability 100–3
    unequal effects 110–12
    see also interference
consumption, daily 50–4, 134
contaminants 292–6
Continental breeding population 156, 355
  midwinter distribution 158–61, 162
  population dynamics 362, 364–77
    passim, 380
  spring migration 173–4
  timing of autumn migration 171–3
cooperation 195–6, 278
copulation 195–6
  extra-pair copulations (*EPCs*) 207–11, 279
  frequency 195, 208–9
crabs 8, 9, 19, 24, 265
Crassostrea gigas 296
Crested Lark (*Galerida cristata*) 231–3
cuckoldry 207–11, 279
culling 309
culture 73–4
Curlew (*Numenius arquata*) 75, 228, 360

daily consumption 50–4, 134
dams, construction of 315–23
Darwin's finches 75
day-time feeding 25–8
deferred maturity 196–7
deformities, physical 138, 139, 140
Delta area 305, 311
  cockle fishery 298, 300–1
  habitat loss 312–24
  migration 161, 163, 182
demographic analysis 328–9 *see also*
    population dynamics; population trends
density dependence
  carrying capacity, mortality and emigration 330–1, 335–6, 347–8
  population dynamics 352–4, 360–4
    habitat change/loss 367–77, 378–9

prey depletion and redistribution of birds
123–7, 127–8, 130
see also bird density
depletion of prey
carrying capacity 337–9, 345–6
patterns of aggregation and 116–31
detection of prey 11–17
low temperatures and 135
searching behaviour and 16–17, 57–9
by sight 12–14, 37
by touch, *see* touch-hunting
diet 264–6
body weight and 148
breadth 264–5
changes in 22–4, 64–7
and distribution 265–6
dominance and 70
growth rate and 239–41
prey species 8, 9, 56–7, 264–5
seasonal changes 23–4, 26, 27, 266
specialization in 56–7, 61–7
see also feeding; prey
digestive bottleneck 52, 53
digestive efficiency 69
digestive pauses 52–4
diplomatist attitude 85, 86
disc equation 107–9
displacement
habitat loss 311–24
interference in feeding 90, 97
distraction displays 271
displays 85, 86, 269–71; *see also* piping
distribution
diet and 265–6
feeding decisions, and *see* feeding
and population trends 354–60
taxonomy and 251–7
disturbance 305–9
divorce 204–7, 279
dogwhelks (*Nucella lapillus*) 8, 20, 268–9
dominance 70, 74, 95–103, 388
behaviour 95–7
and body weight 148
and carrying capacity 332, 339–40, 341
costs and benefits 97–8
global (fighting ability) 100, 101–2, 113–14
intraspecific interactions and 85–7
loss of food/failure to find food 87–90
site-dependent 95–7, 100, 101, 161, 200–2
social careers and competitive ability 100–3
and status signalling 98–100
and survival 98
see also interference
dorsal hammerers/hammering 19, 22, 38, 59, 60, 61
prey selection 21
susceptibility to interference 88
dredging (cockles) 297–8

earthworms (*Lumbricidae*) 8, 9, 141, 214

detection at low temperatures 135
parents feeding chicks 229–30, 231, 236, 237
effective touch area 32, 33
eggs
laying date 212
mortality 365
predation 216–17, 230–3, 273, 282–3
reproductive value 212–14
timing of laying 273
emigration function 330–1, 334, 335–6
see also carrying capacity; migration
encounter rate 43–6
encounters, aggressive 85–7
enemies, natural 149–52; *see also* parasites; predation
energy content of food 236, 237, 238–9
energy requirements, daily 51, 52
cold weather and 134, 180
engineering works (dams) 315–23
episodic events 269
estuaries 7
Eurasian origin 254–7
eutrophication 289–96
evolutionary stable strategies (ESS) 90–1
Exe estuary
carrying capacity of mussel beds 335–45
model and feeding location 112–16
patterns of aggregation and depletion 116–31 passim
extra-pair copulations (*EPCs*) 207–11, 279

fat reserves 115, 332, 334
survival of winter 134, 179–80
feeding 7–29
anatomical adaptations 8–11
chicks, *see* parent-feeding; self-feeding
competition for food, *see* competition
daily consumption and intake rate 50–4
detection/localization of prey, *see* detection of prey
distribution patterns 105–32
autumn and winter 114–15, 123–7
between winters 127–9
build-up phase 113–14, 116–23
Exe estuary model 112–16
ideal free distribution 105–12
in flocks 78–80
food intake rate see intake rate
interference, *see* interference
location and parasite infection 150
night and day 25–8
prey size selection, *see* prey size selection
risks associated with 25
specialization, *see* specialization
supplementary 143–5, 153, 333, 344–5, 347
see also diet; feeding areas/habitats; feeding conditions; feeding time; prey
feeding areas/habitats 7–8, 193–5
habitat loss/change 311–24, 349–50

feeding conditions
  reproductive decisions 212, 214–16
  seasonal changes 140–5
    profitability and rejection thresholds 46–8
  and survival of winter 138–9, 140–5
  see also prey density
feeding techniques 17–22, 148
  changes in 64–7
  cultural transmission 73–4
  and dominance 70
  specialization in 59–67
  see also individual techniques
feeding time
  disturbance and 307–8
  intake rate, daily consumption and 50–4
  seasonal changes 141–3
  severe weather and 135–6
female aggression 195
fields, feeding in (terrestrial feeding) 7–8, 323–4
  chick-rearing 233, 236–9, 240
  survival of winter 141, 142, 143, 143–5
fighting 188, 191
  ability 100, 101–2, 113–14
  queue hypothesis 200–2
fisheries, shell 296–304, 323
fledging period 222
  growth rate and 241, 242
  trade-off with food transport trips 227–44,
    248
fledging success 213, 362
  predation and 241–3
fledgling production
  divorce and 206–7
  population dynamics 361–3, 364–5, 366,
    367, 368
flocking 78–80; *see also* aggregation
following chicks 245–6
food abundance, *see* prey density
food item size 225–6, 228–30, 234–9
food processing rate 50–4
food quality
  energy content 236, 237, 238–9
  feeding distribution 115–16, 116–17, 118,
    136
  and habitat loss 386–7
  see also prey biomass
food shortages 133, 269
food stocks 128–9
food supply
  alternative/supplementary 143–5, 153, 333,
    344–5, 347
  annual fluctuations 341–5
  combined sources 345, 347
  limitation and carrying capacity 345–6
  seasonal variations 330–1, 332
food transport trips 222, 223
  trade-off with fledging period 227–44
foraging 387
  activity and bird density 94–5
  behaviour and detection 16–17
  efficiency

and carrying capacity 331–2, 339–40,
    341, 342
  chick-rearing 236–9, 246
  and survival 148
  see also interference-free intake rate
intake rate as measure 69–70
and kleptoparasitism 91
optimal 39–50, 267–8
severe weather and 135–6
strategy and survival 153
see also detection of prey; intake rate; search
    rate; search time
Fox (*Vulpes vulpes*) 152
France
  hard weather movements 178–9
  hunting Oystercatchers 309–11
frequency-dependent optimization 67–9
Frisian Oystercatchers 169
functional response 107–9, 111–12, 117–23

game theory 68, 90–1, 388
gender differences 260–1, 262, 267
  age of first breeding 189, 193, 277
  bill morphology 66, 67, 69, 262
  and mortality 371, 372
  role differentiation 258–9, 278–9
global dominance (fighting ability) 100, 101–2,
    113–14
Gompertz constant 228
Gondwanaland origin 254–7, 281–2
Goshawks (*Accipiter gentilis*) 151–2
Great Skua (*Stercorarius skua*) 233
Grevelingen estuary 312–15
growth rate 220–1, 222, 223
  between-species comparisons 227–8
  and fledging period 241, 242
  within-species comparisons 239–41
Gulls 82, 84, 227–8

Haag, Den 172–3
habitat
  choice 258–9, 263–4
  feeding specialization and 72
  loss or change 386–7
    carrying capacity 334–6, 349–50
    human activity in coastal zone 311–24
    population dynamics 366–80
    quality of habitat removed 372–5, 379
  saturation 329
    and the social system 187–211
  see also breeding areas/habitats; feeding
      areas/habitats
Haematopus ater 252, 253, 258, 260, 265, 274
  hatching success 282
  piping 270
Haematopus bachmani 252, 253, 258, 260,
    265, 274
  breeding success 282
  brooding 280–1

clutch size 272–3, 276
  population trends 283
  rearing chicks 278–9
Haematopus chathamensis 253, 259, 261, 265, 275, 283
Haematopus fuliginosus 253, 259, 260, 265, 274
Haematopus leucopodus 252, 253, 256, 258, 260, 265, 274
  piping 270
Haematopus longirostris 252–3, 258, 260, 265, 274
Haematopus meadewaldoi 252, 253, 258, 260, 274, 284
Haematopus moquini 252, 253, 258, 260, 265, 274
  anti-predator behaviour 271
  breeding success 282
  diet and distribution 266
  divorce and cuckoldry 279
  habitat choice 263
  investment in eggs 273, 276
  mass mortality 269
  Mytilus galliprovincialis 281
  piping 270
  site fidelity 264
Haematopus ophthalmicus 253, 259, 260, 275
Haematopus ostralegus finschi 253, 259, 261, 265, 275, 283, 284
Haematopus ostralegus longipes 252, 253, 258, 260, 274
Haematopus ostralegus osculans 252, 253, 258, 260, 274
Haematopus ostralegus ostralegus 252, 253, 258, 260, 274, 283, 284
Haematopus palliatus galapagensis 252, 253, 258, 260, 274
Haematopus palliatus palliatus 252, 253, 258, 260, 265, 274
  breeding success 282
  brooding 280–1
  parental investment 273
  piping 270
Haematopus unicolor 253, 257, 259, 260–1, 265, 275
hammering/hammerers 19–20, 38, 144, 267
  body weight 147, 148
  interference 70
  prey size selection 39
  specialization 59–61, 71, 72, 73–4
    prey morphology 61–2
  substrate 22
  see also dorsal hammering; ventral hammering
hand-raking (cockles) 297
handling time 40–1, 41, 42, 48–50
hard weather movements 176–9, 182, 310
Haringvliet–Hollandsch Diep 312, 313
hatching success 213, 282, 362
head: motor patterns 12
helicopters 306

helminths 149–51
Herring Gulls (*Larus argentatus*) 82
higher shore-levels 141, 143, 143–5
hovering ceremony 188, 191
human activities 289–326
  bait-digging 297, 304–5
  disturbance 305–9
  eutrophication and pollution 289–96
  habitat loss 311–24
  shellfishery 296–304
  shooting 309–11

Iceland 169
ideal free distribution (*IFD*) 105–12, 129–30
  equal competitors 107–10
  unequal competitors 110–12
incompatibility hypothesis 204
incubation 274–5, 278, 279–82
inferior phenotype hypothesis 198
inland breeders
  parent- and self-feeding 226–7, 240
  population dynamics 356
    cause of spread inland 356–60
    impact of habitat change/loss 372, 373, 375–7, 379
  seasonal changes in feeding conditions 214–15
  spring migration 173–4
  timing of autumn migration 171–3
  wintering 161–3, 164
inland feeding, *see* fields
intake rate
  dominance and 97, 98
  feeding distribution 106–7, 110–11, 114, 118–20, 130
  interference and 80–2, 87–90, 118–20
  interference-free intake rate (*IFIR*) 97, 114, 118, 331–2
  prey size selection and 39–50
    rejection of prey to increase intake rate 46–50
  processing rate, feeding time and daily consumption 50–4
  severe weather and 135–6
  specialization and differences 68–70
  survival of winter 179–80
intensification of agriculture 359–60
interference 77–104, 141
  avoiding losing mussels 93–5
  carrying capacity 337, 339–40
  causes 80–90
  costs and benefits of kleptoparasitism 90–3
  feeding distribution patterns 109–32 *passim*
  flocking 78–80
  measuring 80
  social careers and competitive ability 100–3
  specialization and 70, 73, 74
  susceptibility to (*STI*) 87–8, 113–14, 332, 339–40
  see also dominance

interference-free intake rate (IFIR) 97, 114, 118, 331–2
interspecific kleptoparasitism 82–4
intertidal flats, feeding on 138–9, 152–3
intraspecific interactions 85–7
intruders, chasing 188, 190

juveniles
    acquisition of specialization 72–3
    carrying capacity 342–3
    and dominance 98, 98–9
    migration 165–6, 183–4
    survival of winter 138, 144–5, 150, 153

kleptoparasitism 60, 87–90, 99
    avoiding 93–5
    costs and benefits of intraspecific 90–3
    interspecific 82–4
Krammer–Volkerak 321–2

Langli/Skallingen 168
Lapwings (*Vanellus vanellus*) 228, 229
large prey 41–3
Larii 222, 223
    see also under individual names
laying date 212
lead intoxication 292–3
leapfrog migration 181–4
leapfrog territory 192–3, 194
    breeding skills hypothesis 198, 199
    chick-rearing 233, 239
        feeding with parents 246–7
        fledging rates 243
        growth rate 239–41
        independence 248
        limpets 234–5
        predation 244
learning specialization 70, 73–4
    see also parental training
life history characteristics 271–6
life history decisions 186–218
    cuckoldry 207–11
    divorce 204–7
    habitat saturation and the social system 187–211
    queue hypothesis 198–204
    reproductive decisions 211–17
    territorial defence 195–6
    territory acquisition 196–9
limpets (*Patella*) 8, 9, 20, 267
    parent-feeding 233, 234–5, 239
    shell shape 268
Littorina (periwinkle) 8, 9, 17, 25
locating prey see detection of prey
longevity 271–2
low cost–low benefit hypothesis 197–8
Lugworms, *see Arenicola*
Lumbricidae, *see* earthworms

Macoma balthica 8, 9, 26, 27, 291
    changes in activity 22–3, 23, 24
    detection 12–13, 14–15, 16
    hammering 20
    and low temperatures 72, 135
    parasite-infected 151
    prey size selection 31, 32, 4–5
        intake rate 49, 50
        rejection threshold 40, 47, 48
        touch area 33
    risks in feeding on 25
    specialization 57, 57–9
        bill morphology 63–4, 65
Marcus Island 282–3
mass mortalities 176, 177, 269
mate change hypothesis 207–11
mate fidelity 195–6, 258–9
maturity, deferred 196–7
meadow birds 358–60
melanism 262
mercury 293–4
Mesodesma donacium 269
methodological problems 378–9
Micrasomacanthus rectacantha 149, 150
migration 155–85
    annual cycle of moult and migration 166–75
    distance 254, 255
    midwinter distribution 157–63
    ring recoveries 155–7
    severe weather in winter 175–84
        hard weather movements 176–9, 182, 310
    summering sites 164–6
    as survival strategy 181–4
    wintering sites 163–4
'mobile nest' 223
monogamy 195–6, 258–9
morphology
    bill 62–7, 71, 74
    characteristics 260–1, 262
    individual variations 69
    prey 21, 61–2
mortality rates
    carrying capacity 330–1, 333, 334, 335–6, 347–8
    Delta area 316–20
    mass mortalities 176, 177, 269
    population dynamics 352–4, 363, 363–4, 365–6, 379
        habitat change/loss 367–75
    severe weather 136–8, 176, 177, 179, 181
moth caterpillars 229–30, 231, 236, 237, 239
moult 294
    annual cycle of migration and 166–75
    location and timing 166–71
mussel fishery 302–4
mussels, *see Mytilus*
Mya arenaria 8, 9, 14
    changes in activity 24
    prey size selection 31, 32, 36–7, 49

rejection threshold 40, 47
   touch area 33
   specialization 57, 58, 63, 64, 65
Mytilus edulis (mussel) 8, 9, 27, 71
   carrying capacity of mussel beds 335–45
   'crashes' in stocks 24
   feeding distribution
      functional response 107–8
      interference function 109, 110
      model for mussel-eating Oystercatchers
         112–31
   hammering 19–20
   handling and comparison of species 267
   mussel-feeders and body weight 147–8
   prey selection
      and feeding technique 21, 38–9
      size selection 32–4, 38–9, 40, 41, 47, 49
   specialization 57, 58, 68, 69
      acquisition of specialization 72–3
      bill morphology 64, 65
      prey morphology 61–2
      use of more than one feeding technique
         60–1
   stabbing 17–19
   upper-level mussel beds and helminths 150
Mytilus galloprovincialis 266, 281

neap–spring tide cycle 114–15, 339–40
Nearctic Oystercatchers 252, 253
Neotropical Oystercatchers 252, 253
Nereis (ragworm) 8, 9, 26, 27, 135
   changes in Oystercatcher diet 22, 23, 24
   detection of 14, 16–17
   specialization 57, 57, 9, 68
      bill morphology 63, 64, 65
Nereis virens (King Ragworm) 297, 304–5
'nest, mobile' 223
nest defence 216–17
Netherlands, The 183
   cockle fishery 298–9, 300–2
   migration 171–3, 173–4
   severity of winters 176, 177, 181
niche width 71
night feeding 25–8, 141
nonbreeding birds 189
   attaining breeding status 203
   distributions 255
North Sea crossings 158–61
Norway 159–61
Nucella lapillus (dogwhelk) 8, 20, 268–9
nutrients 289–92

Oesterdam 316, 317
oil spills 296
ontogenetic trajectories 187
Oosterschelde 168
   dam construction 313, 315, 316–23
   shellfishery 298, 300–1
optical lobes 247

optimal foraging 39–50, 267–8
organics 289–92
Ottenby 171, 172

pair bond 202–3, 204–11, 277
   cuckoldry 207–11, 279
   divorce 204–7, 279
   mate fidelity 195–6, 258–9
   predation during egg-laying 216–17
Palearctic Oystercatchers 252, 253
Paralytic Shellfish Poisoning 269
parasites 25, 54, 69, 149–51, 268–9
parental investment 277–82
parental training 73–4, 247–8
parent-feeding 219–50, 278, 281, 387
   evolutionary stability 248
   predation risk 224–44
      trade-off between fledging period and
         transport trips 227–44
      trade-off between parent- and self-feeding
         224–7
   transition to self-feeding 244–8
Parorchis acanthus 268–9
Parvatrema affinis 25
'passive' prey size selection 30–9
Patella granularis 26, 268, *see also* limpets
pecking 12, 14, 17
pentane extractable lipids (PEL) 295–6
Peregrine Falcon (*Falco peregrinus*) 151–2
pesticides 294–6
Philipsdam 316, 317, 320
piping 85, 86, 95–6, 100, 188
   ceremonies 96
   comparison across species 270
ploughing 12, 17
pollution 289–96, 386–7
Polecat (*Mustella putorius*) 152
polychlorinated biphenyls (PCBs) 294–6
polygyny 195
poor discrimination hypothesis 197
population dynamics 352–83
   demographic consequences of increase
      360–4
   distribution and population trends 354–60
   population model 364–6
   predicting effects of habitat change/loss
      366–77
   total population size 354–5
population trends 283–4, 354–60
'precocial 5' condition 219–20, 221–2, 223
predation 151–2
   anti-predator behaviour 271
   and breeding success 282–3
   eggs 216–17, 230–3, 273, 282–3
   risk and chick-rearing 220–1, 221, 223
      fledging period and food transport trips
         227–44
      trade-off between parent- and self-feeding
         224–7
preening 54

prey
  activity of and changes in prey choice 22–4
  contamination 292–6
  depletion, *see* depletion
  detection, *see* detection of prey
  handling 17–22, 267
  morphology 21, 61–2
  organics' and nutrients' impact 289–92
  parasitized 25, 54, 69, 149–51, 268–9
  risks in feeding 25
  size selection, *see* prey size selection
  species 8, 9, 56–7, 264–5
  see also feeding; food supply; prey biomass;
    prey density; prey size selection
prey biomass 323
  carrying capacity and 337–8, 339, 341–5
  feeding distribution patterns 117–18, 120–5
  organics, nutrients and 290–1
  prey size selection 48–50
  seasonal changes 140–1
prey density
  feeding distribution
    ideal free distribution (*IFD*) 106, 107–12
    patterns 120–5
  parent- and self-feeding 225
  seasonal changes 140–1
prey size selection 30–55, 267–8
  intake rate, processing rate and daily
    consumption 50–4
  observed size selection 34–9
  and optimal foraging 39–50
    large prey 41–3
    predicted 'active' size selection 39–40
    rejection of prey 46–50
    small prey 40–1
    varying encounter rate 43–6
  predicted 'passive' size selection 30–4
primary moult 166–75
probing 247–8
probing depth 31, 44–5
processing rate 50–4
'producing', 'scrounging' and 90–1
  see also kleptoparasitism
profitability 39–40, 50
  large prey 41–3
  and probing depth 44–5
  seasonal variations and rejection thresholds
    46–8
  small prey 40–1, 42, 43
protein reserves 134, 179–80
pseudo-territories 95–7, 101
Psilostomum brevicolle 150

queue hypothesis 101–3, 198–204

ragworms, *see* Nereis
random touch model 30–4, 39
  observed size selection and 34–9
raptors 151–2

rearing chicks, *see* chick-rearing
recovery time 306–7
Recurvirostridae 256–7
Redshank (*Tringa totanus*) 228, 229
rejection threshold 39–40, 46–50, 267–8
repeat clutches 212
representativeness, errors of 378
reproductive decisions 187, 211–17
reproductive rate 347–8, 352–4, 360–3, 363,
    364–5
reproductive value of an egg 212–14
re-scaling coefficients 333
resident territory 192–3, 194, 197
  chick-rearing 233, 248
    feeding with parents 246–7
    fledging rates 243
    food item size 236–9
    growth rate 239–41
    predation 244
risk
  associated with feeding 25
    damage to bill 25, 41, 54, 153
  predation, *see* predation
  robbing, *see* kleptoparasitism
rocky shore 7
Roggenplaat tidal flat 314, 315
role differentiation 258–9, 278–9
ruff 229

sampling territories 202
sandpipers 228–30, 231, 233
scanning rates 93–4
scavenging 60 *see also* kleptoparasitism
Schiermonnikoog 168, 178, 181, 199
Scrobicularia plana 8, 9, 24
  detection 12–13, 16
  functional response 107–8
  prey size selection 31, 32, 35–6
    intake rate 49
    probing depth and encounter rate 44–5
    rejection threshold 40, 47
    touch area 33
  specialization 68
'scrounging', 'producing' and 90–1
  see also kleptoparasitism
searching *see* foraging
search rate 57–9
search time 48–50
  and avoidance 94
seasonal variations
  body weight 145–8
  diet 23–4, 26, 27, 266
  feeding conditions, *see* feeding conditions
  reproductive value of eggs 212–14
sedentary chicks 222–3
self-feeding 281–2, 387
  predation risk and trade-off with parent-
    feeding 224–7
  threshold 226–7
  transition to 244–8

sewing 12, 14, 17
sex roles 258–9, 278–9
  see also gender differences
shellfishery 296–304, 323
shell shape 268
shell thickness 61–2
shooting 309–11
sight-detection of prey 12–14, 37
siphon holes 12–14, 37
site-dependent dominance 95–7, 100, 101, 161,
  200–2
site fidelity 164, 264
size selection of prey, *see* prey size selection
small prey 25
  profitability 40–1, 42, 43
social position 187
  habitat saturation and social system 187–211
  interference and 85–7
  social careers and competitive ability 100–3
  status signalling 98–100
  see also dominance
soil resistance 214–15
specialization 2, 56–76, 264
  acquisition of specializations 72–4
  bill morphology 62–7, 71, 74
  in diet 56–7, 61–7
  ecological consequences of 71–2
  evidence for 56–71
  in feeding technique 59–67
  frequency dependent optimization 67–9
  in other species 75
  prey morphology 61–2
spring migration 173–5
spring–neap tide cycle 114–15, 339–40
St Michel, Baie de Mont 128, 168
stabbing/stabbers 17–19, 21, 21–2, 38–9, 129,
  267
  body weight 147, 148
  interference and 88
  specialization 59, 60, 68, 71, 73–4
    prey morphology 61–2
  supplementary feeding 144–5, 153
starvation
  chicks 241–4, 281
  habitat loss 316–20
state dependency 388
status signalling 98–100
stealing, *see* kleptoparasitism
subordinate chicks 243–4
substrate 22
summer
  length of summer absence 189, 192
  population dynamics 365, 375–7
  sites at different ages 164–6
supplementary feeding 143–5, 153, 333,
  344–5, 347
survival
  chicks 241–3, 246–7
  dominance and 98
  migration as survival strategy 181–4
  population dynamics 367, 368

survival times in severe weather 179–80
  of the winter 133–54
    body condition 145–8
    individuals that die 138–40
    natural enemies 149–52
    seasonal variations in feeding conditions
      140–5
    weather conditions 133–8, 143, 152
survival species 195
susceptibility to interference (*STI*) 87–8,
  113–14, 332, 339–40
  see also dominance
sympatric speciation 257–62

taste 15–16
taxonomy 251–3
Tees estuary 312
'terrestrial' feeders, *see* fields, feeding in
territory
  acquisition 196–204
  competition for, *see* competition
  encounters over 187–8, 190–1
  monogamy and cooperation in defence
    195–6
  site-dependent dominance 95–7, 100, 101,
    161, 200–2
  see also leapfrog territory; resident territory
threshold attack distance 92–3
tidal amplitude 316–20
tidal cycle 22–3, 143
tipulid larvae 8, 9, 214
  chick-rearing 229–30, 231, 236, 237
touch area 31–4
touch-hunting 14–15, 26
  prey size selection 34, 36, 36–7
training, parental 73–4, 247–8
tributyltin (TBT) 296
Turnstones (*Arenaria interpres*) 75, 87, 101–2

Upogebia africana 268
upper flats 141, 143, 143–5

vacancies 202–3
Veersche Gat–Zandkreek 312, 313
ventral hammering/hammerers 19, 21, 22, 38,
  88
  prey morphology 21, 61
  specialization 59–60, 61, 70
vocalizations 269–71

Wadden Sea 311, 356
  eutrophication 291
  habitat loss 312
  hard weather movements 176, 177, 178, 182,
    310
  midwinter distribution 161, 162–3
  moult and migration cycle 169, 170–1

Wadden Sea (*cont.*)
  pollution 294
  severe winter weather 179, 180–1
  shellfishery 298, 301, 302–3
Wash, The 297, 302
weather conditions
  cold spells 71–2
  hard weather movements 176–9, 182, 310
  low temperatures 134–5, 138, 175–6
  severe and migration 175–84
  strong winds 134, 135–6
  and survival of winter 133–8, 143, 152
weight 260–1, 262; *see also* body weight
Westerschelde 319, 323
White-throated Sparrows (*Zonotrichia albicollis*) 100

widowed birds 202–3, 206–7
wing length 260–1
winter
  carrying capacity, *see* carrying capacity
  feeding distribution patterns 114–15, 123–7, 130
    between winters 127–9
  midwinter distribution 157–63
  Oosterschelde winters 316–20
  population dynamics 365, 365–6
    habitat loss/change 369–75, 377–8, 379
  severe weather and migration 175–84
  survival of, *see* survival
  wintering sites 163–4